Topics in
Current Physics

5

Topics in Current Physics Founded by Helmut K. V. Lotsch

Structure and Collisions of Ions and Atoms

Edited by I. A. Sellin

With Contributions by
L. Armstrong, Jr.   J. S. Briggs   S. J. Brodsky
S. Datz   F. Folkmann   P. H. Mokler   P. J. Mohr
I. A. Sellin   N. Stolterfoht   K. Taulbjerg

With 157 Figures

Springer-Verlag Berlin Heidelberg New York 1978

Dr. Ivan A. Sellin

Oak Ridge National Laboratory
Oak Ridge, TN 37830, and
Department of Physics, University of Tennessee,
Knoxville, TN 37916, USA

ISBN-13:978-3-642-81212-5   e-ISBN-13:978-3-642-81210-1
DOI: 10.1007/978-3-642-81210-1

Library of Congress Cataloging in Publication Data. Main entry under title: Structures and
collisions of ions and atoms. (Topics in current physics ; v. 5). 1. Collisions (Nuclear
physics). 2. Ions. 3. Atoms. 4. Atomic structure. I. Sellin, Ivan A. II. Armstrong,
Lloyd, 1940—. III. Series. QC794.6.C6S77   539.7'54 77-26005

2153/3130-543210

Preface

The central subject of this volume is the atomic and molecular physics of heavy particles as investigated with charged particle accelerators. The natural division between atomic structure and ion-atom collision studies, and the similar division between the theoretical and experimental branches of these subjects, are reflected in a parallel subdivision into corresponding chapters. In addition, one chapter is devoted to the important interface between atomic and molecular physics with condensed matter physics. A principal aim of the present volume is to provide a compact description of a number of current interests and trends within the heavy particle structure and collisions field in a sufficiently general, non-specialized way that interested scientists who wish to become acquainted with such interests and trends can do so without becoming bogged down in excessive archival detail. It is, therefore, hoped that the book will be of some use to advanced students who seek a general introduction to these subjects. Numerous, more specialized, archival review articles are frequently referred to in each chapter for the benefit of those who seek more detailed knowledge about particular topics discussed.

The editor wishes to acknowledge the support of two U.S. government agencies: the Office of Naval Research and the National Science Foundation, during the preparation of this volume. Sincere thanks are due Mrs. Betty Thoe for her excellent editorial work on the various manuscripts and Mrs. Deborah Adams for her invaluable help with the preparation of the editor's own contribution. The facilities and assistance provided by the University of Tennessee and the Oak Ridge National Laboratory are gratefully noted. Completion of this volume was accomplished with the kind assistance of the Alexander von Humboldt Foundation, to whom the author is indebted for an Award tenureable at the University of Frankfurt/Main (Fed. Rep. of Germany), and through the gracious hospitality of Professor Dr. K.-O. Groeneveld of that University and Professor Dr. Peter Armbruster of the "Gesellschaft für Schwerionenforschung", Darmstadt (Fed. Rep. of Germany).

Oak Ridge, Tenn.
October 1977   *Ivan A. Sellin*

Contents

List of Contributors

ARMSTRONG, LLOYD, Jr.

   Dept. of Physics, Johns Hopkins University, Baltimore,MD 21218, USA

BRIGGS, JOHN S.

   Atomic Energy Research Establishment, Harwell, Oxfordshire OXII ORA, GB, and
   Inst. of Physics, University of Aarhus, DK-8000 Aarhus C., Denmark

BRODSKY, STANLEY J.

   Stanford Linear Accelerator Center, Stanford, CA 94305, USA

DATZ, SHELDON

   Oak Ridge National Laboratory, Oak Ridge, TN 37830, USA

FOLKMANN, FINN

   Gesellschaft für Schwerionenforschung mbH, D-6100 Darmstadt, Fed. Rep. of Germany

MOHR, PETER J.

   Lawrence Berkeley Laboratory, University of California, Berkeley, CA 94720, USA

MOKLER, PAUL

   Gesellschaft für Schwerionenforschung mbH, D-6100 Darmstadt, Fed. Rep. of Germany

SELLIN, IVAN A.

   Oak Ridge National Laboratory, Oak Ridge, TN 37830, and
   Dept. of Physics, University of Tennessee, Knoxville, TN 37916, USA

STOLTERFOHT, NIKOLAUS

   Hahn-Meitner-Institut, Glienicker Straße 100, D-1000 Berlin 39, Fed. Rep. of
   Germany

TAULBJERG, K.

   Inst. of Physics, University of Aarhus, DK-8000 Aarhus C., Denmark

1. Introduction

I. A. Sellin

The atomic and molecular physics of heavy particles as investigated with charged particle accelerators has become a clearly identifiable field in very recent years. To a great extent this development has paralleled similar growth in the heavy-ion nuclear physics field. The present ability to create almost any ion charge and excitation state of any element at least as heavy as krypton in experimentally useful amounts, and a good many charge and excitation states of substantially heavier elements, has provided the technical means for studies of the structure and lifetimes of few-electron ions, photon and electron spectroscopy on both few- and many-electron excited states, and collisional production and quenching of these ions in both solid and gaseous targets. Many of the states found are multiply excited, have high angular momenta and high excitation energies, and decay not only by allowed processes but also through strong violation of the normal selection rules for radiative and autoionizing processes. Theories of uv, xuv, and x-ray production; ionization; rearrangement in ion-atom collisions; and the penetration of charged particles in condensed matter are all receiving new experimental input from this research. A number of puzzling new phenomena little considered and not accounted for quantitatively by these theories have very recently been discovered. Possibilities for tests of quantum electrodynamics - especially in the domain of so-called supercritical electric fields - in tandem with possibilities for access to the spectroscopy of superheavy atoms and molecules are rapidly becoming actualities. Characterization of even very basic phenomena that will govern production of such systems, e.g., multiple electron ionization and excitation in collisions, has barely begun.

Such experiments will evolve from current research on the physics of ion beams. Before outlining the content of the present volume, it is well to note the breadth of the larger field. Subject matter under study includes fundamental atomic structure, inner and outer shell vacancy production by heavy particles, subsequent emission of radiation (light, x-rays, Auger electrons) during decay of atomic states of the moving ions and the target atoms, the stripping (and often recapture) of electrons from moving ions, radiobiological effects of fast heavy ions, simulation of neutron radiation damage studies, crystal channeling studies of interest in solid-state physics, the study of nuclear lifetimes by the channel blocking technique, the study of hyperfine interactions of recoiling nuclei, the wavelengths and transition probabilities for excited heavy ions in high states of ionization, and collisional production of excited states of heavy ions that may be useful in uv, xuv, and x-ray laser development. Many of the fundamental problems in atomic physics involve so-called forbidden transitions. As always, the objective is to measure the small breakdown of forbiddenness which allows certain states to decay with long but not infinite mean lives. For sufficiently high Z, the breakdown is often found to be appreciable. A measurement leverage advantage arises because the first-order decay rates are zero, so that the measurement accuracy often applies directly to the more subtle interactions that cause departures from the matrix element orthogonalities we commonly refer to as selection rules. A primary objective of much current research is the understanding of the structure, formation, and decay of the simplest heavy ions - those containing just a few electrons - for if these systems are not fundamentally understood, then the behavior of the complicated, multielectron ions will certainly be more poorly understood. Because the rates for many forbidden de-excitation processes involving the excited states of few-electron ions scale with fairly high powers of Z, typically because of stronger magnetic interactions, it has become possible to study the rates of such forbidden processes in accelerator-based lifetime experiments to an accuracy competitive with the best relativistic theoretical structure calculations. The relativistic magnetic interactions are stronger in the higher Z ions of an isoelectronic sequence, because the electrons are closer to each other and to the nucleus, and move at more relativistic speeds.

The natural division between atomic structure studies (including quantum electrodynamics) and atomic collision studies, and between the theoretical and experimental branches of these subjects receives expression in the structure of the present volume. The relation of heavy ion atomic physics to tests of quantum electrodynamics and therefore to our basic understanding of electromagnetism is a principal subject of Chapter 2. Chapter 3 explicitly considers the influence of relativity on atomic structure as manifested in the relativistic behavior of inner electrons of heavy particles. An overview of current ideas and applications of collision theory to ion-atom collisions is a main aim of Chapter 4. The emission of electrons accompanying ion-atom collision processes and their aftermath, particularly atomic Auger electron emission, is described in Chapter 5. It is too rarely recognized that atomic decay by electron emission is relatively more frequent than corresponding decay by photon emission for heavy particles in most ionization-excitation states for much of the periodic table. Chapter 6 discusses many recent experiments in the complementary field of x-ray emission accompanying ion-atom collisions, with emphasis on very heavy particles. Extensions of beam-foil spectroscopy in areas little covered in previous review articles on this basic subject are the topic of Chapter 7. Finally, Chapter 8 addresses itself to the important interface area between atomic collisions and condensed matter physics. The use of heavy particles to probe the structure of solids and to study explicitly solid-state effects accompanying the penetration of heavy particles in matter is an important discipline unto itself.

2. Quantum Electrodynamics in Strong and Supercritical Fields

S. J. Brodsky and P. J. Mohr

With 31 Figures

Quantum electrodynamics (QED), the theory of the interactions of electrons and muons via photons, has now been tested both to high precision - at the ppm level - and to short distances of order 10^{-14} - 10^{-15} cm. The short distance tests, particularly the colliding beam measurements of $e^+e^- \rightarrow \mu^+\mu^-$, $\gamma\gamma$, and e^+e^- [2.1], are essentially tests of QED in the Born approximation. On the other hand, the precision anomalous magnetic moment and atomic physics measurements check the higher-order loop correc- tions and predictions dependent on the renormalization procedure. Despite the extra- ordinary successes, it is still important to investigate the validity of QED in the strong field domain. In particular, high-$Z\alpha$ atomic physics tests, especially the Lamb shift in high-Z hydrogenic atoms, test the QED amplitude in the situation where the fermion propagator is far off the mass shell and cannot be handled in perturba- tion theory in $Z\alpha$, but where the renormalization program for perturbation theory in α must be used. High-Z heavy-ion collisions can be used to probe the Dirac spectrum in the nonperturbative domain of high $Z\alpha$, where spontaneous positron production can occur, and where two different vacuum states must be considered.

Another reason to pursue the high-$Z\alpha$ domain is that the spectrum of radiation emitted when two colliding heavy ions (temporarily) unite can lead to a better under- standing of relativistic molecular physics. This physics is reviewed in the other chapters of this volume. Furthermore, the atomic spectra of the low-lying electron states and outgoing positron continua reflect the nature of the nuclear charge dis- tribution, and could be a useful tool in unraveling the nuclear physics and dynamics of a close heavy-ion collision. Somewhat complementary to these tests are the studies of Delbrück scattering (elastic scattering of photons by a strong Coulomb field) reviewed in [2.2].

One of the intriguing aspects of high field strength quantum electrodynamics is the possibility that it may provide a model for quark dynamics. Present theoretical ideas for the origin of the strong interactions have focused on renormalizable field theories, such as quantum chromodynamics (QCD), where the quarks are the analogues of the leptons, and the gluons - the generalizations of the photon - are themselves charged (non-Abelian Yang-Mills theory). In contrast to QED where the vacuum polari- zation strengthens the charged particle interaction at short distance, in QCD the interactions weaken at short distances and (presumably) become very strong at large separations.

To see the radical possibilites in strong fields, suppose α is large in QED and the first bound state of positronium has binding energy ε > m. The total mass of the atom M is then less than the mass of a free electron M = 2m - ε < m. Consider then an experiment in which an e^+e^- pair is produced near threshold - e.g., via a weak current process. Since an additional virtual pair may be present, the produced pair can spontaneously decay to two positronium atoms in the ground state, each with finite kinetic energy. Thus bound states, and not free fermions, are produced! It is clearly an interesting question whether strong field strength in QED can provide a mechanism analogous to quark confinement in hadron dynamics. The work of WILSON [2.3] and MANDULA [2.4] is especially relevant here. The studies of spontaneous pair production in heavy-ion collisions (see Sec.2.3) provide a simple phenomenological framework in which some of the effects of strong fields can be tested.

It should be noted that our review touches only a limited aspect of high-Zα electrodynamics. We consider only the cases of a fixed or heavy source for a high-Zα Coulomb potential. An important open question concerns the behavior of the Bethe-Salpeter equation for positronium in the large α domain, and in particular, whether the binding energy can become comparable to the mass of the constituents so that M = 2m - ε → 0.

The organization of this Chapter is as follows: We review in detail the recent work on the atomic spectra in high-Z electronic (Sec.2.1) and muonic atoms (Sec.2.2), including muonic helium, with emphasis on the Lamb shift and vacuum polarization corrections which test strong field quantum electrodynamics. The theoretical framework of the QED calculations for strong fields is discussed in Section 2.1.5. The constraints on nonperturbative vacuum polarization modifications and possible scalar particles are presented in Section 2.2.8. A review of recent work on the quantum electrodynamics of heavy-ion collisions, particularly the dynamics of positron production, is presented in Section 2.3. In addition to reviewing the phenomenology and calculational methods (Sec.2.3.2-2.3.4), we also discuss the parameters for possible experiments, with a brief review of vacancy formation (Sec.2.3.6) and background effects (Sec.2.3.7). In our review of heavy-ion collisions, we will also touch on several new topics, including the coherent production of photons in heavy-ion collisions (Sec.2.3.9) and the self-neutralization of charged matter (Sec.2.3.10). We also point out some questions which are not completely resolved, including the relative importance of induced versus adiabatic pair production (Sec.2.3.5) and the nature of radiative corrections in α to spontaneous pair production (Sec.2.3.8).

2.1 The Electrodynamics of High-Z Electronic Atoms

2.1.1 Lamb Shift in Hydrogenlike Ions

At present the most precise and sensitive way to test quantum electrodynamics at high field strength is to compare the theory and measurements of the classic Lamb shift interval, the $2S_{1/2} - 2P_{1/2}$ separation in hydrogenlike ions. In recent work on the Lamb shift, measurements have been extended to hydrogenlike argon (Z=18) by an experiment at the Berkeley SuperHILAC [2.5]. As we shall see, such experiments provide an important test of QED in strong fields. The higher order binding terms in the theory, which are small in hydrogen, become relatively more important at high Z. For example, the terms of order $\alpha(Z\alpha)^6$, which contribute 0.016% of the Lamb shift in hydrogen, give 12% of the Lamb shift in hydrogenlike argon. The theoretical contributions to the Lamb shift are by now well established [2.6,7]. Our purpose here will be to summarize these contributions as an aid to testing the validity of the theory.

The dominant part of the Lamb shift is given by the self-energy and vacuum polarization of order α, corresponding to the Feynman diagrams in Fig.2.1a and b. In the past, most of the theoretical work on the self-energy has been concerned with the evaluation of terms of successively higher order in $Z\alpha$. However, ERICKSON [2.8] has given an analytic approximation which can be used as a guide for the Lamb shift for any Z. This is discussed in detail in [2.9].

More recently, MOHR [2.10] has made a comprehensive numerical evaluation of the $2S_{1/2}$ and $2P_{1/2}$ self-energy to all orders in $Z\alpha$. The method of evaluation is based on the expansion of the bound electron propagation function in terms of the known Coulomb radial Green's functions [2.11] and is described in more detail in Section 2.1.5. To display the results for the order α self-energy contribution $S_{SE}^{(2)}$ to the Lamb shift $S = \Delta E(2S_{1/2}) - \Delta E(2P_{1/2})$, it is convenient to isolate the exactly known low-order terms by writing

$$S_{SE}^{(2)} = \frac{\alpha}{\pi} \frac{(Z\alpha)^4}{6} m \left[\ln(Z\alpha)^{-2} - \ln \frac{K_0(2,0)}{K_0(2,1)} + \frac{11}{24} + \frac{1}{2} \right.$$
$$+ 3\pi\left(1 + \frac{11}{128} - \frac{1}{2} \ln 2\right)(Z\alpha) - \frac{3}{4}(Z\alpha)^2 \ln^2(Z\alpha)^{-2}$$
$$\left. + \left(\frac{299}{240} + 4 \ln 2\right)(Z\alpha)^2 \ln(Z\alpha)^{-2} + (Z\alpha)^2 G_{SE}(Z\alpha)\right] \quad . \qquad (2.1)$$

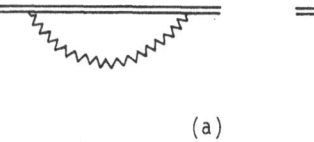 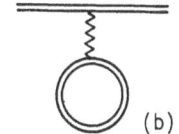

(a) (b)

Fig. 2.1a and b. Feynman diagrams for the lowest order self-energy (a) and vacuum polarization (b). The double line represents an electron in the external Coulomb field

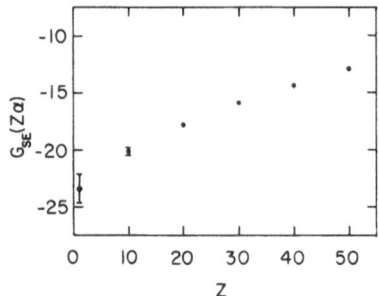

Fig. 2.2. Calculated values of $G_{SE}(Z\alpha)$ for Z=10 to 50 and the extrapolated value at Z=1 (from MOHR [2.10])

We shall always distinguish radiative terms in α from terms in $Z\alpha$ which arise from the nuclear field strength. Values of the remainder $G_{SE}(Z\alpha)$ in (2.1) corresponding to the calculated values of $S_{SE}^{(2)}$ for Z in the range 10 to 50, appear in Fig.2.2. The error bars in that figure represent a conservative estimate of the uncertainty associated with the numerical integration in the evaluation of the self-energy and, at Z = 1, the uncertainty resulting from extrapolation from Z = 10.

Evaluation of the energy level shift associated with the vacuum polarization of order α is facilitated by considering the expansion of the vacuum polarization potential in powers of the external Coulomb potential (see WICHMANN and KROLL [2.12]). Only odd powers of the external potential contribute as a consequence of FURRY's theorem [2.13]. The first term in the expansion gives rise to the Uehling potential [2.14,15]; the associated level shift is easily evaluated numerically. The second nonvanishing term in the expansion is third order in the external potential. The two lowest order contributions to the Lamb shift from this term are given by [2.12,16]

$$\frac{\alpha}{\pi} \frac{(Z\alpha)^6}{6} m \left[\frac{19}{60} - \frac{\pi^2}{36} + \left(\frac{3}{64} - \frac{31\pi^2}{3840} \right) \pi(Z\alpha) \right] \quad . \tag{2.2}$$

A substantial discrepancy between theory and experiment was eliminated when APPELQUIST and BRODSKY [2.17] corrected the fourth order Lamb shift terms by a numerical evaluation. Since then, the terms have been evaluated analytically. The total of the fourth order radiative corrections to the Lamb shift is given by

$$S^{(4)} = \left(\frac{\alpha}{\pi} \right)^2 \frac{(Z\alpha)^4}{6} m \left[\pi^2 \ln 2 - \frac{37\pi^2}{144} - \frac{3767}{1728} - \frac{3}{2} \zeta(3) \right] \quad . \tag{2.3}$$

Recent work on the evaluation of this term is summarized in [2.7]. Note that only the lowest order term in $Z\alpha$ has been evaluated.

The lowest order reduced mass and relativistic recoil contributions to the Lamb shift are given by (see [2.7])

$$S_{RM} = \frac{\alpha}{\pi} \frac{(Z\alpha)^4}{6} m \left(-3 \frac{m}{M} \right) \left[\ln(Z\alpha)^{-2} - \ln \frac{K_0(2,0)}{K_0(2,1)} + \frac{23}{60} \right] \tag{2.4}$$

and

$$S_{RR} = \frac{(Z\alpha)^5}{6\pi} m\left(\frac{m}{M}\right)\left[\frac{1}{4} \ln(Z\alpha)^{-2} - 2\ln\frac{K_0(2,0)}{K_0(2,1)} + \frac{97}{12}\right] \tag{2.5}$$

where M is the nuclear mass.

The finite nuclear size correction to the Lamb shift is given, for Z not too large, by the perturbation theory expression

$$S_{NS} = \left[1 + 1.70(Z\alpha)^2\right] \frac{(Z\alpha)^2}{12} m \left(\frac{Z\alpha R}{\chi}\right)^{2s} \tag{2.6}$$

assuming a nuclear model in which the charge is distributed uniformly inside a sphere, where $s = \sqrt{1-(Z\alpha)^2}$ and R is the rms charge radius of the nucleus. An estimate of the error due to neglected higher order terms in perturbation theory is given in [2.16].

The sum of contributions listed above gives the total Lamb shift S. Values for the individual contributions are listed in Table 2.1 for hydrogenlike argon. Theoretical and experimental values for Z≥3 are compared in Table 2.2. The theoretical values for Z≤30 are listed in [2.16].

Table 2.1. Contributions to the Lamb shift at Z=18, R=3.45(5) fm assumed

Source	Order	Value
Self-energy	$\alpha(Z\alpha)^4[\ln(Z\alpha)^{-2},1,Z\alpha,\ldots]$	40,544(15) GHz
Vacuum polarization	$\alpha(Z\alpha)^4[1,Z\alpha,\ldots]$	-2,598(3)
Fourth order	$\alpha^2(Z\alpha)^4$	11(14)
Reduced mass	$\alpha(Z\alpha)^4 m/M$	-1
Relativistic recoil	$(Z\alpha)^5 m/M$	12(9)
Nuclear size	$(Z\alpha)^4 (R/\chi)^2[1,(Z\alpha)^2\ln(R/\chi),\ldots]$	283(12)
		38,250(25) GHz

Table 2.2. Comparison between theory and measurement of the Lamb shift $E(2S_{1/2})$ - $E(2P_{1/2})$ for Z≥3

	Theory (1σ)	Experiment (1σ)		Ref.
$^6Li^{2+}$	62,737.5(6.6) MHz	62,765(21)	MHz	[2.18]
		62,790(70)	MHz	[2.19]a
		63,031(327)	MHz	[2.20]
$^{12}C^{5+}$	781.99(21) GHz	780.1(8.0)	GHz	[2.21]
$^{16}O^{7+}$	2,196.21(92) GHz	2,215.6(7.5)	GHz	[2.22]
		2,202.7(11.0)	GHz	[2.23]
$^{19}F^{8+}$	3,343.1(1.6) GHz	3,339(35)	GHz	[2.24]a
$^{40}Ar^{17+}$	38.250(25) THz	38.3(2.4)	THz	[2.5]a

a Improved experimental precision is expected.

Most of the experiments listed in Table 2.2 were done by the so-called static field quenching method [2.20]. This method is based on the large difference between the $2S_{1/2}$ and $2P_{1/2}$ lifetimes and the small separation of the levels. The ratio of the lifetimes is roughly $\tau(2S_{1/2})/\tau(2P_{1/2}) \sim 10^8 Z^{-2}$. Atoms in the metastable $2S_{1/2}$ state are passed through an electric field which causes the lifetime of the $2S_{1/2}$ state to decrease by mixing the S and P states. The change in the lifetime as a function of electric field strength leads to a value for the Lamb shift according to the Bethe-Lamb theory. The quenching experiments at higher Z (Z≥6) depend on the electric field in the rest frame of a fast beam of ions passing through a magnetic field to produce the 2S-2P mixing.

The experiments of LEVENTHAL [2.18] and DIETRICH et al. [2.19] with lithium are based on the microwave resonance method. The experiment of KUGEL et al. [2.24] with fluorine measures the frequency of the $2S_{1/2}$-$2P_{3/2}$ separation which is in the infrared range. The Lamb shift is deduced with the aid of the theoretical $2P_{1/2}$-$2P_{3/2}$ splitting which is relatively weakly dependent on QED. In the experiment, one-electron ions of fluorine in the metastable $2S_{1/2}$ state are produced by passing a 64 MeV beam through carbon foils. The metastable atoms are excited to the $2P_{3/2}$ state by a laser beam which crosses the atomic beam, and the x-rays emitted in the transition $2P_{3/2} \rightarrow 1S_{1/2}$ are observed. A novel feature of the experiment is that the resonance curve is swept out by varying the angle between the laser beam and the ion beam which Doppler-tunes the frequency seen by the atoms.

2.1.2 Lamb Shift in Heliumlike Ions

It would be of considerable interest to extend accurate Lamb shift measurements to hydrogenic systems with very high Z in order to test strong field QED. However, it appears unlikely that the hydrogenlike Lamb shift can be measured by the quenching methods in ions with Z≥30 [2.25].

A different possibility for accurate checking of QED at very high Z is the study of two- and three-electron ions with high-Z nuclei. When Z is very large, the electron-nucleus interaction dominates over the electron-electron interaction. Therefore, a theoretical approach which considers non-interacting electrons bound to the nucleus according to the single particle Dirac equation, and treats interactions of the electrons and radiative corrections as perturbations, should be capable of making accurate theoretical predictions [2.26].

As an example, consider the energy separation 2^3P_0 - 2^3S_1 in heliumlike ions. In the high-Z jj-coupling limit, the energy separation is given by $(1s_{1/2}2p_{1/2})0$ - $(1s_{1/2}2s_{1/2})1$, so that if the electron-electron interaction is neglected compared to the electron-nucleus interaction, the absolute energy separation is just the hydrogenic Lamb shift $E(2S_{1/2})$ - $E(2P_{1/2})$. The electron-electron interaction must still be taken into account. The largest term, corresponding to one-photon exchange between the bound electrons, is of the form $\alpha[a(Z\alpha) + b(Z\alpha)^3 + c(Z\alpha)^5 + ...]m$, with the leading term coming from the nonrelativistic Coulomb interaction of the electrons. The dominant energy separation is given by the first two terms which grow more slowly with Z than the Lamb shift $\sim \alpha(Z\alpha)^4$. Hence, the Lamb shift becomes an increasing fraction of the energy separation as Z increases. The ratio of the Lamb shift to the total energy separation is 0.002% for Z=2, 0.8% for Z=18, and 9% for Z=54. At high Z, the main QED corrections in heliumlike ions correspond to Feynman diagrams such as those pictured in Figs.2.3a and b. The energy shift associated with these diagrams is just the hydrogenlike ion Lamb shift. Diagrams with an exchanged photon such as the one in Fig.2.3c are less important (of relative order Z^{-1}), but need to be calculated for a precise comparison with experiment.

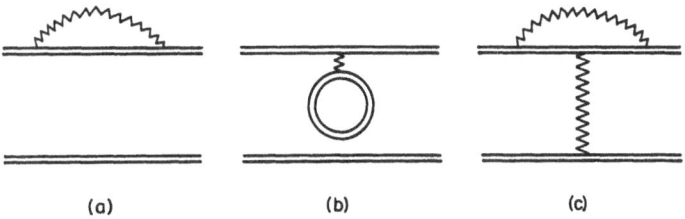

(a)   (b)   (c)

Fig.2.3a-c. Feynman diagrams which contribute to the Lamb shift in heliumlike ions

From the experimental standpoint, the heliumlike Lamb shift has the advantage that both the 2^3P_0 and 2^3S_1 states are long-lived compared to the hydrogenlike 2P states so that the natural width of the states is not the main limitation to the accuracy which may be achieved. In addition, in contrast to the hydrogenlike case, there is no strongly favored decay mode (for zero spin nuclei) to the ground state to depopulate the upper level, which makes direct observation of the decay photons feasible in a beam foil experiment.

Studies of the fine structure in heliumlike argon (Z=18) have been carried out by DAVIS and MARRUS [2.27], who measured the energy of photons emitted in the decays $2^3P_2 \to 2^3S_1$ and $2^3P_0 \to 2^3S_1$ in a beam foil experiment at the Berkeley SuperHILAC. Their results are shown in Table 2.3. In that table, the theoretical values for the QED corrections are the hydrogenlike corrections for Z=18, and are seen to be already tested to the 25% level.

Table 2.3. Fine structure in heliumlike argon, from DAVIS and MARRUS [2.27], in eV

Transition	Self-energy and vacuum polarization	ΔE_{th}	ΔE_{exp}
$2^3P_2 \to 2^3S_1$	-0.15	22.14(3)	22.13(4)
$2^3P_0 \to 2^3S_1$	-0.16	18.73(3)	18.77(3)

GOULD and MARRUS [2.28] have measured the transition rate for the radiative decay $2^3P_0 \rightarrow 2^3S_1$ in heliumlike krypton (Z=36) by observing the x-rays emitted in the sub-sequent M1 decay $2^3S_1 \rightarrow 1^1S_0$. Interestingly, the QED corrections to the $2^3P_0 - 2^3S_1$ energy splitting produce an observable effect in the decay rate. The observed life-time of the 2^3P_0 state is $\tau=1.66(6)$ ns. Assuming that the decay rate is given by the relativistic dipole length formula [2.29]

$$A(2^3P_0 \rightarrow 2^3S_1) = 4/3 \; \alpha\omega^3 \sum_M |<2^3S_1,M \; |\underline{r}_1+\underline{r}_2| \; 2^3P_0,0>|^2 \quad , \tag{2.7}$$

the theoretical value for the lifetime is $\tau=1.59(3)$ ns [$\tau=1.42(3)$ ns] with [with-out] the QED corrections included in the energy separation ω.

2.1.3 Quantum Electrodynamics in High-Z Neutral Atoms

Binding energies of inner electrons in heavy atoms are measured to high accuracy by means of electron spectroscopy of photoelectrons or internal conversion electrons [2.30]. Because of the extraordinary precision of the measurements, surprisingly sensitive tests of QED as well as the many-electron calculations can be made.

Precise calculations of the ground state energies have been given by DESIDERIO and JOHNSON [2.31] and MANN and JOHNSON [2.32]. DESIDERIO and JOHNSON [2.31] have calculated the self-energy level shift of the 1S state in a Dirac-Hartree-Fock po-tential for atoms with Z in the range 70 to 90 (see Sec.2.1.5). They estimated the vacuum polarization correction to the 1S level by employing the Uehling potential contribution for a Coulomb potential reduced by 2% to account for electron screening. MANN and JOHNSON [2.32] have done a calculation of the binding energy of a K elec-tron for W, Hg, Pb, and Rn which takes into account the Dirac-Hartree-Fock eigen-value, the lowest order transverse electron-electron interaction, and an empirical estimate of the correlation energy. The binding energy is taken as the difference between the energy of the atom and the energy of the ion with a 1S vacancy. Their comparison of theory to the experimental values [2.30] corrected for the photoelec-tric work function is shown in Table 2.4. The inclusion of the QED terms dramatical-ly improves the agreement between theory and experiment.

A similar comparison of theory and experiment has been made for Fm (Z=100). FREEDMAN et al. [2.33] and FRICKE et al. [2.34] have calculated the K-electron binding energy in fermium. The results of FREEDMAN et al. are compared to the ex-perimental value obtained by PORTER and FREEDMAN [2.35] in Table 2.5. They used extrapolations of the results for Z=70 to 90 of MANN and JOHNSON for the rearrange-ment energy, and of DESIDERIO and JOHNSON for the QED corrections. If the extra-polated value for the self-energy in that table is replaced by the recently calcu-lated value of CHENG and JOHNSON [2.36], the theoretical energy level is -141.957 keV.

Table 2.4. K-electron energy levels [Ry] from MANN and JOHNSON [2.32]

Element	Self-energy and vacuum polarization[a]	E_{th}	E_{expt}
74^W	8.65	-5110.50	-5110.46 ± 0.02
80^{Hg}	11.28	-6108.52	-6108.39 ± 0.06
82^{Pb}	12.27	-6468.79	-6468.67 ± 0.05
86^{Rn}	14.43	-7233.01	-7233.08 ± 0.90

[a]Calculated by DESIDERIO and JOHNSON [2.31]. These numbers include an estimated correlation energy of -0.08 Ry.

Table 2.5. Calculated K-electron energy level in $_{100}$Fm [in keV] from FREEDMAN et al. [2.33]

Source	Amount
E_{1S} (neutral-atom eigenvalue)	-143.051
Magnetic	+0.709
Retardation	-0.040
Rearrangement	+0.088
Self-energy	+0.484
Vacuum polarization	-0.154
Electron correlation	-0.001
E_{1S} (Z=100)	-141.965±0.025
Experimental value	-141.967±0.013

Extensive calculations of electron binding energies for all the elements in the range $2 \leq Z \leq 106$ have recently been done by HUANG et al. [2.37]. They used relativistic Hartree-Fock-Slater wave functions to calculate the expectation value of the total Hamiltonian. They assumed complete relaxation and included the Breit interaction and vacuum polarization corrections, as well as finite nuclear size effects.

By comparing their results to experiment, it is possible to see the effect of the self-energy radiative corrections to the $2S_{1/2} - 2P_{1/2}$ (L_I-L_{II}) level splitting in heavy atoms. Fig.2.4 shows the relative difference between the theoretical splitting without the self-energy and the experimental values compiled by BEARDEN and BURR [2.30]. The solid line shows theoretical values for the Coulomb self-energy splitting [2.10], and the dashed line shows values modified with a screening correction [2.37].

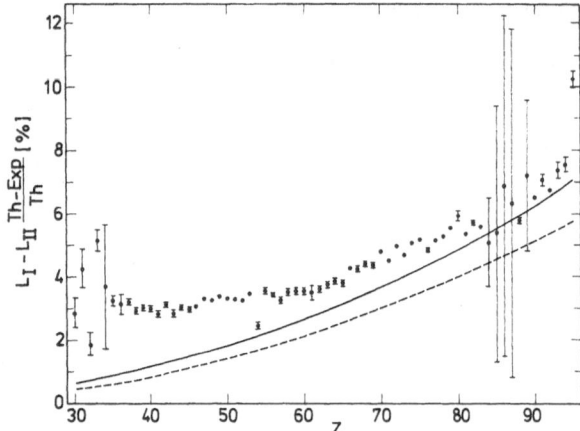

<u>Fig. 2.4.</u> Comparison of theory and experiment for the L_I-L_{II} level splitting in heavy atoms. The error bars give the probable error in the experimental values. Estimated experimental errors smaller than the data points are not shown. See text for explanation

2.1.4 High-Z Atoms and Limits on Nonlinear Modifications of QED

Various reformulations of classical electrodynamics have been proposed which attempt to eliminate the problem of an infinite self-energy of the electron. Among these is the nonlinear theory of BORN and INFELD [2.38,39]. They proposed that the usual Lagrangian L = $1/2(H^2-E^2)$ be replaced by

$$L_{BI} = E_0^2 \left\{ \left[1 + (H^2-E^2)/E_0^2 \right]^{1/2} - 1 \right\} \quad . \tag{2.8}$$

This formulation reduces to the usual form for field strengths much smaller than an "absolute field" E_0. Within the Born-Infeld theory, the electric field of a point charge is given by

$$E_r = \frac{e}{r^2} \left[1 + \left(\frac{e}{r^2} \right)^2 / E_0^2 \right]^{-1/2} \quad . \tag{2.9}$$

The magnitude of E_0 is determined by the condition that the integral of the energy density of the electric field associated with a point charge at rest is just the rest energy of the electron m. This results in a value $E_0 = 1.2 \times 10^{18}$ V cm^{-1} and a characteristic radius $r_0 = 3.5$ fm inside of which the electric field deviates substantially from the ordinary form e/r^2. Due to the large magnitude of E_0, the observable deviations from linear electrodynamics should be most evident in situations involving strong fields.

There has been recent interest in the experimental consequences of the Born-Infeld modification. RAFELSKI et al. [2.40] have found that the critical charge Z_{cr} (see Sec.2.3.1) is increased from about 174 in ordinary electrodynamics to 215 within the Born-Infeld electrodynamics. FREEDMAN et al. [2.33] and FRICKE et al. [2.34] have pointed out that the excellent agreement between the theoretical and experimental 1S binding energies in fermium (Z=100), discussed in Section 2.1.3, is evidence against deviations from the linear theory of electrodynamics. In Fm, the difference in 1S energy eigenvalues between the Born-Infeld theory and ordinary electrodynamics is 3.3 keV, based on a calculation using the Thomas-Fermi electron distribution with a Fermi nuclear charge distribution. This is two orders of magni-tude larger than the combined uncertainty in theory and experiment listed in Table 2.5. Although the other corrections listed in that table might be modified by the Born-Infeld theory, e.g., the self-energy, the linear theory produces agreement with experiment in a case where the effects of possible nonlinearities are large. SOFF et al. [2.41] have found that unless E_0 is greater than 1.7×10^{20} V cm^{-1} which is 140 times the Born-Infeld value, the modification due to L_{BI} (2.8) would disrupt agreement between measured and calculated values for low-n transition ener-gies in muonic lead.

2.1.5 Wichmann-Kroll Approach to Strong-Field Electrodynamics

A common aspect of calculations of strong-field QED effects is the problem of finding a useful representation of the bound interaction (Furry) picture propagator $S_F^e(x_2,x_1)$ for a particle in a strong external potential $A_\mu(x)$. The approaches based on expand-ing $S_F^e(x_2,x_1)$ in powers of either the potential $A_\mu(x)$ or the field strength $\partial_\mu A_\nu(x) - \partial_\nu A_\mu(x)$ suffer from two main drawbacks. First, in the case of the self-energy ra-diative correction, the power series generated in this way converges slowly numer-ically. Second, for both the self-energy and the vacuum polarization, the expressions corresponding to successively higher order terms in the expansion become increas-ingly more complicated and difficult to evaluate.

In their classic study of the vacuum polarization in a strong Coulomb field, WICHMANN and KROLL [2.12] employed an alternative approach to the problem of finding a useful expression for the bound particle propagator. Their method and variations of it have been the basis for studies of strong-field QED effects, so we describe the method in some detail here. We also give a brief survey of calculations of strong-field QED effects based on these methods.

For a time-independent external potential, which we assume has only a nonvanish-ing fourth component $-eA_0(x) = V(\underline{x})$, $\underline{A}(x) = 0$, the bound electron propagation func-tion is

$$
S_F^e(x_2,x_1) = \begin{cases} \sum_{E_n > E_0} \phi_n(\underline{x}_2)\overline{\phi}_n(\underline{x}_1) \exp[-iE_n(t_2-t_1)] & t_2 > t_1 \\ -\sum_{E_n < E_0} \phi_n(\underline{x}_2)\overline{\phi}_n(\underline{x}_1) \exp[-iE_n(t_2-t_1)] & t_2 < t_1 \end{cases} \tag{2.10}
$$

where the $\phi_n(\underline{x})$ are the bound state and continuum solutions of the Dirac equation for the external potential. It has an integral representation given by [2.12,42]

$$S_F^e(x_2,x_1) = \frac{1}{2\pi i} \int_C dz\, G(\underline{x}_2,\underline{x}_1,z)\gamma^0\, e^{-iz(t_2-t_1)} \qquad (2.11)$$

where $G(\underline{x}_2,\underline{x}_1,z)$ is the Green's function for the Dirac equation

$$[-i\underline{\alpha}\cdot\underline{\nabla}_2+V(\underline{x}_2)+\beta m-z]\, G(\underline{x}_2,\underline{x}_1,z) = \delta^3(\underline{x}_2-\underline{x}_1) \qquad (2.12)$$

and the contour C in (2.11) extends continuously from $-\infty$ to $+\infty$ below the real axis for $Re\{z\}<E_0$, through E_0, and above the real axis in the region $Re\{z\}>E_0$. The crossing point E_0 depends on the definition of the vacuum (see Sec.2.3.1). For the Coulomb potential with $(Z\alpha)<1$, it is convenient to choose $E_0=0$. Two possible contours of integration for $(Z\alpha)>1$ are shown in Fig.2.5. In that figure, the branch points of $G(x_2,\underline{x}_1,z)$ at $z=\pm m$ and the bound state poles are also shown.

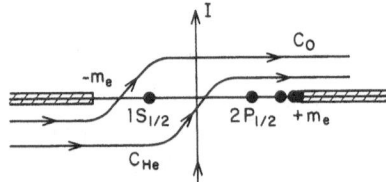

Fig. 2.5. Possible contours of integration in (2.11) for $Z\alpha>1$. The contour labelled C_{He} corresponds to a vacuum state with both 1S levels filled

The Green's function is formally given by the spectral representation

$$G(\underline{x}_2,\underline{x}_1,z) = \sum_E \frac{\phi_E(\underline{x}_2)\phi_E^\dagger(\underline{x}_1)}{E-z} \qquad (2.13)$$

where the sum in (2.13) is over bound state and continuum solutions as in (2.10).

For a spherically symmetric external potential $V(r)$, the Green's function may be written as a sum over eigenfunctions (with eigenvalue $-\kappa$) of the Dirac operator $K = \beta(\underline{\sigma}\cdot\underline{L}+1)$. Each term in the sum can be factorized into a part which depends in a trivial way on the directions of $\underline{x}_2$ and $\underline{x}_1$ and a radial Green's function which contains the nontrivial dependence on r_2 and r_1, the magnitudes of $\underline{x}_2$ and $\underline{x}_1$. The radial Green's function $G_\kappa(r_2,r_1,z)$, written as a 2×2 matrix, satisfies the inhomogeneous radial equation

$$
\begin{bmatrix} V(r_2) + m - z & -\frac{1}{r_2}\frac{d}{dr_2}\, r_2 + \frac{\kappa}{r_2} \\[2ex] \frac{1}{r_2}\frac{d}{dr_2}\, r_2 + \frac{\kappa}{r_2} & V(r_2) - m - z \end{bmatrix} G_\kappa(r_2,r_1,z) = \frac{1}{r_2 r_1}\delta(r_2-r_1) \qquad (2.14)
$$

The utility of this formulation is that the radial Green's functions G_κ can be constructed explicitly from solutions of the homogeneous version of (2.14). Let $A(r)$ and $B(r)$ be the two linearly independent two-component solutions of (2.14) with the right-hand side replaced by 0, where $A(r)$ is regular at $r=0$ and $B(r)$ is regular at $r=\infty$. Then for z in the cut plane (Fig.2.5) and not a bound state eigenvalue, the Green's function G_κ is given by

$$G_\kappa(r_2,r_1,z) = \frac{1}{J(z)}\left[\Theta(r_2-r_1)B(r_2)A^T(r_1)+\Theta(r_1-r_2)A(r_2)B^T(r_1)\right] \qquad (2.15)$$

with the Wronskian $J(z)$ given by [$J(z)$ is independent of r]

$$J(z) = r^2[A_2(r)B_1(r)-A_1(r)B_2(r)] \qquad . \qquad (2.16)$$

In (2.16), 1(2) denotes the upper (lower) component of A or B. Note that the radial Green's function can also be expressed in the form of a spectral representation, in analogy with (2.13), as

$$G_\kappa(r_2,r_1,z) = \sum_E \frac{F_E(r_2)F_E^T(r_1)}{E-z} \qquad (2.17)$$

where $F_E(r)$ is a bound state or continuum solution of the homogeneous radial equation.

In the case of a Coulomb potential, the solutions $A(r)$ and $B(r)$ can be expressed in terms of confluent hypergeometric (or Whittaker) functions [2.12,11]. WICHMANN and KROLL [2.12] employed integral representations for these functions, carried out some of the integrations involved in evaluation of the vacuum polarization, and arrived at relatively compact expressions for the Laplace transform of the vacuum polarization charge density times r^2. Their starting point was the expression for the unrenormalized vacuum polarization charge density of order e

$$\rho_{VP}(r) = -i\frac{em^3}{(2\pi)^2}\sum_{\kappa=1}^{\infty}\kappa\int_C dz\, Tr\{G_\kappa(r,r,z) + G_{-\kappa}(r,r,z)\} \qquad . \qquad (2.18)$$

This expression, which is valid to all orders in $Z\alpha$, may be further expanded in a power series in $Z\alpha$. Many of the calculations relevant to high-Z muonic atoms (see Sec.2.2) are based on results obtained by WICHMANN and KROLL in their extensive study of $\rho_{VP}(r)$.

ARAFUNE [2.43] and BROWN et al. [2.44,45] employed an approximation based on setting m=0 in the radial Green's function G_κ to study finite nuclear size effects on the vacuum polarization in muonic atoms. This approximation considerably simplifies the calculation and corresponds to including only the short-range effect of the vacuum polarization. GYULASSY [2.46-48] constructed Green's functions for a finite nucleus potential in a numerical study of the effect of finite size on the higher order vacuum polarization in muonic atoms and in electronic atoms with Z near the critical value (see Sec.2.3.8). In these studies, it was found that the main correction due to nuclear size arises from the $\kappa=1$ ($j=1/2$) term in (2.18). BROWN et al. [2.49,50] have constructed approximate analytic expressions for the radial Green's functions for a Coulomb potential to estimate the effect of the spatial distribution of the vacuum polarization charge density in muonic atoms.

BROWN et al. [2.51] have developed a method of calculating the 1S self-energy radiative correction for large Z, in which the solutions A(r) and B(r) are generated by numerical integration of a set of coupled differential equations. This method has been generalized to non-Coulomb potentials by DESIDERIO and JOHNSON [2.31] who evaluated the self-energy in a screened Coulomb potential for the 1S state with Z in the range 70 - 90. More recently, CHENG and JOHNSON [2.36] have evaluated the self--energy, with finite nuclear size and electron screening taken into account for Z in the range 70 to 160, and with a Coulomb potential for Z in the range 50 to 130.

MOHR [2.10,11] has evaluated the self-energy radiative correction for the 1S, 2S, and $2P_{1/2}$ states over the range Z=10-110 for a Coulomb potential. In that calculation, the radial Green's functions are evaluated numerically by taking advantage of power series and asymptotic expansions of the explicit expressions for the radial Green's functions in terms of confluent hypergeometric functions. In terms of the radial Green's functions, the (unrenormalized) self-energy has the form

$$\Delta E_{SE} = -\frac{i\alpha}{2\pi} \int_C dz \int_0^\infty dr_2 r_2^2 \int_0^\infty dr_1 r_1^2 \times \sum_\kappa \sum_{i,j=1}^2 \left[f_i(r_2) G_\kappa^{ij}(r_2,r_1,z) f_j(r_1) A_\kappa(r_2,r_1) \right.$$
$$\left. - f_{\bar{i}}(r_2) G_\kappa^{ij}(r_2,r_1,z) f_{\bar{j}}(r_1) A_\kappa^{ij}(r_2,r_1) \right] \tag{2.19}$$

where $\bar{i} = 3-i$, $\bar{j} = 3-j$; $f_i(r)$, i=1,2 are the large and small components of the Dirac radial wave functions, and the A's are functions associated with the angular momentum expansion of the photon propagator and consist of spherical Bessel and Hankel functions. In the numerical evaluation of (2.19), particular care is required in isolating the mass renormalization term [2.11].

2.2 The Electrodynamics of High-Z Muonic Atoms

2.2.1 General Features

Muons, impinging on a solid target, can become trapped in bound states in the target atoms [2.52]. Because the Bohr radius of a particle in a Coulomb potential scales as the inverse of the mass of the particle, the radii of the muon orbits are 1/207 times the radii of the corresponding electron orbits. Thus the muon and the nucleus form a small high-Z hydrogenlike system inside the atomic electron cloud. Observation of the transition x-rays of the muon yields the energy level spacings of the system. The lowest levels of the muon, which have radii comparable to the radius of the nucleus, are sensitive to properties of the nucleus such as charge distribution and polarization effects [2.52]. We are here concerned instead with higher circular orbits of the muon, such as the $4f_{7/2}$ and $5g_{9/2}$ states in lead atoms, which have the property

nuclear radius << muon Bohr radius << electron Bohr radius .

For these states, the effect of the structure of the nucleus and of the bound atomic electrons is small. Hence, precise theoretical predictions for the energy levels can be made and, in comparison with the experimental transition energies, provide a means of testing the effects of QED. In particular, the effect of electron vacuum polarization, which is large for muon levels, is tested to better than 1% with present-day experimental precision.

Experimental determination of the 3d→2p transition energy in muonic phosphorus by KOSLOV et al. [2.53] showed the effect of the lowest order vacuum polarization. More recently, with the use of lithium drifted germanium detectors to measure the x-ray energies, which are typically in the range 100 to 500 keV for the transitions considered here, experiments have become sufficiently accurate to be sensitive to higher order vacuum polarization effects [2.54-59]. The experiments of DIXIT et al. [2.55] and of WALTER et al. [2.56] reported in 1971-72 showed a significant discrepancy with theory; however, more recent experiments of TAUSCHER et al. [2.57], of DIXIT et al. [2.58], and of VUILLEUMIER et al. [2.59] reported in 1975-76 are in agreement with theory for the muonic transition energies. The accurate experiments, and particularly the apparent discrepancy with theory, led to a considerable amount of work on the theory of muonic energy levels. In the following discussion, we describe the present status of the theory, with attention focused on the well-studied transition $5g_{9/2} \rightarrow 4f_{7/2}$ in muonic [208]Pb. Numerical values for the various contributions to the energy levels are collected in Table 2.6 of Section 2.2.6.

The main contribution to the energy levels is the Dirac energy of a muon in a Coulomb potential. A small correction must be added to account for the finite charge radius of the nucleus. This can be calculated either by first order perturbation theory, or by numerical integration of the Dirac equation with a finite nuclear potential. The latter procedure is necessary for low n states where the finite size correction is large. For high n circular states, the correction is small and insensitive to the details of the nuclear charge distribution. For the $5g_{9/2} - 4f_{7/2}$ transition in lead, the correction is -4 eV compared to the reduced mass Coulomb energy

difference of 429,344 eV. The other small non-QED corrections from electron screening, and nuclear polarization and motion are discussed in subsequent sections.

   The largest correction to the Dirac Coulomb energy levels is the effect of electron vacuum polarization which is discussed in the following section. In the remainder of this section, we make some general remarks about the magnitude of the radiative corrections in muonic atoms.

   If we restrict our attention to interactions of photons with electrons and muons,
the QED corrections to the energy levels of a bound muon, to lowest order in α, are
given by the Feynman diagrams in Fig.2.6. In that figure, the double lines represent
electrons or muons in the static field of the nucleus. The diagrams (a), (b), and
(c) represent the muon self-energy, the muon vacuum polarization, and the electron
vacuum polarization, respectively.

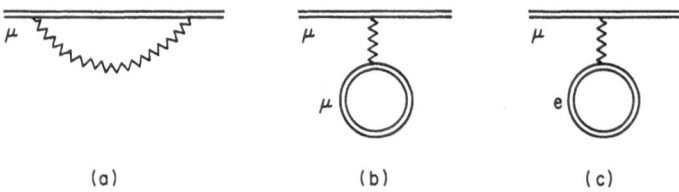

(a)   (b)   (c)

Fig. 2.6a-c. Lowest order QED corrections to the energy levels of a bound muon

   It is of interest to compare the QED corrections to muon levels to the corresponding corrections to electron levels. The lowest order diagrams for a bound electron are given by the diagrams in Fig.2.6 with the μ's and e's interchanged. For a
point nucleus, the electron diagrams corresponding to (a) and (b) give exactly the
same corrections, relative to the electron Dirac energy, as (a) and (b) give, relative to the muon Dirac energy. On the other hand, diagram (c) gives the large vacuum
polarization correction in muonic atoms, while its analogue with μ and e interchanged,
is negligible in electron atoms.

   The relatively greater effect of the electron vacuum polarization in muonic atoms
is due to the short-range nature of the vacuum polarization potential. The leading
(Uehling) term of the potential falls off exponentially in distance from the nucleus
with a characteristic length of $\lambda_e/2$. Hence, the overlap of the vacuum polarization
potential with the muon wave function, which has a radius of $0.2\lambda_e$ for the n=5 state
in lead, is much greater than the overlap of the potential with the electron wave
function, which has a radius of about $550\ \lambda_e$ in the n=2 state of hydrogen.

   The difference in scale between muon and electron atoms has another consequence.
The short-ranged muon wave function is sensitive to the short-range behavior of the
electron vacuum polarization potential, while the long-range electron wave function
is sensitive only to the zero and first radial moments of the potential. Hence, while

the hydrogen Lamb shift, with presently measured precision [2.60,61], tests the vacuum polarization to 0.1%, it is sensitive to a different aspect of the vacuum polarization than the muonic atom tests.

A further difference between muon and electron atoms is that the high-Z muonic atom measurements test higher order than Uehling potential contributions to the vacuum polarization, which are negligible in the hydrogen Lamb shift [2.12].

2.2.2 Vacuum Polarization

The electron vacuum polarization of lowest order in α and all orders in $Z\alpha$ is represented by the Feynman diagram in Fig.2.6c. For a stationary nuclear field corresponding to the charge density $\rho_N(r)$, the effect of the vacuum polarization is equivalent to the interaction of the bound muon with an induced charge distribution given by (-e is the charge of the electron) [2.12,62]

$$\rho_{VP}(\underline{r}) = <0|j_0(\underline{r},t)|0> = \frac{e}{2}\left[\sum_{E>0}|\phi_E(\underline{r})|^2 - \sum_{E<0}|\phi_E(\underline{r})|^2\right]$$

$$= \frac{e}{2\pi i}\int_C dz TrG(\underline{r},\underline{r}',z)\Big|_{\underline{r}'\to\underline{r}} \qquad (2.20)$$

[see also (2.18)], where G is the Green's function for the external field Dirac equation discussed in Section 2.1.5. (The vacuum $|0>$ is the state corresponding to no electrons or positrons in the external potential; $j_\mu(x) = -e/2[\bar{\Psi}(x),\gamma_\mu\Psi(x)]$ has a vanishing vacuum expectation value only in the limit $Z\alpha\to0$). The three expressions for the charge density in (2.20) are formal expressions and require regularization and charge renormalization in order to be well defined. A practical method of regularization is the Pauli-Villars scheme with two auxiliary masses [2.63]. The sum--over-states formula for the charge density in (2.20) is related to the last expression in (2.20) by choosing a suitable contour of integration C and evaluating the residue of the pole in the spectral representation, (2.13), of G [2.12].

To facilitate the evaluation of the charge density (2.20), it is convenient to expand it in powers of the external field. The Feynman diagrams corresponding to this expansion are shown in Fig.2.7. The x's in Fig.2.7 represent interaction with

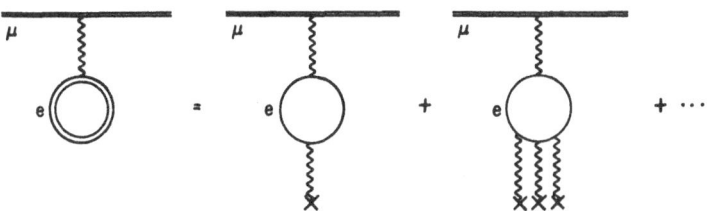

Fig. 2.7. Expansion of the vacuum polarization in powers of the external field

the external nuclear field. Only odd powers of the external field contribute due to FURRY's theorem [2.13]. The expansion in powers of the external field in Fig.2.7 corresponds to the Neumann series generated by iteration of the integral equation for the Green's function

$$G(\underline{r},\underline{r}',z) = G^0(\underline{r},\underline{r}',z) - \int d^3\underline{r}'' \, G^0(\underline{r},\underline{r}'',z) \, V(\underline{r}'') \, G(\underline{r}'',\underline{r}',z) \quad . \tag{2.21}$$

In (2.21), $G^0(\underline{r},\underline{r}',z)$ is the Green's function in the absence of an external potential and $V(\underline{r})$ is the potential energy of the electron in the nuclear field. The term in the expansion of $G(\underline{r},\underline{r}',z)$ linear in $V(\underline{r})$, when substituted for G in (2.20), gives, after charge renormalization, the charge density associated with the Uehling potential [2.15]

$$V_{11}^\rho(\underline{r}) = -\frac{\alpha e}{\pi} \int_1^\infty dt(t^2-1)^{1/2} \left(\frac{2}{3t^2}+\frac{1}{3t^4}\right) \int d^3\underline{r}' e^{-2m_e|\underline{r}-\underline{r}'|t} \frac{\rho_N(\underline{r}')}{|\underline{r}-\underline{r}'|} \quad . \tag{2.22}$$

In (2.22), the charge distribution of the nucleus is normalized such that $\int d^3\underline{r}\rho_N(\underline{r})=Ze$, and the subscripts on V refer to the order of the vacuum polarization, i.e., $V_{nm} = O[\alpha^n(Z\alpha)^m]$. The effect of V_{11}^ρ on a muon energy level is accurately taken into account by adding V_{11}^ρ to the external nuclear potential V in the Dirac equation

$$[\underline{\alpha} \cdot \underline{p} + V(\underline{r}) + V_{11}^\rho(\underline{r}) + \beta m_\mu - E_n]\phi_n(\underline{r}) = 0 \tag{2.23}$$

and solving for the bound state energy E_n numerically. This procedure is equivalent to summing over the higher order reducible contributions of the Uehling potential; Fig.2.8 shows the first three terms in this sum.

For the high-1 states under consideration here, the Uehling contribution is well approximated by the point charge value V_{11}, obtained by making the replacement $\rho_N(\underline{r}) \rightarrow Ze\delta^3(\underline{r})$ in the right-hand side of (2.22), evaluated in first order perturbation theory with Dirac wave functions for a point nucleus. Only the short distance behavior of the electron vacuum polarization is important ($m_e r \approx 0.2$ for r $\approx$ radius of the n=5 state in lead) [2.64-66]:

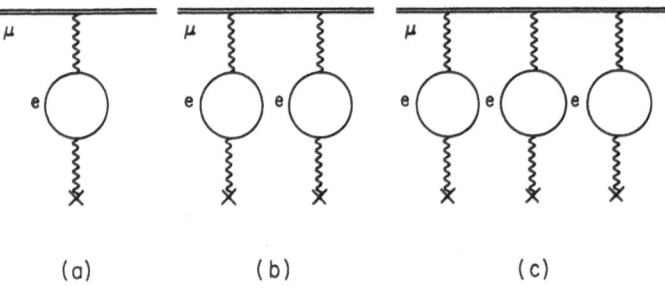

$\qquad$ (a) $\qquad\qquad$ (b) $\qquad\qquad\qquad$ (c)

<u>Fig. 2.8a-c.</u> Sum over all orders in perturbation theory for $V_{11}(r)$

$$V_{11}(r) = \frac{\alpha Z \alpha}{\pi} \left\{ \frac{2}{3r} \left[\ln(m_e r) + \gamma \right] + \frac{5}{9r} - \frac{\pi m_e}{2} + m_e^2 r - \frac{2\pi}{9} m_e^3 r^2 \right.$$

$$\left. + \frac{1}{6} m_e^4 r^3 \left[\ln(m_e r) + \gamma \right] + \frac{7}{18} m_e^4 r^3 + \dots \right\} \tag{2.24}$$

($\gamma = 0.57721 \dots$ is Euler's constant.) There are two non-negligible corrections to V_{11}. The first is the correction due to the finite extent of the nucleus. The small r form of the correction is [2.65]

$$\delta V_{11}(r) = V_{11}^0(r) - V_{11}(r) = \frac{\alpha Z \alpha}{\pi} \left(-\frac{1}{9r^3} <r^2> + \frac{1}{3} \frac{m_e^2}{r} <r^2> - \frac{1}{30r^5} <r^4> + \dots \right) \tag{2.25}$$

where the notation $<>$ denotes an average over the nuclear charge density. The other correction is the second order perturbation correction of the main term corresponding to the diagram in Fig.2.8b

$$\Delta E = \sum_{n \neq 0} <0|V_{11}|n> \frac{1}{E_0 - E_n} <n|V_{11}|0> \quad . \tag{2.26}$$

The energy shifts for the $5g_{9/2} - 4f_{7/2}$ transition arising from these corrections are listed separately in Table 2.6 [2.65].

We next consider the vacuum polarization of order α and third and higher order in $Z\alpha$, corresponding to diagrams with three or more x's in the series in Fig.2.7. The point nucleus approximation is considered first. WICHMANN and KROLL [2.12] obtained an explicit expression for the Laplace transform of r^2 times the vacuum polarization charge density of order $\alpha(Z\alpha)^3$. BLOMQVIST [2.65] has used their result to obtain the vacuum polarization potential $V_{13}(r)$ exactly in coordinate space and found the small r series expansion which is sufficient to evaluate the muon energy shifts

$$V_{13}(r) = \frac{\alpha(Z\alpha)^3}{\pi} \left\{ \left[-\frac{2}{3} \zeta(3) + \frac{1}{6} \pi^2 - \frac{7}{9} \right] \frac{1}{r} + \left[2\pi\zeta(3) - \frac{1}{4} \pi^3 \right] m_e \right.$$

$$+ \left[-6\zeta(3) + \frac{1}{16} \pi^4 + \frac{1}{6} \pi^2 \right] m_e^2 r + \frac{2\pi}{9} m_e^3 r^2 \left[\ln(m_e r) + \gamma \right]$$

$$\left. + \left[\frac{2}{3} \pi\zeta(3) + \frac{4}{9} \pi \ln 2 - \frac{31}{27} \pi \right] m_e^3 r^2 + \dots \right\} \quad . \tag{2.27}$$

This term contributes -43 eV to the $5g_{9/2} - 4f_{7/2}$ transition energy in muonic lead. VOGEL [2.67] has tabulated numerical values of $V_{13}(r)$ as a function of r based on BLOMQVIST's exact expression. Calculations by BELL [2.68] and by SUNDARESAN and WATSON [2.69], based on interpolation of the asymptotic forms of the Laplace transform of the third order vacuum polarization charge density given by WICHMANN and KROLL, are in agreement with the values obtained by BLOMQVIST. An earlier calculation by FRICKE [2.70] had the wrong sign for this term, which accounted for part of the apparent original discrepancy between theory and experiment (see Sec.2.2.1).

The vacuum polarization of order $\alpha(Z\alpha)^5$ and higher can be accounted for by considering the small distance behavior of the induced charge density. For a point nuclear charge density, the effect of the vacuum polarization of third and higher order is to produce a finite change δQ in the magnitude of the charge at the origin and a finite distribution of charge with a mean radius of approximately $0.86\bar{\lambda}_e$ [2.12]. The integral over all space of the induced charge density of order $(Z\alpha)^3$ and higher must, of course, vanish. The induced point charge, which gives rise to a leading term proportional to r^{-1} in the vacuum polarization potential, has the dominant effect on the muon energy. The magnitude of the induced charge was calculated by WICHMANN and KROLL [2.12] to all orders (≥ 3) in $Z\alpha$. Their result has been confirmed by an independent method by BROWN et al. [2.49,71]. WICHMANN and KROLL obtained this result as a special case in a general study of the vacuum polarization, while BROWN et al. were able to simplify the calculation by setting $m_e=0$ from the beginning. That this procedure produces the leading r^{-1} term in third order is seen by inspection of $V_{13}(r)$ in (2.27). The lowest order terms in δQ are given by

$$\delta Q = \frac{e}{3\pi}\left\{\left[2\zeta(3) + \frac{7}{3} - \frac{\pi^2}{2}\right](Z\alpha)^3 - \left[2\zeta(5) + \frac{71}{5}\zeta(3) - \frac{47\pi^4}{240}\right](Z\alpha)^5 + \ldots\right\} \quad .(2.28)$$

The numerical value of the charge to all orders in $Z\alpha$ is displayed by writing

$$\delta Q = -e[0.020940(Z\alpha)^3 + 0.007121(Z\alpha)^5 F_0(Z\alpha)] \qquad (2.29)$$

where $F_0(Z\alpha)$ appears in Fig.2.9. The leading terms of the fifth and seventh order vacuum polarization potential are [2.12,65]

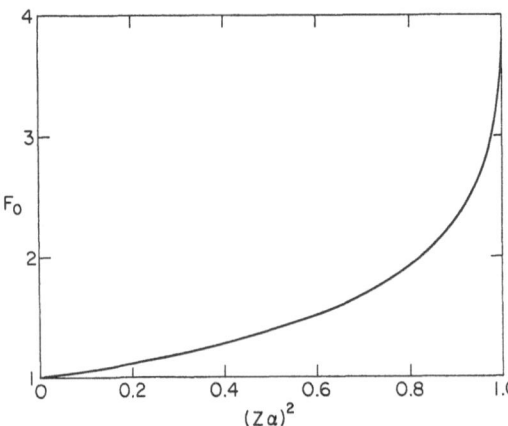

Fig. 2.9. The function $F_0(Z\alpha)$ which describes the charge induced at the nucleus by the higher order vacuum polarization in a Coulomb field (from BROWN et al. [2.49])

$$V_{15}(r) = \frac{\alpha(Z\alpha)^5}{\pi r} \left[\frac{2}{3} \varsigma(5) - \frac{21}{4} \varsigma(4) + \frac{71}{15} \varsigma(3) - \frac{1}{4} \varsigma^2(2) + 0(m_e r) \right]$$

$$V_{17}(r) = \frac{\alpha(Z\alpha)^7}{\pi r} \left[-\frac{2}{3} \varsigma(7) + \frac{445}{24} \varsigma(6) - \frac{286}{21} \varsigma(5) - \frac{1}{4} \varsigma(2)\varsigma(4) \right.$$

$$\left. - \frac{5}{2} \varsigma^2(3) + 0(m_e r) \right] \quad . \tag{2.30}$$

The fifth and seventh order leading terms contribute -7 eV to the $5g_{9/2}$-$4f_{7/2}$ energy separation in lead.

We briefly examine the contribution of terms of higher order in $m_e r$ to the fifth and higher order (in $Z\alpha$) vacuum polarization. The order α potential (excluding the r^{-1} term of the Uehling part) is given by [2.50]

$$V_{11+}(r) = A(Z\alpha) \frac{1}{r} + B(Z\alpha)m_e + C(Z\alpha)m_e(m_e r)^{2\lambda} + D(Z\alpha)m_e^2 r + \ldots \tag{2.31}$$

where $\lambda = [1 - (Z\alpha)^2]^{1/2}$, and the terms omitted from (2.31) are higher order in $m_e r$. The term $A(Z\alpha)r^{-1}$ corresponds to the induced point charge discussed earlier. $B(Z\alpha)$ has not been calculated in fifth or higher order in $Z\alpha$, but gives the same contribution for all states and therefore does not affect the transition energies. The coefficients $C(Z\alpha)$ and $D(Z\alpha)$ have been calculated numerically to all orders in $Z\alpha$ by BROWN et al. [2.50]. Their results show that the part of order fifth and higher in $(Z\alpha)$ in these terms [the third order parts are included in (2.27)] gives a small (of order 1 eV) contribution to the transition energy. A similar conclusion was reached by BELL [2.68].

A correction to the vacuum polarization of third and higher order must be made to account for the finite size of the nucleus. ARAFUNE [2.43] and BROWN et al. [2.44] have independently obtained approximate analytic expressions for the potential corresponding to the finite size correction to the vacuum polarization. ARAFUNE's expression [2.43]

$$\delta V_{11+}(r) \approx -\frac{\alpha(Z\alpha)}{15\pi r} \left[1 - \left(\frac{\pi^2}{3} - 3 - \frac{349}{13,860} \right)(Z\alpha)^2 - \frac{865}{2016}(Z\alpha)^2 \frac{R}{r} \right] \left(\frac{R}{r} \right)^{2\lambda}$$

$$\text{for} \quad r > R \tag{2.32}$$

where $\lambda = [1 - (Z\alpha)^2]^{1/2}$, is based on the following approximations: Terms of relative order $(m_e r)^2$ are neglected, terms of order R^4/r^5 are neglected, higher order terms in $(Z\alpha)^2$ are neglected except in the exponent, and the nucleus is approximated by a uniformly charged sphere of radius R. The effect of the potential inside the radius R is negligible for high-ℓ states. To isolate the contribution of (2.32) to the third and higher order vacuum polarization, it is necessary to subtract from (2.32) the term

$$- \frac{\alpha(Z\alpha)}{15\pi r} \left(\frac{R}{r}\right)^2 \qquad\qquad (2.33)$$

which corresponds to the Uehling potential portion and appears as the first term on the right-hand side of (2.25), $(R^2=5/3<r^2>)$.

BROWN et al. [2.44,45] have done a similar calculation. Their expression allows for an arbitrary nuclear charge distribution and is valid to all orders in $Z\alpha$. The results of these calculations are in excellent agreement and yield a correction of 5 eV for the $5g_{9/2}-4f_{7/2}$ transition in lead.

GYULASSY [2.46,48] has made a numerical study of the effect of finite nuclear size on the higher order vacuum polarization. He was able to calculate the finite size effect with or without the approximations of ARAFUNE and of BROWN et al. The finding was that the approximations introduce a small error of 1 eV, and the finite size correction is 6 eV compared to the 5 eV quoted above. GYULASSY also examined the extent to which the finite size corrections to the third order vacuum polariza- tion are sensitive to the shape of the nuclear charge distribution. The corrections were found to be essentially the same for a uniform spherical distribution and a shell of charge, provided the distributions have the same rms radius.

RINKER and WILETS have evaluated the higher order vacuum polarization correction by a direct numerical evaluation of the sum over eigenfunctions in (2.20). Their early work [2.72], which showed a 16±2 eV finite size correction to the higher order vacuum polarization, compared to 6 eV discussed above, is incorrect due to numerical difficulties [2.73]. More recently, with improved numerical methods, they have eva- luated the higher order vacuum polarization correction for many states and various values of Z in the range 26 to 114 [2.73]. The results in lead are consistent with the work described above.

The fourth order vacuum polarization, of order α^2, corresponding to the Feynman diagrams in Fig.2.10, has been calculated and expressed in momentum space in terms of an integral representation by KÄLLÉN and SABRY [2.74]. The configuration space potential $V_{21}(r)$ derived from the Källén-Sabry representation was obtained by BLOM- QVIST [2.65]. The complete expression for $V_{21}(r)$ is somewhat complicated, so it is

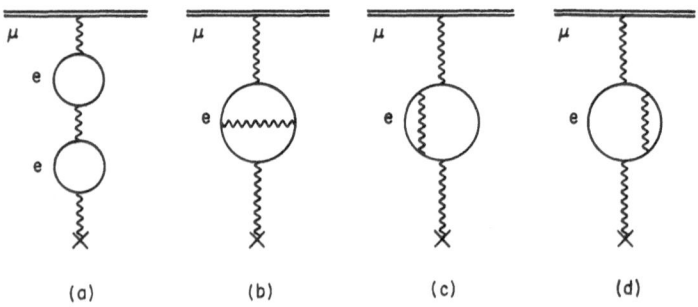

(a)   (b)   (c)   (d)

Fig. 2.10a-d. Fourth order vacuum polarization diagrams

convenient in calculations to employ the first terms in the power series expansion [2.65]

$$V_{21}(r) = \frac{\alpha^2 Z\alpha}{\pi^2} \left\{ - \frac{4}{9r} \left[\ln(m_e r) + \gamma\right]^2 - \frac{13}{54r} \left[\ln(m_e r) + \gamma\right] - \left[\zeta(3) + \frac{\pi^2}{27} + \frac{65}{648}\right] \frac{1}{r} \right.$$

$$\left. + \left(\frac{13}{9}\pi^2 + \frac{32}{9}\pi \ln 2 - \frac{766}{135}\pi\right) m_e + \frac{5}{3} m_e^2 r \left[\ln(m_e r) + \gamma\right] - \frac{65}{18} m_e^2 r + \ldots \right\} \ . $$
(2.34)

The power series represents $V_{21}(r)$ sufficiently well for values of r important for muonic orbits considered here to give accurate values for the energy shifts.

A numerical evaluation $V_{21}(r)$ has been made by VOGEL [2.67], who produced a table of point by point values. FULLERTON and RINKER [2.75] give a numerical approximation scheme to generate the second order potential for a finite sized nucleus based on VOGEL's tabulated values. Earlier estimates of this correction were made by FRICKE [2.70] and by SUNDARESAN and WATSON [2.69]; however, these calculations erroneously counted the diagram in Fig.2.10a twice.

2.2.3 Additional Radiative Corrections

According to the discussion of BARRETT et al. [2.64], it is expected that the self--energy correction to muon energy levels [Fig.2.6a] is reasonably well approximated by the terms of lowest order in $Z\alpha$ [2.26,76]:

$$\Delta E_{SE} = - \frac{4\alpha}{3\pi} \frac{(Z\alpha)^4}{n^3} \left[\ln K_0(n,\ell) + \frac{3}{8} \frac{1}{\kappa(2\ell+1)}\right] m_\mu \quad ; \quad \ell \neq 0 \qquad (2.35)$$

where K_0 is the Bethe average excitation energy, and the second term is due to the anomalous magnetic moment of the muon. For high-ℓ states, the point nucleus values of KLARSFELD and MAQUET [2.77] are used for K_0. This correction contributes -7 eV for the $5g_{9/2}-4f_{7/2}$ transition in lead.

A QED correction of order α^2 which has been the subject of recent interest is shown in Fig.2.11. In that figure, diagrams corresponding to the expansion of the electron loop in powers of the external potential are also shown. The first term in the expansion is the first vacuum polarization correction to the photon propagator. The next three terms correspond to a vacuum polarization correction of order $\alpha^2(Z\alpha)^2$ discussed in the following paragraph.

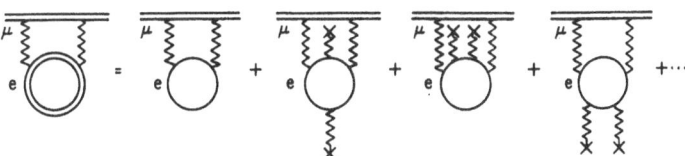

Fig. 2.11. Higher order radiative correction to muon levels

It was suggested by CHEN [2.78] that the contribution of this diagram was larger, relative to similar diagrams, than its nominal order would indicate. He estimated a value of -35 eV for the 5g-4f energy difference in lead. At the same time, WILETS and RINKER [2.73,79] estimated the effect and found that the 4f energy is shifted by an amount in the range 1-3 eV, in conflict with the result of CHEN. Subsequently, FUJIMOTO [2.80] estimated the $\alpha^2(Z\alpha)^2$ correction and found that the energy shift for the $5g_{9/2}-4f_{7/2}$ transition in lead is approximately 0.8 eV which is consistent with the value of WILETS and RINKER. FUJIMOTO simplified the calculation considerably by treating the muon as a static point charge and setting $m_e=0$ in the virtual electron loop. The latter approximation takes advantage of the fact that the distance between the muon and the nucleus is much less than the electron Compton wavelength. The result is then a vacuum polarization modification of the short-range interaction potential between two fixed point charges given by

$$\delta V(r) = - C \frac{\alpha^2(Z\alpha)^2}{r} \qquad (r<<\lambda_e) \tag{2.36}$$

where C=0.028(1). BORIE [2.81] has recently reported an approximate value of 1 eV for the correction.

Additional corrections to the muonic energy levels have been examined and found to be small. SUNDARESAN and WATSON [2.82] have estimated the contributions of hadronic intermediate states in the photon propagator, using a method due to ADLER [2.83]. BORIE has calculated various higher order QED contributions to the muonic atom energy levels, besides the $\alpha^2(Z\alpha)^2$ term just considered, and found them to be negligible compared to the experimental errors [2.84].

2.2.4 Nuclear Effects

Besides the effect of the finite nuclear charge radius which has already been discussed, the effects of nuclear motion and nuclear polarizability must be considered.

The main effect of nuclear motion is taken into account by replacing the muon mass by the reduced mass of the muon-nucleus system in the Dirac expression for the binding energy. This reduced mass correction is exact only in the nonrelativistic limit. The leading relativistic correction for nuclear motion is given by [2.26,85]

$$\delta E = - \frac{m_\mu^2(Z\alpha)^4}{8Mn^4} \tag{2.37}$$

where M is the nuclear mass. The reduced mass correction to the binding energy and relativistic correction contribute -234 eV and 3 eV, respectively, to the $5g_{9/2} - 4f_{7/2}$ transition in lead. The main effects of the nuclear motion are correctly taken into account by using reduced mass wave functions in evaluating the QED corrections, most importantly in the Uehling potential correction.

Up to this point, the nucleus has been treated as a charged object with no structure. There is a small correction to the muon energy levels due to the fact that
the muon can cause virtual excitations of the nucleus. This effect has been considered by COLE [2.86] and by ERICSON and HÜFNER [2.87] for the case of high-ℓ muon
states. The dominant long-range effect is the static dipole polarizability of the
nucleus. It can be roughly described as a separation of the center of charge from
the center of mass of the nucleus induced by the electric field of the muon. The
approximation that the displacement follows the motion of the muon is expected to
be good, because the nuclear frequencies are much higher than the relevant muon
atomic frequencies (5-20 MeV compared to a few hundred keV). The polarization in
this approximation corresponds to an effective potential $V_{E1}(r)$ given by [2.87]

$$V_{E1}(r) = -\alpha_{E1} \frac{e^2}{2r^4} \tag{2.38}$$

where α_{E1} is the static E1 polarizability of the nucleus. The value of α_{E1} can be
obtained from the measured total γ-absorption cross section $\sigma_{E1}(\omega)$ for E1 radiation
in the long wavelength limit by means of the sum rule

$$\alpha_{E1} = \frac{1}{2\pi^2} \int_{\omega_t}^{\infty} d\omega \, \frac{\sigma_{E1}(\omega)}{\omega^2} \quad . \tag{2.39}$$

The energy shifts have been calculated by BLOMQVIST [2.65] using the experimental
photonuclear cross section of HARVEY et al. [2.88] for ^{208}Pb. The result is 4 eV for
the $5g_{9/2}$-$4f_{7/2}$ energy difference, in agreement with COLE's value [2.86].

2.2.5 Electron Screening

In the preceding discussion, the effect of the atomic electrons has been completely
ignored. For the levels of the muonic atom under consideration, it is sufficiently
accurate, to within a few eV, to consider the energy shift of a muon in the potential
due to the charge distribution of the electron density of an atom with nuclear charge
Z-1.

The screening potential is well approximated by a function of the form

$$V_s(r) = V_0 - Cr^K e^{-\beta r} \quad . \tag{2.40}$$

The constant V_0 is relatively large and is approximately equal to the Thomas-Fermi
expression $V_s(0)=0.049 \, Z^{4/3}$ keV. Only the second term in (2.40) contributes to the
energy differences. VOGEL has calculated and tabulated Hartree-Fock-Slater electron
potentials and values for C, K, and β for which (2.40) approximates these potentials
to better than 5% for the range of r relevant to muonic orbits [2.67]. VOGEL finds
that screening contributes -83 eV to the $5g_{9/2}$-$4f_{7/2}$ transition in lead [2.89]. Cal-

culations have also been done by FRICKE [2.90] and by DIXIT (quoted in [2.55,91])
and are in agreement with VOGEL's results and earlier calculations in [2.64] to within a few eV. The approximation, employed in the preceding calculations, of using
the Slater approximation to the exchange potential has been checked by MANN and
RINKER [2.92] and is found to produce a small (1 to 2 eV) error. RAFELSKI et al.
[2.93] discuss the question of how to deal with screening and vacuum polarization
corrections in a consistent way.

A source of uncertainty in the screening calculations is the lack of knowledge
of the extent to which the muonic atom is ionized. During the early stages after
the muon is captured, it cascades in the atom partly by radiative transitions and
partly by Auger transitions. The screening corrections depend on how many electrons
have been ejected by Auger transitions of the muon. This problem has been considered
by VOGEL who finds that the effect of ionization is partly compensated by refilling
of the empty levels, and that the uncertainty in the muon levels is only 1 to 3 eV
[2.94].

2.2.6 Summary and Comparison With Experiment

Numerical values for the corrections described in the preceding sections are listed
in detail for muonic lead in Table 2.6. In Table 2.7 theoretical contributions to
the transition energies for measured transitions with Z in the range 56 to 82 are
listed. The sources of the values are as follows. The point nucleus energy differences are the Dirac values for the muon-nucleus reduced mass $m_\mu M/(m_\mu + M)$. The value
in eV is based on the recent determination of the ratio $m_\mu/m_e = 206.76927(17)$ deduced
from measurement of the muonium hyperfine interval by CASPERSON et al. [2.95] together
with $R_\infty h = 13.605804(36)$ eV recommended by COHEN and TAYLOR [2.96]. (There is a small
change of about 2 eV in the results for the muon energy levels if the value of m_μ/m_p
determined by CROWE et al. [2.97] is used). Numerical values for the contributions
in Table 2.7 are taken from Table 2 of the review by WATSON and SUNDARESAN [2.98]
with the following exceptions. The finite size correction to the higher order vacuum
polarization is evaluated by means of ARAFUNE's formula (with the Uehling term sub-
tracted) in (2.32) and is included in the column labeled $\alpha(Z\alpha)^{3+}$. The $\alpha^2(Z\alpha)^2$ term
is based on the results in [2.79-81]. The self-energy term includes an approximate
error estimate of 30% to account for higher order terms in $Z\alpha$ and finite nuclear
size effects [2.64].

Table 2.8 lists the most recent measurements of muonic x-rays for the transitions
being considered. The 1971-72 experiments show substantial disagreement with theory
whereas the 1975-76 experiments are generally in good agreement with theory, as is
easily seen in Fig.2.12. The apparent agreement of the latest results with theory
provides an impressive confirmation of strong field vacuum polarization effects in
QED.

Table 2.6. Summary of contributions to energy levels in muonic lead ^{208}Pb [eV]

Contribution	Order	$4f_{7/2}$	$5g_{9/2}$
Static external potential			
Dirac Coulomb energy[a]		-1188,314	-758,970
Finite nuclear size		4	0
Vacuum polarization of order α			
Coulomb Uehling potential[a]	$\alpha(Z\alpha)$	-3,652	-1,562
Finite nuclear size corr. to Uehling		-12	-3
Second order perturbation of Uehling		-9	-3
Third order in $Z\alpha$ Coulomb	$\alpha(Z\alpha)^3$	93	50
Fifth order in $Z\alpha$ (leading term)	$\alpha(Z\alpha)^5$	16	10
Seventh order in $Z\alpha$ (leading term)	$\alpha(Z\alpha)^7$	3	2
Finite size corr. to higher order in $Z\alpha$		-8	-3
Vacuum polarization of order α^2			
Coulomb Källén-Sabry potential	$\alpha^2(Z\alpha)$	-25	-11
Self-energy			
Bethe term	$\alpha(Z\alpha)^4$	1	0
Magnetic moment	$\alpha(Z\alpha)^4$	9	3
Other radiative corrections			
Virtual Delbrück diagram	$\alpha^2(Z\alpha)^2$	-1	0
Nuclear motion			
Relativistic reduced mass	$(Z\alpha)^4 m_\mu/M$	-4	-1
Nuclear polarization			
Dipole term	$\alpha(Z\alpha)^4 \alpha_{E1}$	-4	0
Atomic electrons			
Screening correction[b]		-89	-172
Total		-1191,992	-760,660
Transition energy $\cong$ 431,332 eV			

[a] Includes reduced mass correction $m_\mu \rightarrow M m_\mu/(m_\mu+M)$.
[b] Constant term V_0 is not included.

Table 2.7. Theoretical contributions to muonic atom energy separations [eV]

Transition	Pt.nucl.	Finite size	Vacuum polarization				Self-en. $\alpha(Z\alpha)^4$	Rel. rec.	Nuc. pol.	Elec. scr.	Total
			$\alpha(Z\alpha)$	$\alpha(Z\alpha)^{3+}$	$\alpha^2(Z\alpha)$	$\alpha^2(Z\alpha)^2$					
$_{56}$Ba											
$4f_{5/2}-3d_{3/2}$	439,069±1	-146±8	2436	-21±2	17	1	9±3	3	7	-18±1	441,357±9
$4f_{7/2}-3d_{5/2}$	431,654±1	-55±5	2328	-20±2	16	1	-8±2	3	7	-18±1	433,908±6
$5g_{7/2}-4f_{5/2}$	200,544±1	0	761	-9±1	5	0	2±1	1	0	-31±2	201,273±3
$5g_{9/2}-4f_{7/2}$	199,194±1	0	747	-9±1	5	0	-2±1	1	0	-31±2	199,905±3
$_{80}$Hg											
$5g_{7/2}-4f_{5/2}$	414,182±1	-8±1	2047	-42±2	14	1	7±2	2	3	-78±4	416,128±5
$5g_{9/2}-4f_{7/2}$	408,465±1	-2	1972	-40±2	14	1	-6±2	2	3	-79±4	410,330±5
$^{203}_{81}$Tl											
$5g_{7/2}-4f_{5/2}$	424,850±1	-9±1	2117	-44±2	15	1	7±2	2	4	-79±4	426,864±5
$5g_{9/2}-4f_{7/2}$	418,837±1	-3	2039	-43±2	14	1	-7±2	2	4	-81±4	420,763±5
$_{82}$Pb											
$5g_{7/2}-4f_{5/2}$	435,666±1	-10±1	2189	-46±2	15	1	7±2	2	4	-81±4	437,747±5
$5g_{9/2}-4f_{7/2}$	429,344±1	-4	2106	-45±2	15	1	-7±2	2	4	-83±4	431,333±5

Table 2.8. Recent measurements of muonic x-rays [eV], from [2.54-59]

	BACKENSTOSS et al. 1970	DIXIT et al. 1971	WALTER et al. 1972	TAUSCHER et al. 1975[a]	DIXIT et al. 1975[b]	VUILLEUMIER et al. 1976[b]
$_{56}$Ba						
$4f_{5/2}$-$3d_{3/2}$		441,299±21		441,366±13	441,371±12	
$4f_{7/2}$-$3d_{5/2}$		433,829±19		433,916±12	433,910±12	
$5g_{7/2}$-$4f_{5/2}$		201,260±16			201,282± 9	
$5g_{9/2}$-$4f_{7/2}$		199,902±15			199,915± 9	
$_{80}$Hg						
$5g_{7/2}$-$4f_{5/2}$			416,087±23			416,100±28
$5g_{9/2}$-$4f_{7/2}$			410,234±24			410,292±28
$^{203}_{81}$Tl						
$5g_{7/2}$-$4f_{5/2}$			426,828±23			426,851±29
$5g_{9/2}$-$4f_{7/2}$			420,717±23			420,741±29
$_{82}$Pb						
$5g_{7/2}$-$4f_{5/2}$	437,806±40	437,687±20		437,744±16	437,762±13	
$5g_{9/2}$-$4f_{7/2}$	431,410±40	431,285±17		431,353±14	431,341±11	

[a] The new ^{198}Au (412 keV) standard of DESLATTES et al. [2.99] would increase these values by about 10 eV.
[b] Based on the new Au standard.

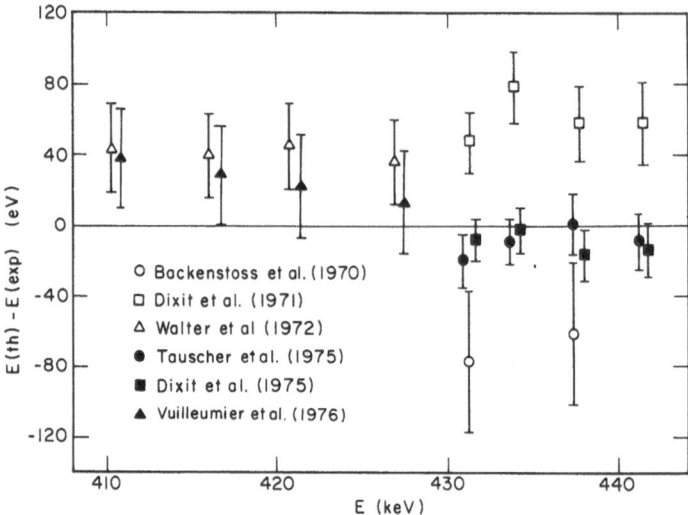

Fig. 2.12. Difference between theory and experiment for muonic atom x-rays plotted as a function of x-ray energy. Experimental values from [2.54-59]

2.2.7 Muonic Helium

Recently, the separation of the $2P_{3/2}$ and $2S_{1/2}$ energy levels in muonic helium $(\mu^4He)^+$ was measured by BERTIN et al. [2.100]. In that experiment, muons were stopped in helium and in some cases formed $(\mu^4He)^+$ in the metastable ($\tau=2\mu s$) 2S state. Transitions to the $2P_{3/2}$ state were induced by a tunable infrared pulsed dye laser, and monitored by observation of the 2P-1S 8.2 keV x-ray. A fit to the resonance curve yielded a line center corresponding to the transition energy

$$\Delta E(exp) = 1527.4(9) \text{ meV} \tag{2.41}$$

The theory of the muonic helium system provides an instructive contrast to the heavy muonic atoms. The relative importance of the various corrections is quite different in the two cases. For example, in muonic lead, the electron vacuum polarization of order $\alpha(Z\alpha)^3$ plays an important role, while it is negligible in muonic helium. On the other hand, the effect of finite nuclear size, which is a small correction to high-ℓ levels in muonic lead, is the major source of uncertainty in the theoretical value of the energy separation $2P_{3/2}-2S_{1/2}$ in muonic helium. In the following, we briefly summarize the contributions to the theoretical value of the $2P_{3/2}-2S_{1/2}$ splitting in $(\mu^4He)^+$. The numerical values are collected in Table 2.9.

The fine structure is qualitatively different from the fine structure of a one-electron atom; the vacuum polarization is the dominant effect in determining the muonic level spacings. The $2S_{1/2}$ level is lowered 1.7 eV by vacuum polarization compared to the Sommerfeld fine structure splitting of 0.1 eV. The finite nuclear size correction is the second largest effect and raises the $2S_{1/2}$ level by 0.3 eV.

Table 2.9. Theoretical contributions to the fine structure in muonic helium [meV]

Source	Lowest order	Value
Fine structure	$(Z\alpha)^4$	145.7
Finite nuclear size	$(Z\alpha)^4 \, m_\mu^2 <r^2>$	$-103.1 <r^2>fm^{-2}$
Electron vacuum polarization Uehling potential	$\alpha(Z\alpha)$	1666.1
Electron vacuum polarization Källén-Sabry term	$\alpha^2(Z\alpha)$	11.6
Self-energy and muon vacuum polarization	$\alpha(Z\alpha)^4 \ln(Z\alpha)^{-2}$	-10.7 ± 1.0
Nuclear polarization		3.1 ± 0.6
Total		1815.8 ± 1.2 $-103.1 <r^2>fm^{-2}$

The starting point for the theoretical contributions is the point nucleus fine structure formula

$$\Delta E_{FS} = \frac{1}{32} (Z\alpha)^4 \frac{m_\mu}{1+m_\mu/M} \left[1 + \frac{5}{8} (Z\alpha)^2 + \ldots \right] \tag{2.42}$$

where M is the nuclear mass. This must be corrected for the finite size of the nucleus. The nuclear charge radius is known only approximately from electron scattering experiments, so it is convenient to parameterize the size contribution to the fine structure in terms of the rms nuclear radius [2.101]

$$\Delta E_{NS} = -103.1 <r^2> meV/fm^2 \quad . \tag{2.43}$$

The value of the sum of the above corrections is in satisfactory numerical agreement with the more recent work of RINKER [2.102].

The largest radiative correction is the electron vacuum polarization of order $\alpha(Z\alpha)$. The value has been calculated by RINKER [2.102], who numerically solves the Dirac equation with a finite-nucleus vacuum polarization potential included (see Sec.2.2.2). The result appears in Table 2.9. The order $\alpha^2(Z\alpha)$ vacuum polarization was calculated by CAMPANI [2.103], by BORIE [2.101], and by RINKER [2.102]; all of the results are in accord.

The point nucleus value for the self-energy and muon vacuum polarization is given by [2.7]

$$\Delta E_{SE} + \Delta E_{VP} = - \frac{\alpha(Z\alpha)^4 m_\mu}{6\pi} \left\{ \frac{1}{(1+m_\mu/M)^3} \left[\frac{19}{30} + \ln(Z\alpha)^{-2} + \ln(1+m_\mu/M) - \ln \frac{K_0(2,0)}{K_0(2,1)} \right] \right.$$

$$\left. - \frac{1}{16} (1+m_\mu/M)^{-2} + 3\pi Z\alpha \left(\frac{427}{384} - \frac{1}{2}\ln 2 \right) + \ldots \right\} \quad . \tag{2.44}$$

The lowest order term may be partially corrected for finite nuclear size effects by replacing the wave function at the origin $|\Psi(0)|^2$ by the expectation value of the nuclear charge density $<\rho_N(\underline{r})>$, as has been done by RINKER [2.102]. An evaluation of the finite-nucleus average excitation energy K_0 would be necessary for a complete evaluation of the effect of finite nuclear size.

A further small correction arises from the effect of the finite nuclear size on the relativistic nuclear recoil terms. RINKER [2.102] estimates a value of 0.3 meV for this correction, using the prescription of FRIAR and NEGLE [2.104] for finite nuclei. This correction is nearly cancelled by the Salpeter recoil term from the noninstantaneous transverse photon exchange of order $(Z\alpha)^5 \, m_\mu^2/M$.

An important effect is nuclear polarization, which has been the subject of some controversy. The simple approximation used for high-ℓ states (see Sec.2.2.4) is not accurate for low-ℓ states in muonic helium. BERNABÉU and JARLSKOG [2.105] calculated a value of 3.1 meV for the nuclear polarizability contribution using photoabsorption cross section measurements as input data. On the other hand, HENLEY et al. [2.106] obtained a value of 7.0 meV based on a harmonic oscillator model for the nucleus. This value agrees with an earlier result of JOACHAIN [2.107]. However, in a subsequent analysis of the discrepancy, BERNABÉU and JARLSKOG [2.108] point out that the harmonic oscillator model predicts a value for the electric polarizability of the nucleus α_{E1} which is in substantial disagreement with the value deduced from existing measurements of the photoabsorption cross section [see (2.39)]. A subsequent calculation by RINKER [2.102] confirms the conclusions of BERNABÉU and JARLSKOG and also yields a value of 3.1 meV for the nuclear polarizability contribution.

The total theoretical value for the $2P_{3/2}$-$2S_{1/2}$ energy separation is given by (see Table 2.9)

$$\Delta E(th) = 1815.8 \pm 1.2 \text{ meV} - 103.1 \, <r^2> \text{ meV/fm}^2 \quad . \tag{2.45}$$

Using a weighted average of the results of electron scattering data for the ^{4}He charge radius ($<r^2>^{1/2}$=1.650±0.025 fm) [2.100] the theoretical energy separation is

$$\Delta E(th) = 1535(9) \text{ meV} \quad , \tag{2.46}$$

in agreement with the experimental result. On the other hand, assuming that the theory is correct, one can equate (2.45) and (2.41) to obtain a measured value for the charge radius

$$<r^2>^{1/2} = 1.673(4) \text{ fm} \quad . \tag{2.47}$$

2.2.8 Nonperturbative Vacuum Polarization Modification and Possible Scalar Particles

A possible deviation of QED from the ordinary perturbation theory predictions might be through a nonperturbative modification of the vacuum polarization. The corresponding change in the vacuum polarization potential would be of the form

$$\delta V(r) = -\frac{\alpha Z\alpha}{3\pi r} \int_{4m_e^2}^{\infty} dt \; t^{-1} \; \delta\rho(t) \; e^{-\sqrt{t} \; r} \tag{2.48}$$

where $\delta\rho(t)$ is a nonperturbative change in the vacuum-polarization spectral function. The change $\delta\rho(t)$ excludes the ordinary electron and muon vacuum polarization contributions of order α and α^2, but might be substantially larger than would normally be expected from perturbation theory terms of order α^3 and higher.

Phenomenological analyses of such a deviation have been given by ADLER [2.83], ADLER et al. [2.109], and BARBIERI [2.110] with particular emphasis on constraints on such a deviation imposed by various comparisons of theory and experiment. ADLER finds, with the technical assumption that $\delta\rho(t)$ increases monotonically with t, that if the vacuum polarization deviation is large enough to produce a change in the muonic atom transition energies of the magnitude of the difference between ordinary QED predictions for high-Z muonic atoms and the disagreeing 1971-72 experimental values, then a) the theoretical value of the muon magnetic moment anomaly $a_\mu = 1/2(g_\mu - 2)$ would be reduced by at least 96×10^{-9}, and b) there would be a reduction of order 27 meV in the theoretical value of the $2P_{3/2} - 2S_{1/2}$ transition energy in muonic helium. Prediction a) would introduce a 2σ difference between theory and experiment in the recent results for a_μ [2.111,112]:

$$a_\mu(\text{exp}) = 1165,895(27) \times 10^{-9}$$
$$a_\mu(\text{th}) \; = 1165,918(10) \times 10^{-9} \quad .$$

Prediction b) appears to be incompatible with the results for muonic helium discussed in Section 2.2.7. However, such modifications of vacuum polarization at a level $\sim$ 3 times smaller have not been ruled out.

A second proposed explanation for the 1971-72 discrepancy between muonic atom measurements and theory is the existence of a light, weakly-coupled scalar boson ϕ. Such particles are predicted by unified gauge theories of weak and electromagnetic interactions, but the mass is not determined. It was pointed out by JACKIW and WEINBERG [2.113] and by SUNDARESAN and WATSON [2.69] that if the mass of the ϕ meson were small enough, then its effect on muonic atom energy levels could account for the discrepancy. The coupling produced by a ϕ-exchange between a muon and a nucleus of mass number A is of the Yukawa form

$$V_\phi(r) = - \frac{g_{\phi\mu\bar{\mu}} g_{\phi N\bar{N}}}{4\pi} A \frac{e^{-M_\phi r}}{r} \tag{2.49}$$

where $g_{\phi\mu\bar{\mu}}$ and $g_{\phi N\bar{N}}$ are the ϕ-muon and ϕ-nucleon couplings, respectively, and M_ϕ is the mass of the ϕ. In gauge models, the ϕ-electron coupling is expected to be of order $(m_e/m_\mu)g_{\phi\mu\bar{\mu}}$ so the effect of such a potential could be observable in muon experiments without affecting the electron g_e-2 or Lamb shift experiments [2.83].

WATSON and SUNDARESAN [2.98] found that the values $g_{\phi\mu\bar{\mu}}g_{\phi N\bar{N}}/(4\pi)=-8\times10^{-7}$ and $M_\phi=12$ MeV would explain the early muonic atom discrepancy (the sign of the coupling is changed here according to ADLER [2.83]). ADLER [2.83] found a range of values for the coupling strength and ϕ mass which explains the discrepancy. However, ADLER [2.83] and BARBIERI [2.110] have shown that such a particle with $M_\phi>1$ MeV which could explain the discrepancy would also reduce the theoretical value for the muonic-helium fine structure ΔE $(2P_{3/2}-2S_{1/2})$ by approximately 27 meV. RESNICK et al. [2.114] pointed out that the effect of a ϕ-meson could be observed in a 0^+-0^+ nuclear decay in which the ϕ is emitted and subsequently decays into an e^+e^- pair. A search for e^+e^- pairs in the decays of the ^{16}O (6.05 MeV) and ^{4}He (20.2 MeV) 0^+ levels to corresponding 0^+ ground states was carried out by KOHLER et al. [2.115], who concluded from the negative results that the mass of the ϕ could not be in the range 1.030 to 18.2 MeV. ADLER et al. [2.109] argue that neutron-electron and electron-deuteron scattering data rules out the ϕ-meson explanation for M_ϕ in the range between 0 and 0.6 MeV.

The most serious constraint, however, was derived by BARBIERI and ERICSON [2.116] who show that low energy neutron-nucleus scattering data yields a limit $g_{\phi N\bar{N}}^2 M_\phi^{-4}/(4\pi)$ $\lesssim 3.4\times10^{-11}$ MeV^{-4}. The Weinberg-Salam theory predicts $g_{\phi\mu\bar{\mu}}^2/(4\pi)=G_F m_\mu^2/(\sqrt{2}2\pi)=1.3\times10^{-8}$; hence for $M_\phi=1$ MeV, for example, $|g_{\phi\mu\bar{\mu}}g_{\phi N\bar{N}}/(4\pi)|\lesssim 7\times10^{-10}$ which is orders of magnitude smaller than the value 1.4×10^{-7} [2.83] required to explain the early muonic atom discrepancy.

2.3 Quantum Electrodynamics in Heavy-Ion Collisions and Supercritical Fields

2.3.1 Electrodynamics for $Z\alpha>1$

One of the most fascinating topics in atomic physics and quantum electrodynamics is the question of what happens physically to a bound electron when the strength of the Coulomb potential increases beyond $Z\alpha=1$. This question involves properties of quantum electrodynamics which are presumably beyond the limits of validity of perturbation theory, so it is an area of fundamental interest. Although a completely rigorous field-theoretic formulation of this strong field problem has not been given, it is easy to understand in a qualitative way what happens physically: As $Z\alpha$ increases beyond a critical value, the discrete bound electron state becomes degenerate in energy with a three-particle continuum state (consisting of two bound electrons plus an outgoing positron wave), and a novel type of pair creation can occur [2.117, 118]. Remarkably, as first suggested by GERSHTEIN and ZELDOVICH [2.119], it may be

possible that such "autoionizing" positron production processes of strong field
quantum electrodynamics can be studied experimentally in heavy-ion collisions.

In addition to the spontaneous pair production phenomena, a number of other
questions of fundamental interest also become relevant at high $Z\alpha$:

a) What is the nature of vacuum polarization if a pair can be created without
the requirement of additional energy?

b) Do higher order radiative effects in α from vacuum polarization and self-
energy corrections significantly modify the predicted high-$Z\alpha$ phenomena?

c) How should the vacuum be defined if the gap in energy between the lowest bound
state and the negative continuum states approaches zero?

d) Can we test the nonlinear aspects of QED, e.g., as contained in the Euler-Hei-
senberg Lagrangian [2.120] and the Wichmann-Kroll calculation [2.12]? (The conventional tests of high-Z electrodynamics are discussed in Sec.2.1 and 2.2).

The high-$Z\alpha$ domain is also fascinating in that it provides a theoretical laboratory
for studying the interplay of single-particle Dirac theory and quantum field theory.
A speculative possibility is that it may be of considerable interest as a model for
strong binding and confinement of elementary particles in gauge theories. In the non-
-Abelian theories, such as "quantum chromodynamics" [2.121], the effective coupling
α_s between quarks could well be beyond the critical value. In addition, theoretical
work on the "psion" family of particles ($J/\Psi,\Psi'$, etc.) has focused on a fermion-anti-
fermion potential and various gauge theory models in the strong coupling regime
[2.122].

Perhaps the most practical way to create the strong fields necessary to test the
exotic predictions of high-$Z\alpha$ electrodynamics is in the slow collision of two ions
of high nuclear charge [2.119]. In addition to the spontaneous and induced pair phenomena, a number of interesting atomic physics questions arise concerning, among
other things, the atomic spectra and radiation of the effective high-Z quasi-molecule
momentarily present in the collisions. These topics are reviewed by MOKLER and
FOLKMANN [2.123] in this volume. The high magnetic field aspects are also of interest
(see Sec.2.3.11). Studies of the high-Z exotic phenomena ideally require highly
stripped ions; the physics of vacancy formation (see Sec.2.3.6) and recent experimental progress is discussed by MEYERHOF [2.124] and references therein.

Historically, the first discussions of the strong field problem were concerned
with the solutions of the Dirac equation for an electron in a Coulomb field,

$$[\underline{\alpha} \cdot \underline{p} + \beta m + V(r)]\Psi = E\Psi \quad ; \quad V(r) = -\frac{Z\alpha}{r} \quad . \tag{2.50}$$

This is, of course, a mathematical idealization for $r{\to}0$ since the nucleus has finite
mass and size. (In the case of positronium, V is effectively modified at small r by
vertex corrections and relativistic finite mass corrections implicit in the Bethe-
-Salpeter formalism. We should emphasize that the analysis of positronium for $\alpha>1$
remains an unsolved problem). The spectrum of the Dirac-Coulomb equation is given

by the Dirac-Sommerfeld fine-structure formula; the energy of the electron in the
1S state is

$$E = \sqrt{1-(Z\alpha)^2}\, m \quad . \tag{2.51}$$

E=0 appears to be the lower limit of the discrete spectrum as $Z\alpha \to 1$, and E is imaginary for $Z\alpha>1$. The Dirac Hamiltonian then is apparently not self-adjoint. Actually,
this result is just a mathematical problem associated with a pure Coulomb potential
[2.125-127]. The solutions are well defined when a nuclear finite size is introduced
[2.117,128-133].

Thus, we should consider the "realistic" potentials

$$V(r) = \begin{cases} -\dfrac{Z\alpha}{r} & r>R \\[2mm] -\dfrac{Z\alpha}{R}\, f(r/R) & r<R \end{cases} \tag{2.52}$$

where, for example, $f(\rho)=1/2(3-\rho^2)$ for the case of a uniform charge density. The
energy eigenvalue is then found by matching the solutions for the Dirac wave function at r=R. Early discussions of the bound state problem for $Z\alpha>1$ appear in [2.128-
130]; accurate extensive calculations were given after 1968 by PIEPER and GREINER
[2.117], by REIN [2.131], and by POPOV [2.133]. The energy spectrum for typical
nuclear radii, from [2.117], is shown in Fig.2.13. In Fig.2.14, POPOV's [2.132,133]
result for the dependence of $(Z\alpha)_{cr}$ (the value of $Z\alpha$ for which E=-m) on the nuclear
radius R is shown. It is clear that the "limit point" E=0 of the point nucleus case
is artificial: at sufficiently large $Z\alpha$, E reaches -m, the upper limit of the negative energy continuum. The critical Z for an extended superheavy nucleus with
$R=1.2\ A^{1/3}$ fm is $Z\approx170$, 185, and 245 for the $1S_{1/2}$, $2P_{1/2}$, and $2S_{1/2}$ levels, respectively [2.134]. The possibility of simulating such a nuclear state with heavy-ion
collisions is discussed in the next section.

It should be noted that the physical situation is already quite unusual if E<0,
let alone when E reaches the negative continuum. If $Z\gtrsim150$ and $E\lesssim0$, then the combined

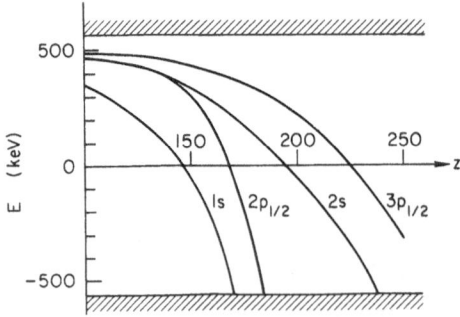

Fig. 2.13. Energy levels for the
Dirac-Coulomb equation as a function
of Z. A uniform charge density with
$R=1.2\ A^{1/3}$ fm is assumed (from PIE-
PER and GREINER [2.117])

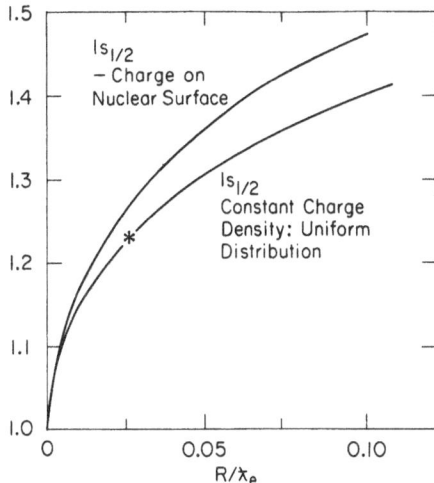

Fig. 2.14. Critical value of $Z\alpha$ in the Dirac-Coulomb equation for $E_{1S}=-m$ as a function of nuclear radius (in units of $\lambda_e=386$ fm) (from POPOV [2.133])

energy of the nucleus and one or two electrons bound in the 1S state is lower than the energy of the nucleus alone! Of course, since charge is conserved, an isolated nucleus of charge $Z\gtrsim150$ cannot "spontaneously decay" to this lower energy state.

However, the situation becomes more intriguing if Z can be increased beyond the critical value $Z_{cr}\sim170$ where E "dives" below -m (see Fig.2.13). In this case, the total energy of a state with a bound electron and an unbound slow positron (with $E_{positron}\sim m$)

$$E_{nucl} + E_{1S} + E_{positron} < E_{nucl} \qquad (2.53)$$

is less than that of the nucleus alone, and an isolated nucleus may decay to that state. In fact, for $Z\gtrsim170$, the nucleus will emit two positrons and fill both 1S levels. Clearly the physics is that of a multiparticle state and we must leave the confines of the single particle Dirac equation. However, in these first two sections we will ignore the higher order QED effects from electron self-energy corrections and vacuum polarization. (This can always be done mathematically - if we envision taking α small with $Z\alpha$ fixed [2.135].) We return to the question of radiative corrections in Section 2.3.8. In the remainder of this section we discuss a qualitative interpretation in terms of a new vacuum state. Quantitative results are discussed in the following sections.

The vacuum state, as originally interpreted by Dirac, is the state with all negative energy eigenstates of the wave equation occupied. Thus, for fermions

$$a_{(+)n}|0> = 0 \quad , \quad a^\dagger_{(-)n}|0> = b_{(+)n}|0> = 0 \qquad (2.54)$$

where $a_{(+)}$ $(a_{(-)})$ are the anticommuting annihilation operators for the positive (negative) energy single-particle states. The operators $b^{\dagger}_{(+)}=a_{(-)}$ can be interpreted as the positron creation (= negative energy electron annihilation) operators. Normally, the $E_n<0$ states are continuum eigenstates. Then, up to a constant, the total energy is

$$H = \sum_{E_n>0} a^{\dagger}_{(+)n} a_{(+)n} E_n + \sum_{E_n<0} b^{\dagger}_{(+)n} b_{(+)n} |E_n| \quad . \tag{2.55}$$

(This is just normal ordering the Hamiltonian - i.e., placing the annihilation operators to the right).

However, in the case of nuclear Coulomb potentials with $Z>Z_0\sim150$, at least one bound state solution of the Dirac equation has negative energy (see Fig.2.13). Thus it is evident that as soon as the field is strong enough to yield bound eigenstates of negative energy, one gains energy by filling these states. For example, imagine that there are two separated nuclei with charge Z and -Z, the latter made of antinucleons! If the charge of both nuclei were increased adiabatically beyond $Z=Z_0$ then there would be spontaneous decay of the nuclear system, to the state where two electrons in the 1S state are bound to the nucleus and two positrons are bound to the antinucleus.

Notice, incidentally, that charge conjugation symmetry is always preserved and one does not have "spontaneous symmetry breaking" in the vacuum decay. This is contrary to the claim of [2.134].

It is thus clear that when $Z>Z_0$, the state where the negative energy bound states are filled represents the natural choice as reference state for excitations [2.136-139]. Accordingly for $Z>Z_0$, we define the "new" Dirac vacuum [2.136]

$$|0_{new}> = a^{\dagger}_{1S}(\uparrow)\, a^{\dagger}_{1S}(\downarrow)\, |0_{old}>$$

$$\equiv b_{1S}(\downarrow)\, b_{1S}(\uparrow)\, |0_{old}> \quad , \tag{2.56}$$

i.e.,

$$|0_{old}> = b^{\dagger}_{1S}(\uparrow)\, b^{\dagger}_{1S}(\downarrow)\, |0_{new}> \quad , \tag{2.57}$$

where we suppose the spin up and down 1S states are the only bound states with negative energy. The charge of the new vacuum is $Q_{new}=Q_{old}-2$. Notice that the operator $b^{\dagger}_{1S}(\uparrow)\equiv a_{1S}(\downarrow)$ creates a hole with respect to the new vacuum, and thus effectively creates a bound positron state with positive energy $E_{pos}=|E_{1S}|$. The old vacuum appears as an excited state of the system; namely, two positrons are bound with positive total energy if $Z<Z_{cr}\sim170$. However, if Z is raised above Z_{cr}, the positrons become unbound. Thus from the standpoint of the new picture, the phenomenon of the instability of the (old) vacuum at $Z=Z_{cr}$ is reinterpreted by the statement that the positron wave function becomes unbound for this value of the charge (see Fig.2.15).

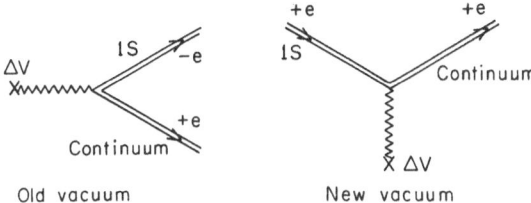

Fig. 2.15. Spontaneous positron production from the viewpoint of the old and new vacuum. The old vacuum is unstable due to the adiabatic introduction of $\Delta V \propto Z - Z_{cr}$

The bound negative energy one-electron state may be written

$$a_{1S}^{\dagger}(\uparrow)\ |0_{old}> \equiv a_{1S}^{\dagger}(\uparrow)\ b_{1S}^{\dagger}(\uparrow)\ b_{1S}^{\dagger}(\downarrow)\ |0_{new}>$$

$$= b_{1S}(\downarrow)\ b_{1S}^{\dagger}(\uparrow)\ b_{1S}^{\dagger}(\downarrow)\ |0_{new}> = -b_{1S}^{\dagger}(\uparrow)\ |0_{new}> \quad , \qquad (2.58)$$

i.e., it is equivalent (with respect to the new vacuum) to a bound positron for $Z<Z_{cr}$ and a continuum positron for $Z>Z_{cr}$. The effective potential $U_{eff}(r)$ in the relativistic Coulomb problem (see Sec.2.3.4) behaves like $-EZ\alpha/r$ as $r\to\infty$, i.e., is attractive for $E>0$ and repulsive for $E<0$ (at sufficiently large distances from the nucleus) [2.133,138]. Thus, as shown by ZELDOVICH and POPOV [2.138], the bound positron moves in a non-monotonic effective potential which becomes shallower as Z increases (see Fig.2.16), until at $Z=Z_{cr}$ it becomes unbound. A plot of the average radius for the bound states as a function of E as computed by POPOV [2.133] is shown in Fig.2.17. The process involved in spontaneous positron production for $Z>Z_{cr}$ is then simply

$$b_{1S}^{\dagger}(\uparrow)\ |0_{new}> \to b_{kS}^{\dagger}(\uparrow)\ |0_{new}> \quad . \qquad (2.59)$$

Formally, from (2.56-58), this is equivalent, in terms of the conventional vacuum, to

$$a_{1S}^{\dagger}(\uparrow)\ |0_{old}> \to b_{kS}^{\dagger}(\uparrow)\ a_{1S}^{\dagger}(\downarrow)\ a_{1S}^{\dagger}(\uparrow)\ |0_{old}> \quad , \qquad (2.60)$$

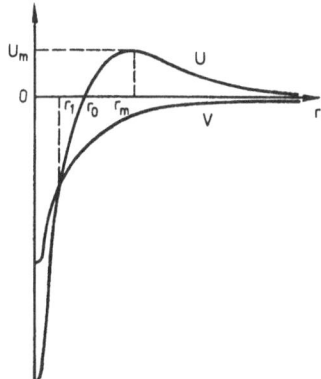

Fig. 2.16. The effective potential $U_{eff}(r)$ of the effective Schrödinger equation (2.71) for $Z\to Z_{cr}$, $E\to -m$. The potential $V(r)$ is the potential in the Dirac equation (from ZELDOVICH and POPOV [2.138])

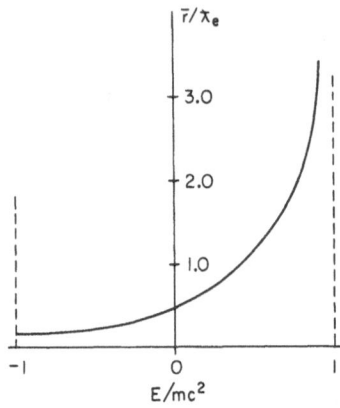

Fig. 2.17. The mean radius $\bar{r}$ of the ground state as a function of its energy E. The radius contracts to $\bar{r}=0.13 \, \lambda_e$ at E=-m (from POPOV [2.133])

corresponding to the degeneracy of the discrete bound electron state with a three-particle continuum. The old vacuum is, however, inappropriate for the description of the system for $Z>Z_{cr}$ simply because it is unstable. The description of high-Z electrodynamics in terms of the new vacuum thus has the advantage of displaying the continuity of the physics at $Z=Z_{cr}$. As we discuss in Section 2.3.8, the vacuum polarization problem is also clarified. The formal aspects of positron autoionization are discussed in more detail in [2.139].

2.3.2 Spontaneous Pair Production in Heavy-Ion Collisions

It would be very interesting if the physical realization of an electron bound to a strong field with Z greater than $Z_{cr} \sim 170$ could be attained experimentally. It is not excluded that nuclei with $Z \sim Z_{cr}$ will eventually be synthesized, but at present this possibility seems remote. Suggestions of positron production in overcritical nuclei were discussed in 1969 by PIEPER and GREINER [2.117] and by GERSHTEIN and ZELDOVICH [2.118]. In the same year, GERSHTEIN and ZELDOVICH [2.119] proposed that the critical field condition could be attained in the close approach of two heavy ions with $Z_1+Z_2>Z_{cr}$. If the velocity of collision is assumed to be sufficient to approach the Coulomb barrier, then, at least in the adiabatic approximation ($V_{ion}/V_{electron} \lesssim 1/10$), a ground state electron sees an effectively coalesced nuclear potential. This pioneering paper by GERSHTEIN and ZELDOVICH and the early papers of POPOV [2.132,133,140] and GREINER et al. [2.40, 141-145] contain many of the fundamental physical ideas which have been subsequently discussed in more quantitative fashion over the past seven years [2.134,139,141,144-154].

In collisions of heavy ions with several-MeV/nucleon kinetic energy, the typical collision time of the ions inside the K-shell of an electron is $T_c \sim 10^{-19}$ s whereas the "orbiting" time of the electron is $T_e \sim 10^{-20}$ s. Thus, roughly speaking, the molecular electronic states have time to adjust to the varying distance $\underline{R}=\underline{R}_2-\underline{R}_1$ between the nuclei. One can then consider an approximate adiabatic treatment of the two--center Dirac equation

$$[\underline{\alpha} \cdot \underline{p} + \beta m + V(\underline{r}-\underline{R}_1) + V(\underline{r}-\underline{R}_2)]\Psi = E\Psi \qquad (2.61)$$

(assuming one electron is present). An extensive discussion of this problem and numerical solutions for the molecular spectra of "intermediate super-heavy molecules" are given in [2.123] and [2.134].

Let us suppose that only one ground state electron is present. For $Z_1+Z_2>Z_{cr}$, there will be a critical distance R_{cr} between the two nuclei for which the electron is bound with an energy -m. Then as the ions collide with $R<R_{cr}$, the lowest one-electron state becomes mixed with the $|e^-e^-e^+>$ continuum level (spontaneous pair production); see Fig.2.18. As the ions recede, we are left with two electrons in the 1S level plus an outgoing positron. Note that double pair production with two outgoing positrons can occur if no ground state electron is initially present. Pair production, however, is suppressed by the Pauli principle if the 1S levels are full, so preionization or stripping is necessary. The energy for the spontaneous pair production is compensated by a decrease in the kinetic energy of the outgoing nuclei. Additional pairs can in principle be produced when the 2P levels in turn reach their critical energy $E_{2P}=-m$. However, according to the calculations of RAFELSKI et al. [2.139] for $_{92}U-_{92}U$ molecular orbitals, the $2P_{1/2}$ level reaches the negative energy continuum at a distance R which may be too close to the Coulomb barrier for experiments to be feasible (see Fig.2.18).

In a series of comprehensive papers, POPOV, GREINER and others have presented detailed analytic and numerical calculations of the spontaneous positron production process in heavy-ion collisions. We shall review the main points of this work which are particularly relevant to practical experiments and refer the reader to the original papers for more details.

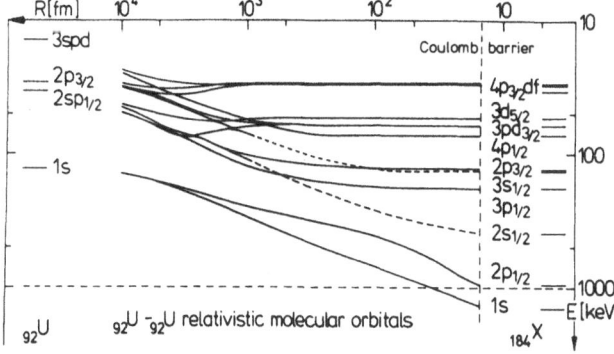

Fig. 2.18. The combined atom relativistic molecular states for $_{92}U-_{92}U$ collisions as a function of the internuclear distance R (from MÜLLER et al. [2.134]). The lowest state reaches the negative continuum at $R_{cr}\sim34$ fm

For simplicity, we consider a beam of completely stripped nuclei Z_1 incident on an ordinary target with nuclear charge Z_2. (The nonstripped case will be discussed in Sec.2.3.6). If $Z_1 > Z_2$, then the K-shell of the combined atom will generally be vacant as a result of the behavior of the molecular terms in the adiabatic approach of the nuclei [2.155]. In fact, the cross section for positron production turns out to be only slightly smaller than in the idealized case of the collision of two bare nuclei (see [2.156]). For definiteness we will usually consider U-U collisions, for which the combined Coulomb field $Z = Z_1 + Z_2 = 184$ is beyond the critical charge $Z_{cr} \cong 170$ necessary for spontaneous positron production.

As shown by MÜLLER et al. [2.157] (see Sec.2.3.3), the critical distance for positron production in U-U collisions (where the energy of the two-center atom reaches the negative continuum) is $R_{cr} \cong 34$ fm=0.088 λ_e. The calculation of R_{cr} can be carried out precisely and requires an analysis of the two-center Dirac problem; we return to this in Section 2.3.3. Thus the lab kinetic energy of the beam at the positron production threshold is

$$E_T = \frac{1}{2} m v_L^2 = \frac{2 Z_1 Z_2 \alpha}{R_{cr}} = 717 \text{ MeV} \tag{2.62}$$

or $\sim$ 3 MeV/nucleon for U-U collisions. The lab velocity is $v_L = 0.08$ which is clearly in the adiabatic nonrelativistic domain, and small compared to the velocity of a K-shell electron.

The classical orbits and Rutherford cross section of nonrelativistic charged particles are, of course, well understood. Detailed discussions and kinematics are given in [2.145,158,159]. It is convenient to define $\eta = E_{LAB}/E_T$ which is also related to the distance of closest approach R_0 for backward scattering of the nuclei: $\eta = R_{cr}/R_0$. Spontaneous positron production is possible only if $\eta > 1$. If θ is the CM angle of the scattering then the requirement that the nuclei are sufficiently close, $R_{min} < R_{cr}$, where R_{min} is the distance of closest approach, is

$$\eta = \frac{E_{LAB}}{E_T} > 1/2 (1 + \operatorname{cosec} \frac{\theta}{2}) = \eta_{min}(\theta) \quad . \tag{2.63}$$

The variation of the positron production cross section as a function of η and θ gives a simple tool for testing the positron production calculations.

For nuclei approaching each other, the two-center Dirac equation can be solved in the adiabatic approximation with nuclear separation $R=R(t)$, assuming $v \ll c$. Each value $R < R_{cr}$ then gives a corresponding (complex) energy level near the lower continuum: $E = E_0 + i \Gamma/2$ [2.153]. (The unusual sign of the imaginary part is discussed in Sec.2.3.4). The real part of the energy level is identified as the produced positron's kinetic energy $T = |E_0| - m$; the positron production rate is $dw/dt = \Gamma$. Integration of dw/dt over the ion's Coulomb orbit then gives the probability of positron production in the collision. Note that here we consider only spontaneous production in the adiabatic approximation; induced production will be discussed in Section 2.3.5.

Clearly the maximum cross section for e^+ production is the geometric limit [2.151], $\sigma_{geom} = \pi R_{cr}^2 (1-\eta^{-1}) = 36b(1-\eta^{-1})$ for U-U collisions, which would be obtained if the Coulomb field of the nuclei succeeded in producing a positron in each collision for which $R_{min} < R_{cr}$. (The corresponding impact parameter at $R_{min} = R_{cr}$ is $\rho = R_{cr}(1-\eta^{-1})^{1/2}$ and $\sigma_{geom} = \pi \rho^2$). In fact, the actual cross section calculated by MARINOV and POPOV [2.159] is exponentially damped at threshold ($\eta \to 1$), rising to a fraction of 0.1% of σ_{geom} for $\eta \gtrsim 2$ and increasing slowly thereafter. The ratio $W_{av} = \sigma/\sigma_{geom}$ averaged over the positron spectrum is shown in Fig.2.19. The maximum energy such that there is no nuclear interaction in the collision is determined by (r_N is the nuclear radius) [2.159]

$$\eta_{max} \sim \frac{R_{cr}}{2r_N + \Delta r} \sim 2.8 \quad , \tag{2.64}$$

allowing $\Delta r \sim 5$ fm for the diffuseness of the nuclear boundary. The background process of e^+e^- production due to Coulomb excitation of the nucleus is discussed in Section 2.3.7.

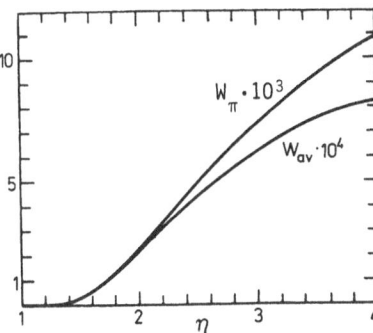

Fig. 2.19. The probability of spontaneous positron production for scattering of uranium nuclei at 180^0 (W_π) and the probability averaged over all angles ($W_{av} = \sigma/\sigma_{geom}$) (from MARINOV and POPOV [2.159]). These curves should be multiplied by the factor 0.54 for $R_{cr} = 34$ fm (see Sec.2.3.3)

Hence for U-U collisions, with $\eta \sim 2$, $E_{LAB} \sim 1.4$ GeV, $v_L \sim 0.1$, we have $\sigma_{geom} \sim 18b$ and $\sigma \sim 2$ mb, i.e., spontaneous positron production occurs roughly in one out of nine thousand nuclear collisions in which the distance $R < R_{cr}$ is reached and the 1S level is unoccupied. A recent semiclassical calculation by JAKUBASSA and KLÉBER [2.160] based on a one-center analysis yields a spontaneous production cross section of similar value.

The ratio of the differential cross section with positron production to the Rutherford cross section $(d\sigma/d\Omega)_R = R_{cr}^2/(4\eta)^2 \sin^4 \frac{\Theta}{2}$ at $\Theta = \pi$ (backward scattering) is given by an integral over the Coulomb path [2.159]

$$W_\pi(\eta) = \frac{d\sigma/d\Omega}{(d\sigma/d\Omega)_R} \bigg|_{\Theta=\pi} = C \int_{\frac{1}{\eta}}^{1} dx \left(\frac{x}{\eta x - 1}\right)^{1/2} \Gamma(x) \tag{2.65}$$

where the positron width is written as a function of $x=R/R_{cr}$. Since $r_N<<(R,R_{cr})<<R_K$ (R_K is the electronic K-shell radius for $Z\alpha\sim1$), the energy E_0 and width Γ are insensitive to the detailed two-center situation; as shown by POPOV [2.153], these quantities are dependent only on the ratio R/R_{cr}. Thus the only critical parameter in the two-center problem is the actual value of R_{cr}. The constant C in (2.65) is proportional to $R_{cr}^{3/2}$ and $[(Z_1+Z_2)\alpha]^{-1/2}$ and is equal to 2.3 for U-U collisions. The result of a numerical calculation for W_π is also shown in Fig.2.19. The corresponding positron spectrum w(T), proportional to the integrand of (2.65), normalized to $\int_0^{T_{max}} dT\ w(T)=1$, is shown in Fig.2.20 for $\Theta_{cm}=\pi$. The maximum positron kinetic energy is

$$T_{max}(\eta,\Theta) = |E_0(R_{min}/R_{cr})| - m \quad . \tag{2.66}$$

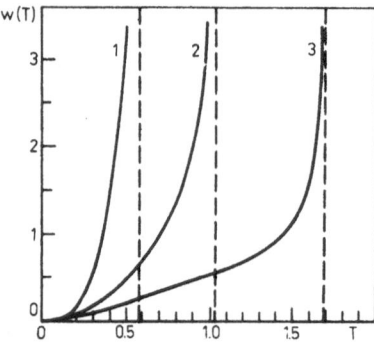

Fig. 2.20. The energy spectrum of spontaneously produced positrons for backward ion scattering ($\Theta=180°$). The curves labelled 1,2,3 refer to $\eta=2$, 2.8, and 4, respectively (from MARINOV and POPOV [2.159])

The distribution in T peaks sharply at T_{max} since the ion spends the most time at the point of closest approach, and $\Gamma(R/R_{cr})$ is largest there. Near the maximum energy, $w(T)\sim(T_{max}-T)^{-1/2}$. For $\eta\sim2$, approximately 60% of the positrons have an energy above $0.9\ T_{max}$. The square-root singularity is due to the fact that the radial velocity vanishes at $R=R_{min}$. The strong peaking and maximal effect at $\Theta=\pi$ are clearly favorable for experiments. For $\Theta=\pi/2$, w is smaller by an order of magnitude. If the angle of the scattered ion is not measured, the peak in w(T) is considerably washed out. The small value for the probability of single positron production $W_{(1)}=W_{av}\lesssim10^{-3}$ means that the double spontaneous positron production probability $W_{(2)}$ (filling both 1S levels with electrons) is only of order $W_{(1)}^2\lesssim10^{-6}$, and is probably not a useful signal for spontaneous production.

2.3.3 Calculation of the Critical Internuclear Distance

An important numerical parameter for phenomena involving a supercritical Coulomb field in heavy-ion collisions is the value of the internuclear distance $R=R_{cr}$ at which the energy of the ground state of the quasi-molecule (Z_1,Z_2,e) crosses the boundary of the lower continuum. For low-velocity collisions with $Z_1=Z_2=Z$, it is sufficient to calculate the energy of the two-center Dirac equation with the potential $V(\underline{r})=-Z\alpha/r_1-Z\alpha/r_2$, where $r_i=|\underline{r}-\underline{R}_i|$ (adiabatic approximation). Since R_{cr} is substantially greater than the nucleon size, the nucleon finite size effects can be estimated from perturbation theory [2.156]. MARINOV and POPOV [2.156] and MARINOV et al. [2.161] calculate R_{cr} using a variational method in which each component of the trial wave function of the two-center Dirac equation is written as a sum of terms with the correct singularity behavior at large and small distances, e.g., near the nuclei the Dirac wave function has the singularity

$$\Psi(\underline{r}) \sim (\xi^2-\eta^2)^{-1+\sqrt{1-(Z\alpha)^2}} \tag{2.67}$$

where $\xi=(r_1+r_2)/R$, $\eta=(r_1-r_2)/R$ are elliptical coordinates. This method converges quickly with just a few terms. As shown by MARINOV et al. [2.156,161], the variational method gives a lower limit on R_{cr}.

A similar variational calculation was also performed for U-U collisions by MÜLLER et al. [2.157,162], giving the result $R_{cr}\sim36$ fm compared to the lower limit, $R_{cr}\gtrsim38$ fm, obtained by MARINOV et al. [2.163]. (It should be noted that the published numerical results beyond the (1,0) approximation given in [2.156,161] need to be revised because of a recently discovered calculational error [2.163].) The small difference between the MÜLLER et al. and MARINOV et al. results could be accounted for by the absence of the relativistic Coulomb singularity of (2.67) in the 100-term MÜLLER et al. trial wave function.

2.3.4 Calculation of the Spontaneous Positron Production Rate

An exact calculation of spontaneous positron production by two colliding nuclei would be extremely difficult. The calculations which have been done for the two-center problem have employed a variety of approximations, which should be carefully considered. In general terms, the analytic analysis of POPOV and co-workers is based on two separated point nuclei where the combined Z is above the threshold $Z_{cr}\sim170$ for positron production. An essential feature of the analysis of GREINER and co-workers is the assumption that the functional dependence of the decay width Γ on the positron energy E in the two-center problem is the same as the dependence in the one-center problem for a finite-size nucleus. A recent semi-classical calculation by JAKUBASSA and KLEBER [2.160] assumes that the width Γ for the two-body problem may be simulated by the width for a one-center atom with an effective nuclear charge Z(R) which depends on the nuclear separation. The GREINER et al. and JAKUBASSA et al. analyses both depend on the nuclear size in the single-center problem which of course cannot accurately represent the effective size of the two-nucleus system.

In the work of POPOV and co-workers, as discussed in Section 2.3.2, the positron production cross section is dependent on the imaginary part Γ of the energy of the level $E=E_0+i\,\Gamma/2$ which for nuclear separation $R<R_{cr}$ is in the lower continuum. In principle the full complexity of the adiabatic two-center problem is required to determine E for the quasi-molecule. However, as shown by POPOV [2.133,146,149,151, 153], E_0 and Γ depend, to a good approximation, only on R/R_{cr}, and their values can be found by matching the "inner" solution with the singularity given in (2.67) to the "outer" solution for the one-center problem: $V(\underline{r})=-(Z_1+Z_2)\alpha/r$. The result is the relation (valid for R_{cr} small compared to $\bar{\lambda}_e$)

$$\ln \frac{R_{cr}}{R} = \Psi\left(-\frac{E}{\lambda}\right) + \ln \frac{\lambda}{m} + \frac{m+E}{m+E+\lambda} \qquad (2.68)$$

where $\lambda=-i(E^2-m^2)^{1/2}$ for $R<R_{cr}$ and $\Psi(z)$ is the derivative of the logarithm of the gamma function. Eq.(2.68) can then be solved numerically for the real and imaginary parts of E. These results are used for the calculation of the positron production cross section and spectrum given in Section 2.3.2.

It is interesting to note that the solution to (2.68) gives a positive imaginary part to E, rather than the familiar position of the pole on the second sheet for a decaying state. This is due to the fact that we are dealing with the solutions of the Dirac equation in non-second-quantized form. The quasi-discrete level $E=E_0+i\Gamma/2$ corresponds physically to a resonance for positron-scattering on a supercritical nucleus at the energy $E_{pos}=-E=|E_0|-i\Gamma/2$. This has the correct sign for the imaginary part [2.153].

An important feature of the solution of (2.68) is the dependence of the imaginary part of E near threshold:

$$\Gamma(v) \sim \gamma_0 \exp[-2\pi(Z_1+Z_2)\,\alpha/v] \qquad (2.69)$$

where v is the positron velocity and $\gamma_0\approx(6/5)\pi m$. This can also be written in terms of the internuclear distance [2.151]

$$\Gamma(R/R_{cr}) \sim \gamma_0 \exp[-b(1-R/R_{cr})^{-1/2}] \qquad (2.70)$$

where $b\approx5.7(Z_1+Z_2)\alpha$. Thus, the positron probability is exponentially small at threshold and increases rapidly with increasing $\eta=E/E_T$.

The threshold dependence of Γ is characteristic of problems involving the penetration of a Coulomb barrier. We can see the origin of the barrier by writing the equation for the large component $G(r)=rg(r)$ of the Dirac-Coulomb equation in the Schrödinger form [2.138,151]

$$\frac{1}{2m} \chi'' + [E_{eff} - U_{eff}(r)]\chi = 0 \qquad (2.71)$$

where $\chi=[m + E -V(r)]^{-1/2}G(r)$, $E_{eff}=1/2(E^2-m^2)$, and U_{eff} (for a finite nuclear size) is shown in Fig.2.16 for $Z \sim Z_{cr}$ and $E \sim -m$. A striking feature of U_{eff} is its large distance behavior: $U_{eff}(r) = E/m \ V(r)$, i.e., it becomes a *repulsive* potential for $E<0$. In the case of a Coulomb potential $V(r) = -Z\alpha/r$,

$$U_{eff}(r) \approx \frac{Z\alpha}{r} - \frac{(Z\alpha)^2 - 1}{2mr^2} \tag{2.72}$$

for the 1S state at $E \sim -m$. At short distances, the relativistic attractive r^{-2} term dominates in U_{eff}. Note that U_{eff} has a maximum at $r = r_m = \textrm{\DH}_e [(Z\alpha)^2 - 1]/Z\alpha$, where $U_{eff}(r_m) = 1/2(Z\alpha)^2 m/[(Z\alpha)^2 - 1] \sim 1.4 \ m$ for $Z = 170$.

It is clear that for E just above $-m$ where $E_{eff} < 0$, the effective potential yields a localized discrete state with a radius of order r_m. The Schrödinger wave function attenuates at infinity in the form [2.138]

$$\chi(r) \sim r^{1/4} \ \exp(-\sqrt{8Z\alpha r/\textrm{\DH}_e}) \quad . \tag{2.73}$$

The radius of the localized ground state is shown in Fig.2.17. For $E_{eff} > 0$, i.e., when E dives into the lower continuum, the wave function extends beyond the barrier, and the width of the quasi-static level is given by the barrier penetrability [2.151, 164]

$$\Gamma = \gamma_0 \ \exp\left\{ -2\pi \left[Z\alpha/v - \sqrt{(Z\alpha)^2 - 1} \right] \right\} \tag{2.74}$$

where

$$\frac{\gamma_0}{m} = \frac{3(Z\alpha)^2 (1 - e^{-2\pi v})}{[2(Z\alpha)^2 + 3] v} \rightarrow \begin{cases} 6\pi/5 \quad , \quad Z\alpha = 1 \\ \\ 3/4(Z\alpha)^{-1}, \quad Z\alpha \gg 1 \end{cases} \tag{2.75}$$

and $v = \sqrt{1 - m^2/E^2}$ is the velocity of the outgoing positron; and in (2.75) $v = 2\sqrt{(Z\alpha)^2 - 1}$. The simplest interpretation is that given in Section 2.3.1: If $E < -m$, then the one-electron state becomes degenerate with the two-electron plus one-positron state. The electron-positron pair is created near the nucleus where $V(r) > 2m$; the form of (2.74) reflects the probability that the positron can penetrate the barrier. Because Γ is small compared to m, the single particle analysis of the Dirac equation used to describe the supercritical atom can be justified [2.164].

One can now proceed to the calculation of the positron production probability by integrating over the classical ion trajectory, which is familiar from treatments by BATES and MASSEY [2.165] for autoionization in a slow collision and the level crossing formula of Landau and Zener for the probability of near adiabatic transitions between discrete levels [2.166]. POPOV's result for the positron production cross section is [2.151]

$$\sigma = \frac{4\pi}{V_L} \int_{R_0}^{R_{cr}} dR\ R^{3/2}(R-R_0)^{1/2}\ \Gamma(R) \tag{2.76}$$

where $1/2 M v_L^2$ is the lab kinetic energy of the incident nucleus and R_0 is the distance of closest approach at zero impact parameter. [If the probability $P=\sigma/\sigma_{geom}$ is not small, P should be replaced by $(1-e^{-P})$]. The cross section for producing a positron with kinetic energy T is [2.151]

$$\frac{d\sigma}{dT} = \frac{4\pi}{V_L}\ \Gamma(R)R^{3/2}\ (R-R_0)^{1/2}\left(\frac{dE_0}{dR}\right)^{-1} \tag{2.77}$$

where $T=|E_0(R/R_{cr})|-m$. Since Γ is small compared to $|E|$, the positron energy can be identified with a given internuclear distance R, and σ is correctly obtained by integrating over the complete ion orbit.

The calculation of GREINER and co-workers differs substantially in approach from that of POPOV et al. The first step is the computation of the width for positron production in a supercritical atom using the Fano autoionization method. The second step is to use these results to calculate the spontaneous and induced positron production rate, assuming that the relation between Γ and the bound state energy is the same in the one- and two-center problems.

The method used by GREINER and co-workers [2.145,158] to evaluate the positron width in the one-center problem is based on the analogy of the supercritical nucleus to that of autoionization in nuclear and atomic physics, where a bound state is imbedded in a continuum, and FANO's formalism [2.167] should be applicable. One begins by assuming that the nuclear potential is at criticality so that $|\phi_{cr}\rangle$, the single particle bound-state solution of the Dirac equation, has E=-m. Let $|\psi_E\rangle$ denote the s-wave negative-energy continuum solutions to the same equation with E<-m. If V is increased above criticality, $\Delta V=V-V_{cr}<0$, then to first order in ΔV

$$\Delta E_{cr} = \langle\phi_{cr}|\Delta V|\phi_{cr}\rangle \tag{2.78}$$

and

$$\Gamma = \tau_{pos}^{-1} = 2\pi|\langle\psi_E|\Delta V|\phi_{cr}\rangle|^2\ . \tag{2.79}$$

Calculations [2.143] show that $\Delta E_{cr}\sim-30\ (Z_1+Z_2-Z_{cr})$ keV; see Fig.2.21. Thus the bound state $|\phi_{cr}\rangle$ dives into the negative continuum with an energy shift roughly linear in $\Delta Z=Z_1+Z_2-Z_{cr}$ and a monotonically increasing width. If one defines the negative energy solutions $|\psi_E\rangle$ for the Dirac equation with $V_{cr}+\Delta V$, then

$$|\langle\psi_E|\phi_{cr}\rangle|^2 = \frac{1}{2\pi}\ \frac{\Gamma}{[E-(E_{cr}+\Delta E_{cr})]^2+\Gamma^2/4} \tag{2.80}$$

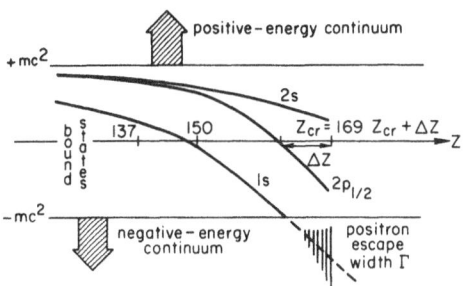

Fig. 2.21. Dependence of the atomic levels on nuclear charge. The positron escape width, which gives the rate for spontaneous decay, increases monotonically with $\Delta Z = Z - Z_{cr}$ (adapted from MÜLLER et al. [2.143])

is the probability that the bound electron is promoted to $|\psi_E\rangle$ when ΔV is added adiabatically. (This Breit-Wigner form for the admixture probability neglects an extra energy shift from the energy variation of Γ). This treatment thus far is, in principle, complementary to that given earlier in this section.

There may, however, be difficulties with using the autoionization method and a perturbation expansion near Z_{cr} for calculating the spontaneous decay width. In the papers of MÜLLER et al. [2.134,145], the estimate

$$\Gamma = 2\pi |\langle \phi_{cr} | \Delta V | \psi_E \rangle|^2 \sim (\Delta Z)^2 \; 50 \text{ eV} \tag{2.81}$$

is reported to be a good approximation to the exact one-center Dirac equation width, at least for $\Delta Z = Z_1 + Z_2 - Z_{cr} \gtrsim 3$. [For example, in the more recent work of SMITH et al. [2.134,168] a curve is given showing numerical results based on a one-center Dirac equation calculation for the function $\gamma(T) = m\Gamma(T)/T^2$ where T is the positron kinetic energy. For $180 < Z < 210$, R > 15 fm, and 0.4 MeV < T < 1.2 MeV, $\gamma(T)$ is roughly constant at 0.015, giving $4.7 < \Gamma(T) < 42$ keV which agrees within a factor of two with (2.81)]. However, as noted by MÜLLER et al. [2.145] and POPOV [2.153], the threshold behavior characteristic of the Coulomb barrier leads to strong exponential damping at zero positron momentum; for $\Delta Z \rightarrow 0$, POPOV [2.153] obtains $\Gamma \propto \exp[-b(Z_{cr}/\Delta Z)^{1/2}]$. Thus, (2.81) should not be considered a reliable approximation for small ΔZ. The equation $\Gamma = 2\pi |\langle \phi_{cr} | \Delta V | \psi_E \rangle|^2$ is evidently invalid near threshold since it requires an expansion in ΔZ about the critical value. Furthermore, the continuum solution $|\psi_E\rangle$ which is defined to obey the Dirac equation for $V = V_{cr}$ does not see the Coulomb barrier in the effective potential U_{eff} appropriate to $V = V_{cr} + \Delta V$ and the level energy E_0. [In the paper by POPOV and MUR [2.164], it is argued that $\Gamma/m \lesssim e^{-2\pi Z\alpha} \ll 1$ even for $|E_0| \gg m$ in a one-center problem because the barrier in U_{eff} increases in height as $E_0 \rightarrow \infty$. In fact, the width of the level can increase with $|E_0|$ despite the increasing barrier height because the tunneling distance becomes smaller [2.160]. This is also consistent with the barrier penetration formula (2.74) which gives $\Gamma \rightarrow \gamma_0 e^{-\pi/(Z\alpha)}$ in the limit $Z \rightarrow \infty$, $v \rightarrow 1$].

The results of a recent one-center calculation of Γ by JAKUBASSA and KLEBER
[2.160] (based on a semi-classical method) are shown in Fig.2.22. We have also indicated the values calculated from (2.81). The width computed by JAKUBASSA and KLE-
BER is about three times as large as that given by (2.81) at $Z{\sim}200$.

It should be noted that the corresponding calculations of Γ/E_0 as a function of
R/R_{cr} by POPOV [2.153] for the *two-center* problem yield much smaller widths, with
$\Gamma/E_0 < 0.012$, i.e., an order of magnitude below those of the one-center problem results
of JAKUBASSA and KLEBER. This may indicate that the one-center values, which are
based on the nuclear size - and not on the actual nuclear separation - give an over-
estimate of the decay width.

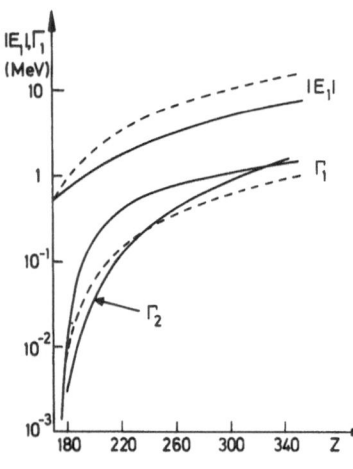

Fig. 2.22. Energy and width of the supercritical
IS state calculated by JAKUBASSA and KLEBER
[2.160]. The solid lines labeled $|E_1|$ and Γ_1
give the energy and width in the WKB approximation; the dashed lines give the corresponding
results when the effective potential is modified
by adding a centrifugal term. The results of
MÜLLER et al. [2.134,145], labeled Γ_2, for the
width of the 1S state (2.81) have been added to
the figure for comparison (adapted from JAKUBASSA
and KLEBER [2.160])

2.3.5 Induced Versus Adiabatic Pair Production

One of the controversial questions concerning positron production in heavy-ion collisions is the relative importance of pair production induced by the changing Coulomb
field. In the review of ZELDOVICH and POPOV [2.138] and the later papers of POPOV
[2.151,153], arguments are given that this mechanism can be neglected since 1) the
frequency of collision ω_c (equal to the inverse of the collision time τ_c) is a small
fraction of 2m, and 2) the characteristic electron time is much shorter than the
collision time, so that the electron state can adiabatically adjust to the changing
Coulomb potential. However, as emphasized in the papers by SMITH et al. [2.154,168],
the energy required for pair production during the collision is just the (narrow)
gap between the 1S level and the negative continuum. Thus the changing Coulomb field
can induce a transition (pair production) even at very low velocities. In typical
inelastic atom-atom collisions, an appreciable cross section for transitions occurs
when [2.169]

$$\Delta E \lesssim 2\pi \frac{v}{a} \tag{2.82}$$

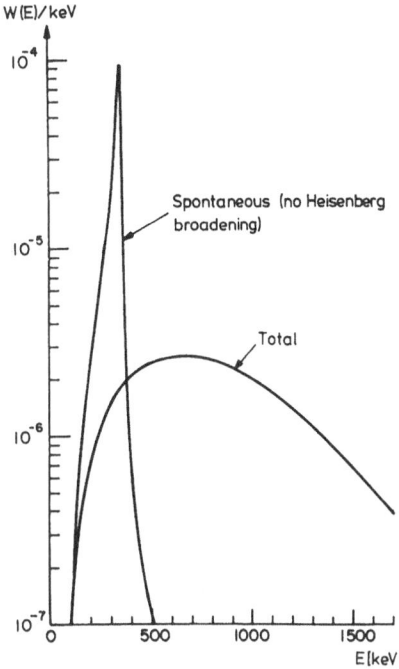

W(E)/keV

10^{-4}

10^{-5}

Spontaneous (no Heisenberg broadening)

Total

10^{-6}

10^{-7}

0   500   1000   1500

E[keV]

Fig. 2.23. Probability W(E) for production of a positron with energy E per ΔE= 1 keV. Comparison of the spontaneous and spontaneous-plus-induced spectrum as calculated by SMITH et al.(from the review of RAFELSKI and KLEIN [2.170])

where v is the relative velocity and a is a length characteristic of the inducing potential. Taking a~50 fm, v~0.05, gives ΔE~1 MeV. Thus, induced pair production with a continuum positron could well be an important process even for collisions in which diving does not occur. In fact, SMITH et al. [2.168] find that in typical U-U collisions, induced positron production is two orders of magnitude larger than the spontaneous cross section alone! Obviously, the induced contribution also will spread the kinetic energy spectrum of the positron, with substantial contributions occurring at kinetic energies ~ 1 MeV beyond the kinematic limit for spontaneous production. The results of SMITH et al. [2.168] compared with the spontaneous positron production spectrum calculated by PEITZ et al. [2.158] are shown in Fig.2.23. The cross section calculated by SMITH et al. for positron production in central U-U collisions at 812.5 MeV center-of-mass energy is shown in Fig.2.24. The curve denoted A is the contribution "during diving", i.e., integrated over the times when the 1S level joins the negative continuum. Curve B denotes the contribution before and after diving. The coherent sum is also shown. The Rutherford cross section for U-U scattering and the ionization probability L_0 have been divided out. By integrating over energy, one sees that roughly 5% of all the collisions with a 1S vacancy will produce a positron by the induced process. The positron production cross section for different CM kinetic energies is shown in Fig.2.25.

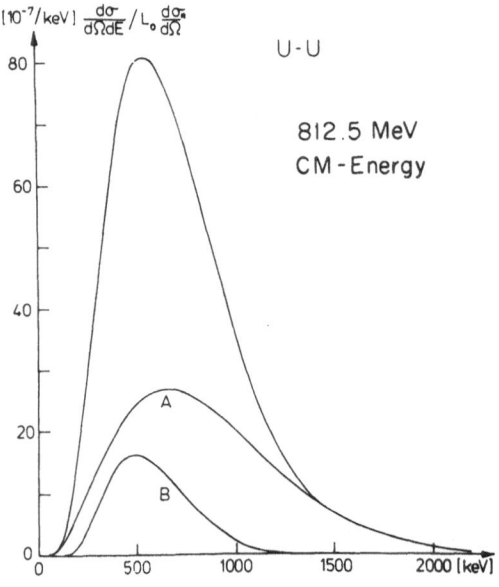

Fig. 2.24. Cross section for positron production as a function of positron energy, divided by the ionization probability L_0 and Rutherford cross section for U-U central collisions at 812.5 MeV, as calculated by SMITH et al. (from the review of RAFELSKI and KLEIN [2.170])

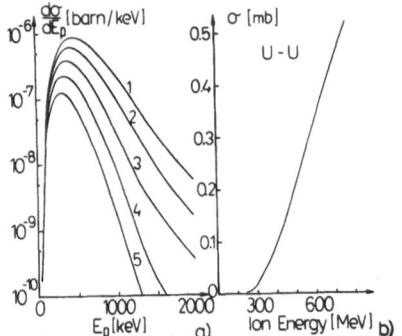

Fig. 2.25. (a) The positron cross sections calculated for U-U central collisions, with L_0 set equal to 10^{-2}. The ion center-of-mass kinetic energies are: [1] 815.5 MeV (distance of closest approach, 15 fm); [2] 609.4 MeV (20 fm); [3] 478.5 MeV (25 fm); [4] 706.3 MeV (30 fm); [5] 398.2 MeV (35 fm). The vertical scale here is corrected according to the relevant footnote in [2.134]. (b) The total positron cross section dependence on the ion CM energy (from SMITH et al. [2.168])

The during-diving positron production amplitude computed by SMITH et al. [2.154, 168] takes the form

$$C_D = i \int_{-t_{cr}}^{t_{cr}} dt \ V_E(t) \ \exp\left\{\int_{-\infty}^{t} dt' [iE - iE_{1S}(t') - 1/2\Gamma(t')]\right\} \tag{2.83}$$

where E is the positron energy level and $\Gamma(t')$ gives the positron resonance width at time t'. The perturbing potential is taken to be

$$V_E(t) = \langle\psi_E|\Delta V[R(t)]|\phi_{cr}\rangle \tag{2.84}$$

as in Section 2.3.4.

As we have remarked in Section 2.3.4, this expression for $V_E(t)$ and the definition $\Gamma=2\pi|V_E(t)|^2$ are in apparent conflict with the analytic results of POPOV [2.153] for the resonance width near threshold. Hence, it is important that calculations of the induced process which avoid expansions in $\Delta V=V(R)-V(R_{cr})$ about the diving point be done.

Recently, JAKUBASSA and KLEBER [2.160] have also presented a method of evaluation of induced positron production in heavy-ion collisions within the WKB semi-classical approximation. Their results for both spontaneous and induced production are about ten times larger than those calculated by SMITH et al. [2.168]. In particular, JAKUBASSA and KLEBER find a total cross section $\sigma=4b$ for $v_L=0.15$ and $\sigma=1b$ for $v_L=0.1$ in U-U collisions with $R_{cr}=34$ fm.

It should be noted that so far all the calculations of the induced production rate are based on the single-center results for the widths and transition matrix elements.

We also note that the induced process may make positron production experimentally practical even for medium-Z heavy-ion collisions. Diving is not critical. Further, induced pair production where both the positron and electron are in the continuum may be feasible in U-U collisions even without ionization. In fact, pair production requires not much more energy transfer than induced 1S hole production when the energy is near critical.

From a general point of view, induced and spontaneous positron production in heavy-ion collisions can be identified with the Feynman diagram shown in Fig.2.26.

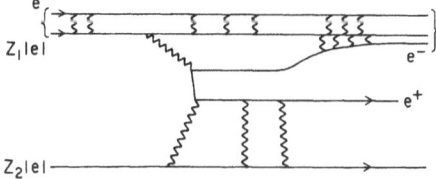

Fig. 2.26. Feynman diagram for positron production in ion-ion collisions. The produced electron becomes bound to the nucleus with charge $Z_1|e|$

The physical process, however, goes beyond perturbation theory in that the production occurs as a result of the coherent energy of the Coulomb interaction in the strong field relativistic domain.

2.3.6 Vacancy Formation in Heavy-Ion Collisions

The physics of inner-shell vacancy formation is currently a subject of active experimental [2.124] and theoretical interest [2.171,172], and our discussion here
can only be very brief. According to the extrapolation of BLASCHE et al. [2.173],
completely ionized U atoms are possible at beam energies of order 300 MeV/nucleon
(which should be achievable within the next five years at the LBL Bevalac). However,
as we have seen, the optimum kinetic energy of the ion for the positron production
experiments is in the few MeV/nucleon range. Ions with this kinetic energy could be
achieved either by ion deceleration, or more ingeniously (as suggested by GREINER
and SCHMELZER [2.171]) by arranging low relative velocity collisions between colliding beams in storage rings, as in Fig.2.27. A similar configuration could be attained
using the configuration of the CERN-ISR, with both ion beams circulating in the same
direction.

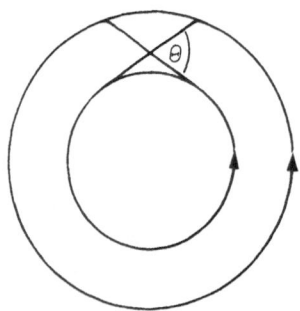

Fig. 2.27. Schematic diagram of a heavy-ion storage
ring configuration, suggested in [2.171], arranged
to obtain low relative velocity collisions. By changing Θ, the relative velocity can be adjusted (from
GREINER [2.171])

At lower and more practical energies one must rely on the formation of the 1S
vacancy of the combined quasi-molecule which occurs in the same atomic collision
which produces the pair. (The lifetime of the vacancy is too short at high Z for
collisions involving more than one target atom to be important [2.124]). For U-U
collisions, estimates of the vacancy formation probability L_0 range from 0.2 for
$E_{LAB}=1600$ MeV [2.174] (using an "atomic" model in which the time variation of the
Coulomb field causes energy to be transferred to the electron, which is ejected)
to values between 10^{-4} and 10^{-6} depending on projectile energy [2.175,176] (using
a "molecular" model in which the collision is assumed slow enough to allow the
electrons to adjust themselves to the diatomic molecular levels; transitions caused
by the varying Coulomb fields then produce vacancies).

However, very recently, BETZ et al. [2.177] have performed an approximate calculation of the ground state vacancy production probability in U-U collisions. The
vacancies are produced by the Coulomb field variation in the two-center Dirac equation. They find the vacancy production probability at 1600 MeV and zero impact parameter to be larger than 0.08 - much larger than was anticipated and very encouraging
for the experiments discussed here.

Since 1S state vacancy production followed by positron production in a single collision will be accompanied by direct pair production associated with the time variation of the fields in the collision, and the latter process may have an amplitude of comparable magnitude, the cross section for atom+atom→atom+atom+e^-+e^+ should be calculated from the coherent sum of both processes. In fact, all processes which produce a pair must, of course, be considered together. This includes summation over pair production processes which fill higher n vacancies. Although pair production which fills higher n vacancies may be less likely, there could be a partial compensation due to a larger probability of vacancy formation in higher n states. The relative importance of these processes has not yet been estimated for the case of collisions between neutral atoms.

2.3.7 Nuclear Excitation and Other Background Effects

There are general background effects which can complicate the experimental observation of positron production associated with the overcritical Coulomb field. We will closely follow the discussion of POPOV [2.151], OKUN [2.178], and OBERACKER et al. [2.179,180] here.

When heavy particles collide, e^+e^- pairs can be produced by hard photon bremsstrahlung and pair conversion. The cross section is small [2.151,178] because the motion of the nuclei is nonrelativistic. A typical cross section for Z=92, R_0=40 fm is $\sigma < 10^{-16}$b. For identical nuclei U-U, this is suppressed by a few more orders of magnitude since the cross section for dipole radiation, proportional to $(Z_1A_2-Z_2A_1)^2$, vanishes [2.151].

The most important background is the production of e^+e^- pairs by pair conversion in transitions resulting from Coulomb excitation of nuclei. An estimate given by POPOV [2.153] for U-U collisions gives $\sigma_C^{e^+e^-} \sim 10^{-4}$b which is somewhat smaller than the estimates for the cross section for spontaneous positron production.

Extensive calculations of the nuclear and Coulomb excitation cross sections have been recently performed by OBERACKER et al. [2.179,180]. The calculated differential cross section (dashed lines) for ^{238}U-^{238}U collisions at the Coulomb barrier E_{cm}^{kin} =800 MeV, as a function of the ion scattering angle Θ_{ion}, is shown in Fig.2.28. The two dashed lines correspond to two different models for the nuclear states. The associated cross section, calculated by PEITZ (quoted in [2.179,180]), via spontaneous and induced decay (assuming R_{cr}=35 fm, and the K-vacancy probability L_0=10^{-2}) is given by the solid line. Representative total nuclear and Coulomb excitation production cross sections calculated by OBERACKER et al. [2.179,180] for U-U collisions range from $\sigma_C^{e^+e^-} \sim 1.25 \times 10^{-4}$b to $\sigma_C^{e^+e^-} \sim 2.28 \times 10^{-4}$b depending on the model for the nuclear states. This is in good agreement with the estimate of POPOV. As noted by OBERACKER et al., the nuclear background is suppressed in the backward and forward directions for symmetric systems ^{238}U-^{238}U. In addition, the nuclear positron

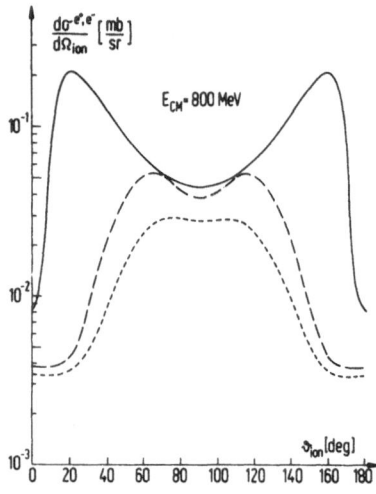

Fig. 2.28. Differential pair production cross sections (CM) as a function of ion angle for $^{238}U-^{238}U$ collisions. Spontaneous and induced positron production cross section (solid line) and pair production from Coulomb and nuclear excitation cross sections (dashed lines corresponding to two nuclear models) are shown (from OBERACKER et al. [2.179])

spectrum terminates at $E_p \sim 800$ keV while the induced positron spectrum extends to much higher energies. Both of these characteristics should aid in separating out the background positrons. More complicated, but negligible backgrounds, involving conversion of gamma rays from nuclear transitions where the electron occupies the vacant ground state are also estimated by OBERACKER et al.

2.3.8 Radiative Corrections in Critical Fields

There is considerable theoretical interest in the question of whether radiative corrections could modify or even eliminate the predictions discussed here for pair production at $Z > Z_{cr}$. As we have noted in Section 2.3.1, the radiative corrections are controlled by α rather than $Z\alpha$ so they are, in principle, independently controllable in their physical effects, and thus one would not expect dramatic changes in the previous description. One also would not expect that calculations based on a Feynman diagram treatment indicated by Fig.2.26 could be much affected by effects of order α. However, since virtual pairs may be produced with an arbitrarily small expenditure of energy as $Z \to Z_{cr}$, the smallness of α is not necessarily decisive. In the following, the tractable model of a single nucleus of charge Z is examined. The results for a heavy-ion collision are expected to be qualitatively similar.

The situation is well understood in the case of the order α vacuum polarization corrections; the modifications turn out to be small. For r at the maximum of the near-critical 1S probability distribution, $r_0 \sim 0.1 \, \lambdabar_e$, the Coulomb Uehling potential has the value

$$V_{11}(r_0) = -\frac{2\alpha}{3\pi} \frac{Z\alpha}{r_0} \left(\ln \frac{\lambdabar_e}{r_0} - \gamma - \frac{5}{6} + \ldots \right) \approx -9 \text{ keV} \quad . \tag{2.85}$$

Calculation of the 1S Uehling energy shift at $Z \approx Z_{cr}$ by SOFF et al. [2.181] gives
$\Delta E = -11.8$ keV for Z=171, in good agreement with an extrapolation of the results, for
Z≤160, of PIEPER and GREINER [2.117] and with the order-of-magnitude estimate in
(2.85). The corresponding shift ΔZ_{cr} in Z_{cr} is found, with the aid of [2.47,144,145]

$$\left. \frac{dE_{1S}}{dZ} \right|_{Z_{cr}} = -27 \text{ keV} \quad , \tag{2.86}$$

to be $\Delta Z_{cr} = -0.4$, i.e., the critical charge is reduced by less than one unit. The
result of POPOV [2.133], $\Delta Z_{cr} = 0(10^{-3})$, appears to be an underestimate.

The higher order correction to the Uehling potential of the Wichmann-Kroll type
(from the Coulomb interactions of the electron-positron pair) is an order of magnitude smaller. Arguments of POPOV [2.133] and of MÜLLER et al. [2.144,145] suggest
that the higher order corrections are small. A calculation has been done by GYULASSY
[2.47,48], who found $\Delta E_{1S} = 1.2$ keV which is negligible compared to the Uehling term.
GYULASSY [2.47] has also shown numerically that the vacuum polarization charge density associated with the charged vacuum, discussed by FULCHER and KLEIN [2.136],
varies smoothly as Z passes through Z_{cr}. The vacuum polarization associated with
the charged vacuum is formally related to the ordinary vacuum polarization by a
shift in the contour of integration in the bound electron propagator as discussed
in Section 2.1.5 [2.47].

In the case of the self-energy corrections to the electron level, a simple heuristic argument is that a fraction α of the lepton charge is spread out over a
Compton radius of the electron $\bar{\lambda}_e$ [modulo a logarithmic tail out to the Bohr radius
$(Z\alpha m)^{-1}$ associated with the Bethe sum]. Such a distribution convoluted with the
nuclear size distribution could only change Z_{cr} by a small amount. Also, since the
determination of the nuclear radius R derives from electron scattering experiments,
the influence of radiative corrections is already partially included. The situation
for the self-energy is more obscure at very high Z where higher order terms in $Z\alpha$
are important; quantitative calculation is necessary.

Results of calculations of the self-energy of the 1S state for large Z are shown
in Fig.2.29. The self-energy for a Coulomb potential appears to become infinite as
$Z\alpha \rightarrow 1$. This is clearly an anomaly due to the point charge singularity. CHENG and
JOHNSON [2.36] have extended calculations to Z=160 for a finite nucleus. By extrapolation, they find that the 1S self-energy is at least 1% of the binding energy
at $Z = Z_{cr}$. Although it seems unlikely, we note that if the self-energy were to increase sufficiently rapidly as $Z \rightarrow Z_{cr}$, there might be no diving phenomenon, and further analysis would be necessary. In any case, induced positron production would
still be possible (see Sec.2.3.5).

Calculation of radiative corrections to the positron emission rate would be very
complicated, because the self-energy graph will include the long-range radiative

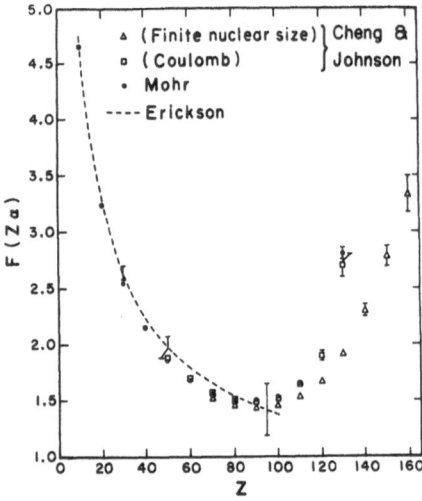

Fig. 2.29. Results of various calculations of the 1S self-energy at high Z. The energy shift is given by $\Delta E = (\alpha/\pi)(Z\alpha)^4 F(Z\alpha)m$. The results are from [2.8,11,36]. Error estimates smaller than 2% are not shown (from CHENG and JOHNSON [2.36])

correction associated with the outgoing charged particle. The effect of photon emission would have to be separated from the energy shift.

2.3.9 Coherent Production of Photons in Heavy-Ion Collisions

Another intriguing, possibly feasible test of strong field electrodynamics utilizing high-Z ion collisions is single or multiple hard photon production. The quantum electrodynamic process is a variation of Delbrück, or light-by-light scattering (see Fig.2.30). The photons are created by the coherent energy of the ions' Coulomb field. Unlike bremsstrahlung processes, the spectrum of the photon peaks in the electron mass (MeV) range. The production cross section for n photons should be of order α^n times the Rutherford cross section for collisions in which the potential energy at the distances of closest approach significantly exceeds the total photon energy. Such photons should be distinguishable from nuclear excitation photons and combined-atom x-rays by their (calculable) spectrum and angular distribution, and by coincidence (correlation) measurements.

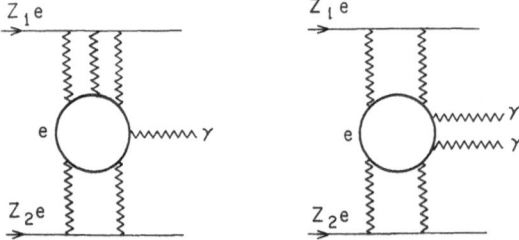

Fig. 2.30. Feynman diagrams for production of photons by vacuum polarization in high-Z collisions

Note that this photon production process occurs at ion energies and charges well below those required for spontaneous pair production. Conversely, if the photons have energies beyond 1 MeV, they provide a background for positron production from internal pair conversion, or by conversion in a nearby atom.

2.3.10 Self-Neutralization of Matter

The possibility of spontaneous pair production at high Coulomb field strength leads to a rather novel self-neutralization mechanism of ionized matter. Suppose that one could arrange a contained plasma of completely stripped uranium ions (no electrons present). For any finite temperature there will occasionally be ion-ion collisions at sufficient velocity such that the distance of closest approach is less than $R_{cr} \sim 35$ fm, where diving of the lowest electronic level of the two-center Dirac system begins. Eventually, all the bound electron atomic levels which dive will be filled by the pair production process and - assuming the continuum positrons are allowed to escape - the ionic system will be partially neutralized. Although the process can occur in principle at any finite temperature via the Maxwell velocity distribution, the positron production probability becomes large only at high temperatures, $kT \sim O(1$ GeV$)$, i.e., $T \sim O(10^{13}$ K$)$. At still higher temperatures, other pair production mechanisms become important; however, the spontaneous pair production is the lowest energy mechanism.

2.3.11 Very Strong Magnetic Field Effects

RAFELSKI and MÜLLER [2.182] have suggested looking at heavy-ion collisions as a means of testing the behavior of matter in strong magnetic fields. Such tests would be sensitive to possible anomalous higher order effects of strong fields. In aeheavy--ion collision, the magnetic fields are produced by the motion of the charged nuclei, with the corresponding vector potential given by

$$e\underline{A}(\underline{r}) = - Z\alpha \frac{\underline{V}_1}{|\underline{r}-\underline{R}_1|} - Z\alpha \frac{\underline{V}_2}{|\underline{r}-\underline{R}_2|} \tag{2.87}$$

where $\underline{V}_i$ are the nuclear velocities and $\underline{R}_i$ are the position vectors of the nuclei. In a sub-Coulomb barrier heavy-ion collision, the magnetic field created in the vicinity of the colliding nuclei is of the order of 10^{14} Gauss over a small volume [2.182]. The magnetic fields give rise to a splitting of the lowest quasi-molecular states through interaction with the electron spin. RAFELSKI and MÜLLER have calculated the magnetic splitting which can be expected in various heavy-ion collisions. Fig.2.31 shows the energy separation of the spin-up and spin-down states relative to the binding energy for various systems. For example, the separation of the $1s\sigma$ state in a U-U collision with T=9 MeV/nucleon and impact parameter b=20 fm is calculated to be approximately 0.1 m=50 keV. This splitting corresponds to an average

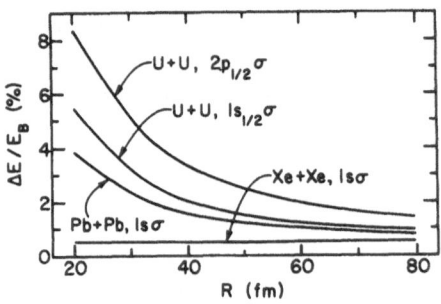

Fig. 2.31. Relative magnetic splitting $(E_\downarrow - E_\uparrow)/E_B$ (E_B is the binding energy) for selected quasi-molecular states. Collision parameters are $E_{LAB}=9$ MeV/nucleon and impact parameter $b=13$ fm (from RAFELSKI and MÜLLER [2.182])

magnetic field $\bar{B} \sim 4 \times 10^{12}$ Gauss. The magnetic splitting results in a difference of 3.3 fm in R_{cr} for the two states.

We note that any model for strong magnetic field anomalies which could be evident in heavy-ion collisions would be constrained by existing fine structure measurements in exotic atoms. The determination of the magnetic moment of the anti-proton to 1% accuracy by the fine-structure measurement in $\bar{p}$Pb by HU et al. [2.183] has yielded a value which is in excellent agreement with the proton magnetic moment (in accord with the TCP theorem). The fine structure in the lower level ($n=10, \ell=9$) in that experiment arises from interaction of the anti-proton moment with an average magnetic field of order 10^{14} Gauss. A similar test in somewhat stronger magnetic fields is made by muonic atoms. The measured fine structure splitting in the 2P state of muonic lead [2.184] agrees with theory, and the average magnetic field seen by the muon is of order 10^{16} Gauss.

2.4 Conclusion

All of the tests of high-$Z\alpha$ quantum electrodynamics which we have discussed in this review probe in various ways the Furry bound state interaction picture description of the bound leptons. In the strong field domain where $Z\alpha$ is not small, a natural question is whether this generalization of weak field perturbation theory continues to be applicable if the binding strength is not small compared to the mass of the bound particle.

Thus far the tests of high field strength QED involving the spectra of bound electrons and muons are in extraordinary agreement with predictions, ruling out anomalous nonlinear interactions, low-mass scalar particles with certain couplings, and anomalous modifications of vacuum polarization at momentum transfers $\sim Z\alpha m_\mu$. The high-$Z\alpha$ spectra also test electromagnetic interactions in the strong magnetic field regime, where the effective fields reach 10^{16} Gauss. Tests of relativistic bremsstrahlung in high magnetic fields are reviewed in [2.185]. Further tests of the

Furry picture of bound leptons and their radiative corrections are possible by measurements of the bound state gyromagnetic ratio via Zeeman interactions and by photon scattering from high-Z atoms. The Lamb shift measurements in heavy atoms confirm the calculations of radiative corrections for highly off-shell electrons. The measurements in high-Z few-electron ions provide a means of testing QED in strong fields with multiparticle systems for which the theory is still tractable. Precision measurements in muonic atoms are now beginning to confirm higher order vacuum polarization corrections of order $\alpha(Z\alpha)^3$, ruling out broad classes of anomalous muon-nuclear interactions [2.186].

Although the basic predictions for positron production in heavy-ion collisions appear to be understood from a fundamental point of view, there are many quantitative questions which have not been completely settled. As we have noted, it is difficult to compare details of results based on different calculations because of the wide range of models employed. This is particularly critical in the questions concerning the absolute magnitude of both the spontaneous and induced positron production rates.

The dynamical tests of high-$Z\alpha$ QED, especially positron production (and possibly anomalous photon production) in heavy-ion collisions, are particularly interesting because they require an extension of the theory to a domain which is otherwise unexplored. For example, when the binding becomes critical, the ordinary vacuum is effectively unstable, and a new vacuum reference state is required.

There are many issues of fundamental interest which still need to be resolved. These include a complete field theoretic treatment of the positron production problem which considers the effects of radiative corrections; the problem of the Klein-Gordon equation for $Z\alpha > 1$, where Bose condensation can occur [2.138,187]; and the nature of positronium when $\alpha > 1$, and in particular, whether there is a mechanism (possibly recoil corrections) which can moderate the singular Coulomb problem.

Acknowledgments. We are grateful to R. Anholt, L. Madansky, J. Mandula, M. Marinov, R. Marrus, W. Meyerhof, and M. Weinstein for helpful conversations.

This work was supported in part by the U.S. Energy Research and Development Administration.

References

2.1   R. Hofstadter: Proceedings of the 1975 International Symposium on Lepton and Photon Interactions at High Energies, ed. by W.T. Kirk (SLAC, Stanford 1975) p. 869
2.2   P. Papatzacos, K. Mork: Phys. Reports 21, 81 (1975)
2.3   K. Wilson: Cornell preprint CLNS 356 (1977)
2.4   J. Mandula: MIT preprint (1976)

2.5 H. Gould, R. Marrus: Reported at the 5th International Conference on Atomic Physics, Berkeley, July 1976, and private communication
2.6 S.J. Brodsky, S.D. Drell: Ann. Rev. Nucl. Sci. 20, 147 (1970)
2.7 B.E. Lautrup, A. Peterman, E. de Rafael: Phys. Reports 3, 193 (1972). This paper gives complete references to the original work.
2.8 G.W. Erickson: Phys. Rev. Lett. 27, 780 (1971)
2.9 G.W. Erickson: U. C. Davis preprint (1976)
2.10 P.J. Mohr: Phys. Rev. Lett. 34, 1050 (1975)
2.11 P.J. Mohr: Ann. Phys. (N.Y.) 88, 26, 52 (1974)
2.12 E.H. Wichmann, N.M. Kroll: Phys. Rev. 96, 232 (1954); 101, 843 (1956)
2.13 W.H. Furry: Phys. Rev. 51, 125 (1937)
2.14 R. Serber: Phys. Rev. 48, 49 (1935)
2.15 E.A. Uehling: Phys. Rev. 48, 55 (1935)
2.16 P.J. Mohr: In Beam-Foil Spectroscopy, ed. by I.A. Sellin and D.J. Pegg (Plenum Press, New York 1976) p. 89
2.17 T. Appelquist, S.J. Brodsky: Phys. Rev. Lett. 24, 562 (1970); Phys. Rev. A 2, 2293 (1970)
2.18 M. Leventhal: Phys. Rev. A 11, 427 (1975)
2.19 D. Dietrich, P. Lebow, R. deZafra, H. Metcalf: Bull. Amer. Phys. Soc. 21, 625 (1976)
2.20 C.Y. Fan, M. Garcia-Munoz, I.A. Sellin: Phys. Rev. 161, 6 (1967)
2.21 H.W. Kugel, M. Leventhal, D.E. Murnick: Phys. Rev. A 6, 1306 (1972)
2.22 G.P. Lawrence, C.Y. Fan, S. Bashkin: Phys. Rev. Lett. 28, 1612 (1972)
2.23 M. Leventhal, D.E. Murnick, H.W. Kugel: Phys. Rev. Lett. 28, 1609 (1972)
2.24 H.W. Kugel, M. Leventhal, D.E. Murnick, C.K.N. Patel, O.R. Wood, II: Phys. Rev. Lett. 35, 647 (1975)
2.25 R. Marrus: In Beam-Foil Spectroscopy, ed. by S. Bashkin, Topics in Current Physics, Vol. 1 (Springer, Berlin, Heidelberg, New York 1976) p. 209
2.26 H.A. Bethe, E.E. Salpeter: Quantum Mechanics of One- and Two-Electron Atoms (Berlin-Göttingen-Heidelberg: Springer 1957)
2.27 W.A. Davis, R. Marrus: Phys. Rev. A (to be published)
2.28 H. Gould, R. Marrus: In Beam-Foil Spectroscopy, ed. by I.A. Sellin and D.J. Pegg (Plenum Press, New York 1976) p. 305
2.29 H. Gould, R. Marrus, P.J. Mohr: Phys. Rev. Lett. 33, 676 (1974)
2.30 J.A. Bearden, A.F. Burr: Rev. Mod. Phys. 39, 125 (1967); Atomic Energy Levels (U.S. Atomic Energy Commission, Oak Ridge 1965)
2.31 A.M. Desiderio, W.R. Johnson: Phys. Rev. A 3, 1267 (1971)
2.32 J.B. Mann, W.R. Johnson: Phys. Rev. A 4, 41 (1971)
2.33 M.S. Freedman, F.T. Porter, J.B. Mann: Phys. Rev. Lett. 28, 711 (1972)
2.34 B. Fricke, J.P. Desclaux, J.T. Waber: Phys. Rev. Lett. 28, 714 (1972)
2.35 F.T. Porter, M.S. Freedman: Phys. Rev. Lett. 27, 293 (1971)
2.36 K.T. Cheng, W.R. Johnson: Phys. Rev. A 14, 1943 (1976)
2.37 K. Huang, M. Aoyagi, M.H. Chen, B. Crasemann, H. Mark: At. Data and Nucl. Data Tables 18, 243 (1976)
2.38 M. Born, L. Infeld: Proc. Roy. Soc. (London) A 144, 425 (1934)
2.39 M. Born: Ann. Inst. Henri Poincaré 7, 155 (1937)
2.40 J. Rafelski, L.P. Fulcher, W. Greiner: Phys. Rev. Lett. 27, 958 (1971)
2.41 G. Soff, J. Rafelski, W. Greiner: Phys. Rev. A 7, 903 (1973)
2.42 G.E. Brown, G.W. Schaefer: Proc. Roy. Soc. (London) A 233, 527 (1956)
2.43 J. Arafune: Phys. Rev. Lett. 32, 560 (1974)
2.44 L.S. Brown, R.N. Cahn, L.D. McLerran: Phys. Rev. Lett. 32, 562 (1974)
2.45 L.S. Brown, R.N. Cahn, L.D. McLerran: Phys. Rev. D 12, 609 (1975)
2.46 M. Gyulassy: Phys. Rev. Lett. 32, 1393 (1974)
2.47 M. Gyulassy: Phys. Rev. Lett. 33, 921 (1974)
2.48 M. Gyulassy: Nucl. Phys. A244, 497 (1975)
2.49 L.S. Brown, R.N. Cahn, L.D. McLerran: Phys. Rev. D 12, 581 (1975)
2.50 L.S. Brown, R.N. Cahn, L.D. McLerran: Phys. Rev. D 12, 596 (1975)
2.51 G.E. Brown, J.S. Langer, G.W. Schaefer: Proc. Roy. Soc. (London) A 251, 92 (1959)
2.52 C.S. Wu, L. Wilets: Ann. Rev. Nucl. Sci. 19, 527 (1969)
2.53 S. Koslov, V. Fitch, J. Rainwater: Phys. Rev. 95, 291 (1954)
2.54 G. Backenstoss, S. Charalambus, H. Daniel, Ch. Von der Malsburg, G. Poelz, H.P. Povel, H. Schmitt, L. Tauscher: Phys. Lett. 31B, 233 (1970)

2.55 M.S. Dixit, H.L. Anderson, C.K. Hargrove, R.J. McKee, D. Kessler, H. Mes, A.C. Thompson: Phys. Rev. Lett. 27, 878 (1971)

2.56 H.K. Walter, J.H. Vuilleumier, H. Backe, F. Boehm, R. Engfer, A.H. v. Gunten, R. Link, R. Michaelson, G. Petitjean, L. Schellenberg, H. Schneuwly, W.U. Schröder, A. Zehnder: Phys. Lett. 40B, 197 (1972)

2.57 L. Tauscher, G. Backenstoss, K. Fransson, H. Koch, A. Nilsson, J. De Raedt: Phys. Rev. Lett. 35, 410 (1975)

2.58 M.S. Dixit, A.L. Carter, E.P. Hincks, D. Kessler, J.S. Wadden, C.K. Hargrove, R.J. McKee, H. Mes, H.L. Anderson: Phys. Rev. Lett. 35, 1633 (1975)

2.59 J.L. Vuilleumier, W. Dey, R. Engfer, H. Schneuwly, H.K. Walter, A. Zehnder: Z. Physik A 278, 109 (1976)

2.60 S.R. Lundeen, F.M. Pipkin: Phys. Rev. Lett. 34, 1368 (1975)

2.61 D.A. Andrews, G. Newton: Phys. Rev. Lett. 37, 1254 (1976)

2.62 J. Schwinger: Phys. Rev. 82, 664 (1951)

2.63 W. Pauli, F. Villars: Rev. Mod. Phys. 21, 434 (1949)

2.64 R.C. Barrett, S.J. Brodsky, G.W. Erickson, M.H. Goldhaber: Phys. Rev. 166, 1589 (1968)

2.65 J. Blomqvist: Nucl. Phys. B48, 95 (1972)

2.66 K. Huang: Phys. Rev. A 14, 1311 (1976)

2.67 P. Vogel: At. Data and Nucl. Data Tables 14, 599 (1974)

2.68 T.L. Bell: Phys. Rev. A 7, 1480 (1973)

2.69 M.K. Sundaresan, P.J.S. Watson: Phys. Rev. Lett. 29, 15 (1972)

2.70 B. Fricke: Z. Physik 218, 495 (1969)

2.71 L.S. Brown, R.N. Cahn, L.D. McLerran: Phys. Rev. Lett. 33, 1591 (1974)

2.72 G.A. Rinker, Jr., L. Wilets: Phys. Rev. Lett. 31, 1559 (1973)

2.73 G.A. Rinker, Jr., L. Wilets: Phys. Rev. A 12, 748 (1975)

2.74 G. Källén, A. Sabry: Dan. Mat. Fys. Medd. 29, no. 17 (1955)

2.75 L.W. Fullerton, G.A. Rinker, Jr.: Phys. Rev. A 13, 1283 (1976)

2.76 H.A. Bethe, J.W. Negele: Nucl. Phys. A117, 575 (1968)

2.77 S. Klarsfeld, A. Maquet: Phys. Lett. 43B, 201 (1973)

2.78 M. Chen: Phys. Rev. Lett. 34, 341 (1975)

2.79 L. Wilets, G.A. Rinker, Jr.: Phys. Rev. Lett. 34, 339 (1975)

2.80 D.H. Fujimoto: Phys. Rev. Lett. 35, 341 (1975)

2.81 E. Borie: Bull. Amer. Phys. Soc. 21, 625 (1976)

2.82 M.K. Sundaresan, P.J.S. Watson: Phys. Rev. D 11, 230 (1975)

2.83 S.L. Adler: Phys. Rev. D 10, 3714 (1974)

2.84 E. Borie: Helv. Phys. Acta 48, 671 (1975)

2.85 H. Grotch, D.R. Yennie: Rev. Mod. Phys. 41, 350 (1969)

2.86 R.K. Cole, Jr.: Phys. Lett. 25B, 178 (1967)

2.87 T.E.O. Ericson, J. Hüfner: Nucl. Phys. B47, 205 (1972)

2.88 R.R. Harvey, J.T. Caldwell, R.L. Bramblett, S.C. Fultz: Phys. Rev. 136, B126 (1964)

2.89 P. Vogel: Phys. Rev. A 7, 63 (1973)

2.90 B. Fricke: Lett. Nuovo Cimento 2, 859 (1969)

2.91 H.L. Anderson: Proc. of the 3rd Int. Conf. on High Energy Physics and Nuclear Structure, New York, September 1969, ed. by S. Devons (Plenum Press, New York 1970) p. 640

2.92 J.B. Mann, G.A. Rinker, Jr.: Phys. Rev. A 11, 385 (1975)

2.93 J. Rafelski, B. Müller, G. Soff, W. Greiner: Ann. Phys. (N.Y.) 88, 419 (1974)

2.94 P. Vogel: Phys. Rev. A 8, 2292 (1973)

2.95 D.E. Casperson, T.W. Crane, V.W. Hughes, P.A. Souder, R.D. Stambaugh, P.A. Thompson, H. Orth, G. zu Putlitz, H.F. Kaspar, H.W. Reist, A.B. Denison: Phys. Lett. 59B, 397 (1975)

2.96 E.R. Cohen, B.N. Taylor: J. Phys. Chem. Ref. Data 2, 663 (1973)

2.97 K.M. Crowe, J.F. Hague, J.E. Rothberg, A. Schenck, D. L. Williams, R.W. Williams, K.K. Young: Phys. Rev. D 5, 2145 (1972)

2.98 P.J.S. Watson, M.K. Sundaresan: Can. J. Phys. 52, 2037 (1974)

2.99 R.D. Deslattes, E.G. Kessler, W.C. Sauder, A. Henins: *Atomic Masses and Fundamental Constants 5*, ed. by J.H. Sanders and A.H. Wapstra (Plenum Press, New York 1976) p. 48

2.100 A. Bertin, G. Carboni, J. Duclos, U. Gastaldi, G. Gorini, G. Neri, J. Picard, O. Pitzurra, A. Placci, E. Polacco, G. Torelli, A. Vitale, E. Zavattini: Phys. Lett. 55B, 411 (1975); Nuovo Cimento 34A, 493 (1976)

2.101 E. Borie: Z. Physik A 275, 347 (1975)
2.102 G.A. Rinker: Phys. Rev. A 14, 18 (1976)
2.103 E. Campani: Lett. Nuovo Cimento 4, 982 (1970)
2.104 J.L. Friar, J.W. Negele: Phys. Lett. 46B, 5 (1973)
2.105 J. Bernabéu, C. Jarlskog: Nucl. Phys. B75, 59 (1974)
2.106 E.M. Henley, F.R. Krejs, L. Wilets: Nucl. Phys. A256, 349 (1976)
2.107 C. Joachain: Nucl. Phys. 25, 317 (1961)
2.108 J. Bernabéu, C. Jarlskog: Phys. Lett. 60B, 197 (1976)
2.109 S.L. Adler, R.F. Dashen, S.B. Treiman: Phys. Rev. D 10, 3728 (1974)
2.110 R. Barbieri: Phys. Lett. 56B, 266 (1975)
2.111 J. Bailey, K. Borer, F. Combley, H. Drumm, C. Eck, F.J.M. Farley, J.H. Field,
      W. Flegel, P.M. Hattersley, F. Krienen, F. Lange, G. Petrucci, E. Picasso,
      H.I. Pizer, O. Runolfsson, R.W. Williams, S. Wojcicki: Phys. Lett. 55B, 420
      (1975)
2.112 M.A. Samuel, C. Chlouber: Phys. Rev. Lett. 36, 442 (1976)
2.113 R. Jackiw, S. Weinberg: Phys. Rev. D 5, 2396 (1972)
2.114 L. Resnick, M.K. Sundaresan, P.J.S. Watson: Phys. Rev. D 8, 172 (1973)
2.115 D. Kohler, B.A. Watson, J.A. Becker: Phys. Rev. Lett. 33, 1628 (1974)
2.116 R. Barbieri, T.E.O. Ericson: Phys. Lett. 57B, 270 (1975)
2.117 W. Pieper, W. Greiner: Z. Physik 218, 327 (1969)
2.118 S.S. Gershtein, Ya.B. Zeldovich: Lett. Nuovo Cimento 1, 835 (1969)
2.119 S.S. Gershtein, Ya.B. Zeldovich: Zh. Eksper. I. Teor. Fiz. 57, 654 (1969);
      [Sov. Phys. JETP 30, 358 (1970)]
2.120 W. Heisenberg, H. Euler: Z. Physik 98, 714 (1936); V. Weisskopf, Kgl. Danske,
      Videnskab. Selskab, Mat. Fys. Medd. 14, no. 6 (1936)
2.121 R. Dashen: Proceedings of the 1975 International Symposium on Lepton and Pho-
      ton Interactions at High Energies, ed. by W.T. Kirk (SLAC, Stanford 1975)
      p. 981
2.122 H. Harari: Ref. [2.121], p. 317
2.123 P. Mokler, F. Folkmann: Chapter 6 this volume
2.124 W.E. Meyerhof: Science 193, 839 (1976)
2.125 K.M. Case: Phys. Rev. 80, 797 (1950)
2.126 A.M. Perelomov, V.S. Popov: Teor. Mat. Fiz. 4, 48 (1970); [Theor. Math. Phys.
      4, 664 (1970)]
2.127 C. Schwartz: J. Math. Phys. 17, 863 (1976)
2.128 L.I. Schiff, H. Snyder, J. Weinberg: Phys. Rev. 57, 315 (1940)
2.129 I. Pomeranchuk, J. Smorodinsky: J. Phys. (USSR) 9, 97 (1945)
2.130 F.G. Werner, J.A. Wheeler: Phys. Rev. 109, 126 (1958)
2.131 D. Rein: Z. Physik 221, 423 (1969)
2.132 V.S. Popov: Zh. ETF Pis. Red. 11, 254 (1970); [JETP Lett. 11, 162 (1970)]
2.133 V.S. Popov: Yad. Fiz. 12, 429 (1970); [Sov. J. Nucl. Phys. 12, 235 (1971)]
2.134 B. Müller, R.K. Smith, W. Greiner: Atomic Physics 4, ed. by G. zu Putlitz,
      E.W. Weber and A. Winnacker (Plenum Press, New York 1975) p. 209
2.135 S.J. Brodsky: Comments Atom. Mol. Phys. 4, 109 (1973)
2.136 L. Fulcher, A. Klein: Phys. Rev. D 8, 2455 (1973)
2.137 J.S. Greenberg, C.K. Davis, B. Müller, W. Greiner: Proceedings of the Inter-
      national Conference on Reactions Between Complex Nuclei, Vol. 2, ed. by R.L.
      Robinson, June 1974, at Nashville, Tennessee (American Elsevier Publ. Co.,
      New York 1974) p. 67
2.138 Ya.B. Zeldovich, V.S. Popov: Usp. Fiz. Nauk. 105, 403 (1971); [Sov. Phys. Usp.
      14, 673 (1972)]
2.139 J. Rafelski, B. Müller, W. Greiner: Nucl. Phys. B68, 585 (1974)
2.140 V.S. Popov: Zh. Eksper. I. Teor. Fiz. 59, 965 (1970); [Sov. Phys. JETP 32,
      526 (1971)]
2.141 L.P. Fulcher, W. Greiner: Lett. Nuovo Cimento 2, 279 (1971)
2.142 P.G. Reinhard, W. Greiner, H. Arenhövel: Nucl. Phys. A166, 173 (1971)
2.143 B. Müller, H. Peitz, J. Rafelski, W. Greiner: Phys. Rev. Lett. 28, 1235 (1972)
2.144 B. Müller, J. Rafelski, W. Greiner: Z. Physik 257, 62 (1972)
2.145 B. Müller, J. Rafelski, W. Greiner: Z. Physik 257, 183 (1972)
2.146 V.S. Popov: Yad. Fiz. 14, 458 (1971); [Sov. J. Nucl. Phys. 14, 257 (1972)]
2.147 V.S. Popov: Yad. Fiz. 15, 1069 (1972); [Sov. J. Nucl. Phys. 15, 595 (1972)]
2.148 V.S. Popov: Zh. ETF Pis. Red. 16, 355 (1972); [JETP Lett. 16, 251 (1972)]
2.149 V.S. Popov: Yad. Fiz. 17, 621 (1973); [Sov. J. Nucl. Phys. 17, 322 (1973)]
2.150 J. Rafelski, W. Greiner, L.P. Fulcher: Nuovo Cimento 13B, 135 (1973)

2.151 V.S. Popov: Zh. Eksper. I. Teor. Fiz. 65, 35 (1973); [Sov. Phys. JETP 38, 18 (1974)]
2.152 B. Müller, J. Rafelski, W. Greiner: Nuovo Cimento 18A, 551 (1973)
2.153 V.S. Popov: Yad. Fiz. 19, 155 (1974); [Sov. J. Nucl. Phys. 19, 81 (1974)]
2.154 K. Smith, B. Müller, W. Greiner: J. Physik B 8, 75 (1975)
2.155 S.S. Gershtein, V.S. Popov: Lett. Nuovo Cimento 6, 593 (1973)
2.156 M.S. Marinov, V.S. Popov: Zh. Eksper. I. Teor. Fiz. 68, 421 (1975); [Sov. Phys. JETP 41, 205 (1975)]
2.157 B. Müller, J. Rafelski, W. Greiner: Phys. Lett. 47B, 5 (1973)
2.158 H. Peitz, B. Müller, J. Rafelski, W. Greiner: Lett. Nuovo Cimento 8, 37 (1973)
2.159 M.S. Marinov, V.S. Popov: Yad. Fiz. 20, 1223 (1974); [Sov. J. Nucl. Phys. 20, 641 (1975)]
2.160 D.H. Jakubassa, M. Kleber: Z. Physik A 277, 41 (1976)
2.161 M.S. Marinov, V.S. Popov, V.L. Stolin: J. Comp. Phys. 19, 241 (1975)
2.162 B. Müller, R.K. Smith, W. Greiner: Phys. Lett. 53B, 401 (1975)
2.163 M.S. Marinov: private communication (1976)
2.164 V.S. Popov, V.D. Mur: Yad. Fiz. 18, 684 (1973); [Sov. J. Nucl. Phys. 18, 350 (1974)]
2.165 D.R. Bates, H.S.W. Massey: Phil. Mag. 45, 111 (1954); D.R. Bates: In *Atomic and Molecular Processes*, ed. by D.R. Bates (Academic Press, New York 1962) p. 613
2.166 D.R. Bates: Ref. [2.165], p. 608
2.167 U. Fano: Phys. Rev. 124, 1866 (1961)
2.168 K. Smith, H. Peitz, B. Müller, W. Greiner: Phys. Rev. Lett. 32, 554 (1974)
2.169 M.S.W. Massey: Rep. Prog. Phys. 12, 248 (1949)
2.170 J. Rafelski, A. Klein: Proceedings of the International Conference on Reactions Between Complex Nuclei, Vol. 2, ed. by R.L. Robinson, June 1974, at Nashville, Tennessee (American Elsevier Publ. Co., New York 1974) p. 397
2.171 W. Greiner: Proceedings of the Third General European Physical Society Conference (European Physical Society, Petit-Lancy, Switzerland 1976) p. 257
2.172 J.S. Briggs: Rep. Prog. Phys. 39, 217 (1976)
2.173 Blasche, Franke, Ch. Schmelzer: quoted in Ref. [2.171]
2.174 D. Burch, W.B. Ingalls, H. Wieman, R. Vandenbosch: Phys. Rev. A 10, 1245 (1974)
2.175 W.E. Meyerhof: Phys. Rev. A 10, 1005 (1974)
2.176 C. Foster, T. Hoogkamer, P. Worlee, F.W. Saris: Abstracts submitted to IX International Conference on Physics of Electronic and Atomic Collisions, ed. by J.S. Risley and R. Geballe (University of Washington Press 1975) p. 511
2.177 W. Betz, G. Soff, B. Müller, W. Greiner: Frankfurt/Main preprint (1976)
2.178 L.B. Okun: Dokl. Akad. Nauk. SSSR 89, 883 (1953)
2.179 V. Oberacker, G. Soff, W. Greiner: Phys. Rev. Lett. 36, 1024 (1976)
2.180 V. Oberacker, G. Soff, W. Greiner: Nucl. Phys. A259, 324 (1976)
2.181 G. Soff, B. Müller, J. Rafelski: Z. Naturforsch. A29, 1267 (1974)
2.182 J. Rafelski, B. Müller: Phys. Rev. Lett. 36, 517 (1976)
2.183 E. Hu, Y. Asano, M.Y. Chen, S.C. Cheng, G. Dugan, L. Lidofsky, W. Patton, C.S. Wu, V. Hughes, D. Lu: Nucl. Phys. A254, 403 (1975)
2.184 H.L. Anderson, C.K. Hargrove, E.P. Hincks, J.D. McAndrew, R.J. McKee, R.D. Barton, D. Kessler: Phys. Rev. 187, 1565 (1969)
2.185 T. Erber, D. White, W. Tsai, H.G. Latal: Ann. Phys. (N.Y.) 102, 405 (1976)
2.186 R. Staffin: private communication
2.187 A.B. Migdal: Zh. Eksper. I. Teor. Fiz. 61, 2209 (1972); [Sov. Phys. JETP 34, 1184 (1972)]

3. Relativistic Effects in Highly Ionized Atoms

L. Armstrong Jr.

In this chapter, we investigate how relativity manifests itself in highly ionized atoms. We shall study its effects on both atomic structure and transition probabilities since it is through transitions that one studies the structure. Our emphasis will not be on a review of what has been done but rather on providing all the pertinent concepts and equations necessary for a basic understanding of the subject.

In Section 3.1, we give a brief qualitative picture of relativistic effects in highly ionized atoms. This hopefully will serve to orient the reader by providing a rough outline of the Z dependence and relative magnitudes of various effects. In Section 3.2 we discuss the interactions in atoms from two different views: in Section 3.2.1, we describe the relativistic Hamiltonian; in Section 3.2.2, we consider the nonrelativistic Hamiltonian with relativistic corrections. Relativistic transition operators and associated quantities such as oscillator strengths are discussed in Section 3.3. In Section 3.4, we outline methods of obtaining relativistic wave functions (e.g., Dirac-Hartree-Fock). Finally, in Section 3.5, we review some of the results of relativistic calculations.

3.1 Background

Physicists often speak of "turning on" or "turning off" some basic and obviously unchangeable interaction such as the Coulomb interaction. (Theoreticians are particularly prone to do this, as it allows them to define solvable problems). A rare occasion to effectively do this, that is to change the strength of an interaction, occurs in studies of isoelectronic sequences. By looking at the structure and interactions in a neutral atom and the sequence of ions having the same number of electrons as the atom, but with increasing nuclear charge, one can see the effects of "turning up" the strength of the electron-nucleus interaction while holding all other interactions constant. Since it is the strength of this electron nucleus interaction which determines whether or not the energy of the electron will be relativistic, isoelectronic sequences are obviously a reasonable starting point for our study of relativistic effects in highly ionized atoms.

It is probably useful to have a very rough qualitative picture of what can be expected to happen as one moves up such a sequence before considering the exact forms of the rather complicated relativistic operators and their effects. Such a picture can be obtained by imagining that the electronic wave functions are hydrogenic. The Schrödinger energy of each electron will, in that case, be simply $E_n = -(Z^2/n^2)R$, where R is the Rydberg; that is, the energy increases negatively as Z^2. Obviously, we must also add the spin orbit energy of each electron, $H_{so} = a_{so}\underline{\ell}\cdot\underline{s}$, where

$$a_{so} = - E_n \frac{(Z\alpha)^2}{n\ell(\ell+1/2)(\ell+1)} \quad . \tag{3.1}$$

Thus, the strength of the spin orbit interaction increases as Z^4 as we move along the sequence, and is approximately $(Z\alpha)^2$ as large as E_n. Since $(Z\alpha)^2 \sim (v/c)^2$ for hydrogenic functions, we can obviously label this as a relativistic effect. Evaluating the angular factor $\underline{\ell}\cdot\underline{s}$, one finds that H_{so} is positive for electrons with $j_+ = \ell+1/2$, and negative for electrons with $j_- = \ell-1/2$.

Clearly, the further one moves along the sequence, the larger the energy of the electrons. Thus, one would expect to begin to see familiar relativistic effects such as mass variation of the electron; this contributes an energy

$$E_m = E_n \frac{(Z\alpha)^2}{n^2} \left(\frac{n}{\ell+1/2} - 3/4 \right) \quad (\ell \neq 0) \tag{3.2}$$

to each electron. Another contribution of the same order is the Darwin term, which affects only s electrons:

$$E_s = - E_n \frac{(Z\alpha)^2}{n} \quad . \tag{3.3}$$

Both of these terms are of the same relative strength as the spin orbit interaction, and we can add them all together to give a quantity which we denote as E_R:

$$E_R = E_n \frac{(Z\alpha)^2}{n^2} \left(\frac{n}{j+1/2} - 3/4 \right) \quad . \tag{3.4}$$

The net effect of these three relativistic terms is to increase the energy of each electron in a negative sense beyond the energy predicted by the Schrödinger equation, E_n. In addition, the importance of this additional energy given by E_R increases as $(Z\alpha)^2$, and decreases with increasing n. Note also that the energy of the j_+ electron is increased less than that of the j_-.

We could, of course, have obtained these results by starting directly with the Dirac equation for a hydrogenic electron. The energy in this case is given by the rather more complicated expression [3.1]

$$E_D = mc^2 / \left(1 + \left\{Z\alpha/[n-k+(k^2-Z^2\alpha^2)^{1/2}]\right\}^2\right)^{1/2} \qquad (3.5)$$

where $k=j+1/2$. Expanding E_D to order Z^4, one obtains the rest energy of the electron, mc^2, plus E_n+E_R. This formula is, of course, more correct than that of E_n+E_R, since it contains terms of all orders in $(Z\alpha)^2$. We see, however, that even when the more correct expression is used, the relativistic energy will always be lower (larger negative) than the Schrödinger energy. The complexity of E_D rather precludes its use in obtaining simple physical pictures of what is happening; for that reason, we shall use the simpler nonrelativistic equations for our qualitative expressions of this section.

We have thus far assumed that our wave functions are hydrogenic. This will be true only if there is no interaction between the electrons. This is clearly an unreasonable assumption, and we must take into account the interelectron Coulomb interactions if we are to obtain realistic results. Treating the interaction $H_C = \sum e^2/r_{ij}$ as a perturbation to the electron-nucleus interaciton, one obtains a first order contribution to the energy of the form

$$E_C \sim \left(\Psi_1\Psi_2 \left| \frac{e^2}{r_{12}} \right| \Psi_1\Psi_2 \right) \qquad (3.6)$$

where Ψ_1 and Ψ_2 are hydrogenic wave functions. Of course, E_C also depends on the number of electrons in the atom, but since we are considering only an isoelectronic sequence, this dependence is of no interest. Now, the matrix element of $1/r_{12}$ varies in Z just as the matrix element of $1/r$. Thus, we can immediately say that

$$E_C \sim Z \ R \sim 1/Z \ E_n \qquad .$$

The Z dependence of higher order terms in the perturbation series can also easily be seen: the n-th order term contains a product of n matrix elements such as appear in (3.6) in the numerator and a product of n-1 energy differences proportional to E_n in the denominator. Combining, we have that the n-th order perturbation term varies as $Z^n/Z^{2(n-1)} = Z^{-n+2}$. Thus, the introduction of the interelectronic Coulomb interaction means that the energy has an expansion of the form [3.2]

$$E = E_n(1+Z^{-1}A_1+Z^{-2}A_2+\ldots) \qquad (3.7)$$

exclusive of relativistic effects. This expansion shall be considered in more detail in Section 3.4.1. However, from this rough discussion, one can see that the importance of the interelectronic interaction decreases with increasing Z, in direct contrast to the importance of the relativistic effects, which increases with increasing Z.

One should also note that it is E_C which creates the term structure of the atom. L and S exist as good quantum numbers, of course, only so long as the spin orbit interaction remains weak. Since the spin orbit interaction increases in strength as Z^4, while E_C increases only as Z, the coupling scheme of the electrons will have more and more jj character and less and less LS character as Z increases along the isoelectronic sequence.

The two most important "relativistic" effects which remain are the Breit interaction [3.3], and the Lamb shift [3.4]. The Breit interaction, which we shall discuss in more detail in Section 3.2.1, is an approximation to the interaction between two electrons; it is of order $\alpha(Z\alpha)E_n$. The Lamb shift, which is smaller still, arises from the virtual emission and reabsorption of a photon by an electron; the magnitude of this effect varies roughly as $\alpha(Z\alpha)^2 E_n$. Because of its small size, we shall not consider this term further.

There are, of course, other interactions which can affect the electronic energies and structure. For example, the hyperfine interaction varies as $\mu_I \mu_N 1/r^3$, or roughly as $\alpha(Z\alpha)(m/M_p)E_n$, where M_p is the proton mass. Thus the importance of the hyperfine structure relative to the term structure will increase as Z increases.

In any discussion of atomic energy levels, transition probabilities must play a central role, since it is through transitions that energy levels are measured. The electric dipole transition is generally the dominant transition; the probability per unit time for an electric dipole transition A is proportional to

$$A \sim (\Delta E)^3 |(\Psi_f|r|\Psi_i)|^2 \equiv (\Delta E)^3 S(i,f) \tag{3.8}$$

where ΔE is the energy of the transition, Ψ_i is the initial state, and Ψ_f is the final state. The quantity $S(i,f)$ is called the line strength, and it varies roughly as $1/Z^2$. The Z dependence of ΔE for a $\Delta n \neq 0$ transition can be obtained roughly from E_n, i.e., it is proportional to Z^2. Thus one finds that

$$A \sim Z^4 \quad (\Delta n \neq 0) \quad . \tag{3.9}$$

In fact, the $\Delta n \neq 0$ transitions are often not the most interesting transitions in highly ionized atoms. This is because these transitions rapidly become very energetic with increasing Z. For example, if we take the energy of the electron to be given by E_n, then a transition from a state of n=3 to a state of n=2, which has a wavelength of about 6563 Å at Z=1, will have a wavelength of only 66 Å for Z=10. On the other hand, transitions in which the principal quantum number does not change, $\Delta n=0$, only increase in energy roughly as Z. That is because the energy of the $\Delta n=0$ transitions arises primarily through H_C, whose matrix elements we showed above are proportional to Z. Thus, for a $\Delta n=0$ transition, one finds that

$$A \sim Z \quad (\Delta n=0) \quad . \tag{3.10}$$

Another commonly calculated value is the oscillator strength f of the transition

$$f \sim |\Delta E| S(i,f) \tag{3.11}$$

which, if we use E_n to calculate ΔE, and take $S(i,f)$ to vary as $1/Z^2$, gives

$$f \sim const \quad (\Delta n \neq 0) \quad . \tag{3.12}$$

However, if we take the interelectronic Coulomb interaction into account, both ΔE and $S(i,f)$ will take on a more complicated Z dependence. To obtain the Z dependence of the latter, consider the effect of H_C on the wave functions:

$$\Psi_f \rightarrow \Psi_f + \sum_\alpha \frac{(\Psi_\alpha|H_C|\Psi_f)}{E_f - E_\alpha} \Psi_\alpha + \ldots = \Psi_f + \sum_\alpha Z^{-1} a_\alpha \Psi_\alpha + \ldots \quad . \tag{3.13}$$

Using this expansion to calculate $S(i,f)$, and (3.7) to obtain ΔE, one finds that f can be expanded as

$$f = f_0 + 1/Z\, f_1 + \ldots \quad (\Delta n \neq 0) \quad . \tag{3.14}$$

For $\Delta n = 0$ transitions, one obtains in a like fashion

$$f = 1/Z\, f_0 + \ldots \quad (\Delta n = 0) \quad . \tag{3.15}$$

The introduction of relativistic effects, such as the spin orbit interaction, (3.1), also leads to a perturbation expansion in the wave function:

$$\Psi_f \rightarrow \Psi_f + \sum_\beta \frac{(\Psi_\beta|H_{so}|\Psi_f)}{E_f - E_\beta} \Psi_\beta + \ldots = \Psi_f + \sum (Z\alpha)^2 b_\beta \Psi_\beta + \ldots \quad , \tag{3.16}$$

i.e., an expansion in powers of $(Z\alpha)^2$. If we include both this and the relativistic terms in the energy expansion, as well as the effects of the Coulomb interaction H_C, the expression for the f value becomes a double expansion

$$f = \sum f_{nm} Z^{-n} (Z\alpha)^{2m} \quad . \tag{3.17}$$

This rather uninformative-looking double expansion can be understood somewhat better if we look at $S(i,f)$ separately from the energy. In our studies, we have found that for many cases, a very good approximation to $S(i,f)$ is

$$S(i,f) = S_D(i,f)(1 + A/Z + \ldots) \tag{3.18}$$

where $S_D(i,f)$ is the line strength evaluated with Dirac hydrogenic wave functions. These functions have been studied by YOUNGER and WEISS [3.5]. The relativistic part of $S(i,f)$ is thus to a pretty fair approximation absorbed into $S_D(i,f)$, and the effects of the Coulomb operator are once again manifested in an expansion in powers of $1/Z$. A handy rule of thumb which describes the results of YOUNGER and WEISS [3.5] (and which has been confirmed by our studies of the Li and Be sequences [3.6]) is that relativistic effects on the line strength are stronger for $|\Delta j|=|\Delta n|$ transitions than for $|\Delta j|\neq|\Delta n|$ transitions.

One can also make a few simple arguments which give some indication of where relativistic effects will be strongest in transition energies. For $\Delta n=0$, the energy of the $\Delta j=0$ transition must, in leading order, be proportional to E_C (3.6), that is, to Z, since hydrogenic states of the same j are degenerate, (3.4); for $|\Delta j|=1$ transitions the leading terms arise not only from E_C, but also from the relativistic splitting of states of different j. Thus, for $\Delta n=0$ transitions, the $|\Delta j|=1$ transitions show the greatest influence of relativity, and the transition energy has an expansion of the form

$$\Delta E^0 = Z(E_{1,0}+Z^3\alpha^2 E_{4,2}+\ldots) \quad . \tag{3.19}$$

For $|\Delta n|\neq 0$ transitions on the other hand, the transition energy can be expanded in the form:

$$\Delta E^1 = Z^2(E_{2,0}+Z^2\alpha^2 E_{4,2}+\ldots) \quad . \tag{3.20}$$

i.e., the proportional importance of relativity will not grow as rapidly as for the $\Delta n=0$, $|\Delta j|=1$ transition.

Combining our results, one sees that for $\Delta n=0$, $|\Delta j|=1$ transitions, the effects of relativity on the transition energy will probably be more important than the effects on the line strength. For other transitions, relativity will probably be equally important in both components of the oscillator strength.

One type of transition whose f increases with increasing Z is that which arises due to the breakdown of LS coupling—the intercombination lines. The spin orbit interaction mixes states of different L and S, permitting transitions previously forbidden because of selection rules on L and S. The dependence on Z of the f value associated with this type of transition is easily obtained using the arguments above; one finds

$$f \sim Z^6\alpha^4 \quad (\Delta n\neq 0)$$
$$f \sim Z^5\alpha^4 \quad (\Delta n=0) \quad . \tag{3.21}$$

We have thus far mentioned only electric dipole transitions. As Z increases, so will the probability of transitions which arise from higher order multipoles in the expansion of the electromagnetic fields. The magnetic dipole and electric quadrupole transition operators are of the order of $(Z\alpha)^2$ times the electric dipole operator, etc. Thus, with increasing Z, transitions which violate the nonrelativistic electric dipole selection rules will begin to appear. As we shall see later, relativity also changes the selection rules for electric dipole transitions themselves—again a $(Z\alpha)^2$ effect.

This brief introduction should give some idea of the types of systematic effects which will be observed in moving along an isoelectronic sequence. In the sections which follow, we will go into the specifics of the various effects which we have touched upon here.

3.2 Interactions

To make the discussion of the previous section more quantitative, we need to discuss the Hamiltonians suitable for calculations of highly ionized atoms. We shall first discuss the existing relativistic Hamiltonians. As we shall see, these Hamiltonians are not altogether satisfactory, and at times it may be preferable to start with the nonrelativistic Schrödingers equation and add the Breit-Pauli terms, which provide relativistic corrections of order $(Z\alpha)^2$ to the Schrödinger equation results.

3.2.1 Relativistic Interactions

The Dirac Equation

The starting point for any calculation involving relativistic electrons is naturally the Dirac equation [3.1,7]. We shall discuss only the form of the Dirac equation for an electron in a central potential. A more complete discussion of the general properties of this relativistic wave function can be found in, for example, DIRAC [3.1], MESSIAH [3.8], and BETHE and SALPETER [3.9].

The Hamiltonian operator for a relativistic electron in a central potential $U(r)$ is of the form

$$H_D = c\underline{\alpha} \cdot \underline{p} + \beta mc^2 - e\, U(r) \tag{3.22}$$

where $\underline{\alpha}$ is given in terms of the Pauli 2×2 $\underline{\sigma}$ matrices, and β in terms of a 2×2 unit matrix by

$$\underline{\alpha} = \begin{pmatrix} 0 & \underline{\sigma} \\ \underline{\sigma} & 0 \end{pmatrix} \quad \text{and} \quad \beta = \begin{pmatrix} I & 0 \\ 0 & I \end{pmatrix} \ . \tag{3.23}$$

The charge of the electron is -e.

The Dirac wave equation

$$H_D |\psi) = E|\psi) \tag{3.24}$$

has some properties which are quite different from the Schrödinger equation. One of these is that the solutions to (3.24) appear in pairs. This arises because the relativistic energy equation (neglecting potential energy)

$$E^2 = c^2(p^2+m^2c^2) \tag{3.25}$$

has two solutions given by

$$E = \pm mc^2 \left(1 + \frac{p^2}{m^2c^2}\right)^{1/2} . \tag{3.26}$$

Thus, for a given momentum, there are two solutions, one having an energy $E \approx mc^2$ (if $|p| << mc^2$), the other an energy $E \approx -mc^2$. These two solutions correspond to electrons and positrons, respectively. Addition of a small potential $U(r)$ does not change this situation, and (3.24) will therefore have solutions corresponding to both positive and negative eigenvalues. Obviously, (3.25) and (3.26) imply that any correct relativistic equation must have solutions which appear in pairs in this way.

   The general form of solutions of (3.24) can easily be found. One notices, firt that

$$(H_D,\underline{\ell}) = -1/2(H_D,\underline{\sigma}) = ic(\underline{\alpha}\times\underline{p}) \tag{3.27}$$

where $*\hbar\underline{\ell}=\underline{r}\times\underline{p}$ is the orbital angular momentum. Thus, neither $\underline{\ell}$ nor the spin $\underline{s}=1/2\underline{\sigma}$ commutes with H_D, but their vector sum $\underline{j}=\underline{\ell}+\underline{s}$ will. This, of course, implies that $|\psi)$ of (3.24) must be an eigenfunction of j^2 [with eigenvalue $j(j+1)$] and can be constructed so as to also be an eigenfunction of j_z (with eigenvalue m_j). This in turn suggests that H_D is probably separable into radial and angular equations with the angular eigenfunctions being expressible in terms of eigenfunctions $|1/2\ell jm_j)$ of j^2 and j_z:

$$|1/2\ell jm_j) = \sum (-1)^{1/2-\ell-m_j} [j]^{1/2} \begin{pmatrix} 1/2 & \ell & j \\ m_s & m_\ell & -m_j \end{pmatrix} Y_{\ell m_\ell}(\theta,\varphi) \chi_{m_s}^{(1/2)} \tag{3.28}$$

with

$$\chi_{1/2}^{(1/2)} = \begin{pmatrix} 1 \\ 0 \end{pmatrix} , \quad \chi_{-1/2}^{(1/2)} = \begin{pmatrix} 0 \\ 1 \end{pmatrix} , \quad \text{and } [j] = 2j+1 .$$

$*\hbar=h/2\pi$ (normalized Planck's constant)

In fact, one can show in a straightforward fashion that, if the eigenfunction $|\psi)$ is written in the form

$$|\psi) = \left| \begin{array}{l} \frac{F}{r} \; |1/2 \; \ell jm_j) \\ \frac{iG}{r} \; |1/2 \; \bar{\ell} jm_j) \end{array} \right) \tag{3.29}$$

where $\bar{\ell} = 2j - \ell$, (3.24) can be reduced to two coupled first order differential equations

$$\left(\frac{d}{dr} + \frac{\kappa}{r} \right) F = -\frac{1}{\hbar c} \; [E + mc^2 + e \; U(r)] \; G(r)$$

$$\left(\frac{d}{dr} - \frac{\kappa}{r} \right) G = -\frac{1}{\hbar c} \; [-E + mc^2 - e \; U(r)] \; F(r) \tag{3.30}$$

with $\kappa = (-1)^{\ell+j+1/2} [j]/2$. Normalization of the wave functions leads to the further condition

$$\int (F^2 + G^2) dr = 1 \quad . \tag{3.31}$$

Obviously, F and G are functions of both ℓ and j; as in the case of the corresponding nonrelativistic radial wave function, they can also be classified according to a principal quantum number n which characterizes the number of nodes of these functions. In the case of F, there are $n-\ell-1$ nodes; and for G, there are $n-|\kappa|$ nodes (both numbers excluding the origin and infinity). If we approximate the eigenvalue E by mc^2, we see [using (3.30)] that for electrons, G is of the order of $(\hbar/mcr)F \sim (Z\alpha)F$; as a consequence, the part of $|\psi)$ containing F is called the large component, and the part containing G is called the small component. The situation is just reversed for positrons.

If we take for U(r) the Coulomb potential of the nucleus Ze/r and solve (3.24), we obtain E_D of (3.5). Thus, from the discussion of Section 3.1, we see that H_D contains within it such effects as the mass correction term, the Darwin term, and the spin orbit interaction. This can, of course, be demonstrated directly by a reduction of H_D to nonrelativistic limits; details of such reductions can be dound in, for example, MESSIAH [3.8] or ARMSTRONG and FENEUILLE [3.10]. Such a reduction shows, as might be expected, that for a more general choice of U(r), H_D still contains all of these effects with, however, the values suitable to U(r) rather than to Ze/r. Thus, the Darwin term becomes

$$- \frac{e\hbar^2}{8m^2c^2} \; \nabla^2 U \tag{3.32}$$

and the spin orbit operator is given by

$$\frac{-e\hbar^2}{2m^2c^2} \frac{1}{r} \frac{dU}{dr} (\underline{\ell} \cdot \underline{s}) \qquad . \qquad (3.33)$$

The Dirac equation can be thought of as containing these interactions to all orders of perturbation theory.

Two-Body Interactions

We have thus far considered the interaction of an electron with a central potential $U(r)$. In reality, the atomic electrons do not move in a purely central field, but in the field produced by the nucleus (which is central) and the field produced by all the other electrons (which is not, in general, central). We must therefore explicitly consider the electron-electron interactions. Unfortunately, one cannot at this time write down in Hamiltonian form the electron-electron interaction correct to all orders in $Z\alpha$. The instantaneous Coulomb interaction is, of course, correctly given by the operator $\sum e^2/r_{ij}$; however, the S matrix term for the interchange of a single transverse photon cannot be put into the form of the matrix element of an operator [3.11,12]. Nevertheless, this transverse photon term can be approximated to give a Hamiltonian-like operator which is accurate roughly to order $(Z\alpha)^4$; the electron-electron interaction then becomes:

$$H_e = \sum \frac{e^2}{r_{ij}} + H_B \qquad (3.34)$$

where H_B, the Breit interaction [3.3], is given by

$$H_B = - \sum \frac{e^2}{2r_{ij}} \left[\underline{\alpha}_i \cdot \underline{\alpha}_j + \frac{(\underline{\alpha}_i \cdot \underline{r}_{ij})(\underline{\alpha}_j \cdot \underline{r}_{ij})}{r_{ij}^2} \right] \qquad . \qquad (3.35)$$

The total Hamiltonian for the atomic electrons can then be written as

$$H_R = \sum \left(c\underline{\alpha}_i \cdot \underline{p}_i + \beta_i mc^2 - \frac{Ze^2}{r_i} \right) + \sum \frac{e^2}{r_{ij}} + H_B \equiv H_c + H_B \qquad . \qquad (3.36)$$

The approximation made in obtaining H_B involves, however, the discarding of the electron-positron and positron-positron interactions, keeping only the electron-electron interaction. The immediate consequence of this truncation of the interaction is that H_B can only be introduced into the problem as a perturbation, and then only to first order [3.9]. That is, if one uses H_B in a second (or higher) order expression, the sum over intermediate states will include states of negative energy.

Since H_B does not correctly describe the interaction which changes positive to negative energy particles, these terms will be given incorrectly. Then, if we view
solving the equation

$$H_R|\psi) = E|\psi) \qquad\qquad (3.37)$$

as being formally equivalent to solving

$$H_c|\psi) = E_c|\psi) \qquad\qquad (3.38a)$$

and evaluating H_B to all orders in perturbation theory, the above limitation on the
use of H_B implies that (3.37) is not a valid equation.

The statement which is generally made at this point is that one should therefore
solve (3.38a) and then treat H_B as a first order perturbation. However, there is
reason to believe that there are difficulties involved even in solving (3.38a)[1].
Consider the two-electron calculation in which one first evaluates

$$H_2|\psi) = E_2|\psi)$$

where

$$H_2 = \sum_{i=1}^{2} \left(c\underline{\alpha}_i \cdot \underline{p}_i + \beta_i mc^2 - \frac{Ze^2}{r_i} \right)$$

and then treat the perturbation e^2/r_{12} to all orders. This should be equivalent to
solving (3.38a) in the two-electron case. Now, the simplified eigenvalue equation
involving H_2 must have eigenfunctions $|++)$ corresponding to two positive energy par-
ticles, $|+-)$ corresponding to a positive and a negative energy particle, and $|--)$
corresponding to two negative energy particles; this complete set of states must be
used in evaluating the perturbation produced by e^2/r_{12}. Let us take any pair of bound
positive energy states $|++)$ with a total combined energy of the two states being
E_{++} for our zeroth order wave function. However, for any $|++)$, there is an infinite
set of states $|+-)$ with the combined energy of the positive and negative energy
states being E_{+-} such that

$$E_{++} = E_{+-} \qquad .$$

---

[1] I thank Dr. J. Sucher for the following argument.

Thus, one cannot use simple nondegenerate perturbation theory to evaluate the first order correction to $|++)$; rather, one is faced with using degenerate perturbation theory in a situation of infinite zeroth order degeneracy. In such a case, it is not possible even to show that the problem has bound state solutions.

These problems can be circumvented by replacing e^2/r_{12} by $\Lambda_{++}^{12}(e^2/r_{12})\Lambda_{++}^{12}$, where Λ_{++}^{12} is a projection operator which leaves only positive energy particles:

$$\Lambda_{++}^{12}|++) = |++) \quad , \quad \Lambda_{++}^{12}|+-) = 0 \quad , \quad \text{etc.}$$

If this replacement is made, the state $|++)$ is no longer coupled to the infinity of states $|+-)$, etc., by the perturbation, and bound state solutions can obviously be found. Thus, rather than solving (3.38a), one should first solve

$$H'_c|\psi) = \left[\sum_{i=1}^{n}\left(c\underline{\alpha}_i\cdot\underline{p}_i+\beta_imc^2-Ze^2/r_i\right) + \sum\Lambda_{++}^{ij}\left(\frac{e^2}{r_{ij}}\right)\Lambda_{++}^{ij}\right]|\psi) = E_c|\psi) \qquad (3.38b)$$

and then treat both H_B and the interactions included in (3.38a) but excluded in (3.38b) using perturbation theory.

The requirement that H_B be introduced only as a first order perturbation can be overcome by replacing H_B by $\Lambda_{++}H_B\Lambda_{++}$, where this latter interaction simply implies H_B of (3.35) multiplied on the left and right inside of the summation by Λ_{++}^{ij}. This modified Breit interaction can be used to all orders. Thus by implication, one can try to obtain solutions to

$$H'_R|\psi) = (H'_c+\Lambda_{++}H_B\Lambda_{++})|\psi) = E|\psi) \qquad (3.39)$$

and treat all contributions arising from negative energy states using perturbation theory. In fact, if one takes the $Z^4\alpha^2$ limit of the Bethe-Salpeter equation, one obtains (3.39) rather than (3.37) [3.13].

There are, in fact, cases in which H_B itself can be used in second order perturbation expressions. In particular, it can be used for corrections to matrix elements of the interaction of an electron with an $e\underline{\alpha}\cdot\underline{A}$ (see Sec.3.3), involving one interaction with the field and one interaction with H_B [3.14], i.e.,

$$\sum\frac{(\psi_1|e\underline{\alpha}\cdot\underline{A}|\psi')(\psi'|H_B|\psi_2)}{\Delta E} \quad .$$

Examples of this are found in the calculation of transition matrix elements [3.14], in studies of the Zeeman effect [3.15], and in investigations of hyperfine structure [3.16].

One can, of course, work directly with the complete S matrix expression for exchange of a transverse photon, rather than reducing it to its Hamiltonian-like ap-

proximation H_B. For example, MANN and JOHNSON [3.17] have computed the total energies of the ground states of selected atoms from Z=2 to Z=102 using both the complete
S matrix expression and also H_B. They find that for Z<48, H_B and the complete transverse interaction give energy contributions which agree to within 1 eV; by Pu, however, the difference is over 3 eV. In all cases, H_B overestimates the contribution
to the energy from the transverse photon.

The Central Field Approximation

In reality, (3.38b) cannot be solved exactly; the difficulty arises in this case
from the mathematics, however, rather than from the physics as was the case for
(3.37). There are many ways to find approximate solutions for (3.38b); we shall discuss some of them in detail in Section 3.4. The type of approximate solution to be
used depends on the type of problem which is being considered and the accuracy which
is desired. Most approximate solutions depend on some type of central field approximation. In this approximation, the effects of the operators

$$\sum \left(-\frac{Ze^2}{r_i} \right) + \sum \Lambda_{++}^{ij} \frac{e^2}{r_{ij}} \Lambda_{++}^{ij}$$

are approximated by a central potential term $-e\sum_i U(r_i)$. Then

$$H_R = \sum H_D(i) + H_p \qquad (3.40)$$

where $H_D(i)$ is H_D of (3.22) defined for the i-th electron, and

$$H_p' = e\sum U(r_i) + \sum \Lambda_{++}^{ij} \frac{e^2}{r_{ij}} \Lambda_{++}^{ij} - \sum \frac{Ze^2}{r_i} + \Lambda_{++}H_B\Lambda_{++} \qquad . \qquad (3.41)$$

By making this approximation, one is able to work with zero order wave functions
which, as explained in Section 3.2.1, p. 76, are separable into radial and angular
parts, and for which the radial parts satisfy known differential equations [(3.30)].
Any approach in which the wave functions are not separable leads, of course, to
immense complications.

3.2.2 Nonrelativistic Interactions

Since, as was explained in the previous section, no completely correct Hamiltonian
exists to describe the interaction between two relativistic electrons, it may in
many cases make more sense to begin with the nonrelativistic Schrödinger equation
and add on relativistic correction terms. Thus, one solves the many-electron Schrödinger equation

$$H_S |\psi\rangle = \left[\sum_i \left(\frac{p_i^2}{2m} - \frac{Ze^2}{r_i} \right) + \sum \frac{e^2}{r_{ij}} \right] |\psi\rangle = E|\psi\rangle \tag{3.42}$$

and then uses perturbation theory to treat the "Pauli limit" operators [3.9,12]

$$H_P = H_m + H_s + H'_{so} + H_{BP} \tag{3.43}$$

where H_m is the mass correction term of Section 3.1:

$$H_m = \frac{-p^4}{8m^3c^2} \quad . \tag{3.44}$$

H_s is the Darwin term, also discussed in Section 3.1, which in its form suitable
for the zeroth order Hamiltonian of (3.42) is

$$H_s = \frac{\hbar^2}{8m^2c^2} \left[\sum_{i<j} \left(\nabla_i^2 + \nabla_j^2 \right) \left(\frac{e^2}{r_{ij}} \right) - 2 Ze^2 \sum_i \nabla_i^2 \left(\frac{1}{r_i} \right) \right] \quad . \tag{3.45}$$

The term H'_{so} is

$$H'_{so} = \frac{\hbar}{4m^2c^2} \left[\sum_{i<j} \underline{\sigma}_i \cdot \underline{\nabla}_i \left(\frac{e^2}{r_{ij}} \right) \times \underline{p}_i + \sum_i \frac{Ze^2\hbar}{r_i^3} \underline{\sigma}_i \cdot \underline{\ell}_i \right] \tag{3.46}$$

and includes both the spin orbit interaction and part of the spin-other-orbit interaction. In both (3.45) and (3.46), the first term arises from the electron-electron
Coulomb interaction, and the second from the electron-nucleus interaction. The "spin
orbit" part of H'_{so} is obtained by taking only the electron-nucleus interaction and
that part of the electron-electron interaction which results when only the first
term in the expansion of e^2/r_{ij} is kept. The three terms H_m, H_s, and H'_{so} can be obtained by taking the nonrelativistic limit of H_c (3.36). Details can be found in,
for example, BETHE and SALPETER [3.9], or AKHIEZER and BERESTETSKI [3.12]. In a
sense, then, the relativistic equation

$$H_c |\psi\rangle = E_c |\psi\rangle \tag{3.47}$$

can be said to contain all the effects obtained by solving the nonrelativistic
equation

$$H_S |\psi\rangle = E_S |\psi\rangle \tag{3.48}$$

and treating $H_m + H_s + H'_{so}$ to all orders in perturbation theory.

The remaining term in H_p, H_{BP}, is obtained by reducing the Breit interaction (3.35) to the Pauli limit [3.9,12], i.e., to its lowest nonvanishing expression in a power series in $(Z\alpha)$. One finds in a straightforward way

$$
H_{BP} = \frac{\hbar}{2m^2c^2} \sum_{i<j} \left[\underline{\sigma}_j \cdot \underline{\nabla}_i \left(\frac{e^2}{r_{ij}} \right) \times \underline{p}_i \right] - \frac{e^2}{2m^2c^2} \sum_{i<j} \left[\frac{\underline{p}_i \cdot \underline{p}_j}{r_{ij}} \right.
$$

$$
+ \frac{\underline{r}_{ij}(\underline{r}_{ij} \cdot \underline{p}_i) \cdot \underline{p}_j}{r_{ij}^3} \right] + \frac{e^2\hbar^2}{4m^2c^2} \sum_{i<j} \left[\frac{\underline{\sigma}_i \cdot \underline{\sigma}_j}{r_{ij}^3} - \frac{3(\underline{\sigma}_i \cdot \underline{r}_{ij})(\underline{\sigma}_j \cdot \underline{r}_{ij})}{r_{ij}^5} \right]
$$

$$
- \frac{2e^2\hbar^2}{3m^2c^2} \pi \sum_{i<j} \underline{\sigma}_i \cdot \underline{\sigma}_j \, \delta(\underline{r}_{ij}) \quad , \tag{3.49}
$$

where $\underline{r}_{ij} = \underline{r}_i - \underline{r}_j$. The first of these terms is the remainder of the spin-other-orbit interaction; the second is the orbit-orbit interaction; the third is the spin-spin interaction, and the last is the spin-spin contact term. The operator H_{BP} can be used to all orders in perturbation, since it is a projection of the action of H_B onto positive energy states.

I began this section by saying that since the electron-electron interaction is known only to $(Z\alpha)^2 E_n$, it might in some cases be more sensible to use, instead of the relativistic interactions, the Breit-Pauli operators, which are the $(Z\alpha)^2 E_n$ reductions of the relativistic operators. There are those (e.g., [3.12]) who claim that it always makes more sense to use the Breit-Pauli operators. This argument is based upon the conviction that if any term of order $(Z\alpha)^n$ is to be kept, all terms of this and all lower orders must be kept. Thus, if one can use the $(Z\alpha)^2$ Breit operator only to first order, it makes no sense to treat the $(Z\alpha)^2$ spin orbit operator to higher than first order either. Since, as discussed above, the Dirac equation contains H_{so} to all orders, it then never makes sense to use the Dirac equation.

This argument does not really strike me as convincing for calculations in highly ionized atoms. As noted above, the Breit operator is really of relative order $\alpha(Z\alpha)E_n$, there being no $Z^2\alpha^2 E_n$ contribution to the transverse photon interaction. Thus, although for low Z the Breit interaction is of the same size as the $Z^2\alpha^2 E_n$, electron-nuclear part of the spin orbit interaction, for large Z the latter interaction is obviously considerably larger than the former. In fact, for moderately large Z, the next term in the electron-nucleus interaction, which is of order $Z^4\alpha^4 E_n$, will almost certainly be equal in magnitude to the $Z\alpha^2 E_n$ Breit interaction, etc. Therefore, using the Dirac equation, which describes this electron-nucleus to all orders, would seem to be quite sensible for highly ionized atoms, even if one does not know the electron-electron interaction exactly.

In general, I would expect results obtained using the relativistic equations of the previous section to be more accurate than results obtained using the equations of this section. However, calculations based on the Schrödinger equation are often

simpler to carry out than those based on the Dirac equation, and nonrelativistic
wave functions are certainly much more readily available than are relativistic wave
functions. Thus, when high accuracy is not needed, the equations of this section
may be preferable to those of the previous section.

3.3 Transition Operators

3.3.1 The Interaction

The interaction between a nonrelativistic electron and an external field is obtained
by replacing $\underline{p}$ by the conjugate momentum $\underline{\pi} = \underline{p} + (e/c)\underline{A}$, and E by $E + e\varphi$, where $\underline{A}$ is the
vector potential and φ the scalar potential of the field (we use Gaussian units).
The same replacement can be used in the Dirac equation (3.22) to obtain the interaction between a relativistic electron and an external field. It is these fields
which cause the transition from one relativistic state to another, each represented
by an eigenfunction of (3.22). The probability of such a transition will be proportional to the square of the matrix element of the interaction operator

$$M_{fi} = (\Psi_f | \sum (e\underline{\alpha}_j \cdot \underline{A}_j - e\varphi_j) | \Psi_i) \tag{3.50}$$

where the sum is over all electrons in the atom, and Ψ_i includes both the electronic
and field wave functions.

The vector and scalar potentials which belong to the Lorentz gauge are solutions
of the equations

$$\frac{1}{c^2}\frac{\partial^2 \underline{A}}{\partial t^2} - \nabla^2 \underline{A} = 0 \quad , \quad \frac{1}{c^2}\frac{\partial^2 \varphi}{\partial t^2} - \nabla^2 \varphi = 0 \quad , \quad \frac{1}{c}\frac{\partial \varphi}{\partial t} + \underline{\nabla} \cdot \underline{A} = 0 \quad . \tag{3.51}$$

For our purposes, the spherical wave solutions to (3.51) will be the most appropriate.
These solutions vary in time as $e^{\pm i\omega t}$, and are expressed in terms of spherical tensor
operators. This type of solution describes photon states having definite parity and
angular momentum. Thus, for electric type fields [3.12]

$$\underline{A}^e = \sum_{\omega, L, M} \left\{ a^e_{\omega, L, M} \underline{A}^e_{\omega, L, M}(r,t) + a^{e\dagger}_{\omega, L, M} \left[\underline{A}^e_{\omega, L, M}(r,t) \right]^* \right\} \tag{3.52}$$

where (normalized in the unit sphere)

$$A^e_{-\omega LM}(r,t) = i^{L+1}(4\pi\hbar\omega)^{1/2}\left\{\left(\frac{L}{2L+1}\right)^{1/2}Y_{-L,L+1,M}\,j_{L+1} - \left(\frac{L+1}{2L+1}\right)^{1/2}Y_{-L,L-1,M}\,j_{L-1}\right.$$

$$\left. - G\left[\left(\frac{L+1}{2L+1}\right)^{1/2}Y_{-L,L+1,M}\,j_{L+1} + \left(\frac{L}{2L+1}\right)^{1/2}Y_{-L,L-1,M}\,j_{L-1}\right]\right\}e^{-i\omega t} \tag{3.53}$$

and

$$\varphi^e = \Sigma\left\{a^e_{\omega LM}\,\varphi^e_{\omega LM}(r,t) + a^{e\dagger}_{\omega LM}\left[\varphi^e_{\omega LM}(r,t)\right]^*\right\} \tag{3.54}$$

with

$$\varphi^e_{\omega LM} = Gi^L(4\pi\hbar\omega)^{1/2}Y_{LM}j_L\,e^{-i\omega t} \quad . \tag{3.55}$$

In (3.52) and (3.54), $a^{e\dagger}_{\omega LM}$ and $a^e_{\omega LM}$ are, respectively, creation and annihilation operators of electric-type photons of energy $\hbar\omega$, angular momentum L and Z projection M, with parity $(-1)^L$. The quantity Y_{-JLM} is a vector spherical harmonic, defined by

$$Y_{-J,L,M} = \sum_{Qq} Y_{LQ}\,\varepsilon_q \begin{pmatrix} L & 1 & J \\ Q & q & -M \end{pmatrix}(-1)^{L+1+M}[J]^{1/2} \quad . \tag{3.56}$$

Y_{LQ} is a spherical harmonic, and $\underline{\varepsilon}$ is a tensor of rank one composed of the unit vectors ε_x, ε_y, and ε_z:

$$\varepsilon_0 = \varepsilon_z \quad , \quad \varepsilon_{\pm 1} = \mp 2^{-1/2}(\varepsilon_x \pm i\varepsilon_y) \quad .$$

$j_L \equiv j_L(\omega r/c)$ is a spherical Bessel function. The constant G is completely arbitrary; its choice corresponds to choice of gauge [3.12,18] within the restricted gauge transformation

$$\underline{A} \to \underline{A} + \underline{\nabla}\Lambda \quad ; \quad \varphi \to \varphi - \frac{1}{c}\frac{\partial\Lambda}{\partial t}$$

where $\Box^2\Lambda=0$. The Coulomb gauge corresponds to a vector potential having no longi-tudinal part; this is obtained for the case G=0.

Similarly, the magnetic type fields are given by [3.12]

$$\underline{A}^m = \sum_{\omega,L,M}\left\{a^m_{\omega LM}\,\underline{A}^m_{-\omega LM}(r,t) + a^{m\dagger}_{\omega LM}\left[\underline{A}^m_{-\omega LM}(r,t)\right]^*\right\} \tag{3.57}$$

where

$$\underline{A}^m_{-\omega LM}(r,t) = (4\pi\hbar\omega)^{1/2}i^L j_L Y_{-L,L,M}e^{-i\omega t} \tag{3.58}$$

and $\varphi^m = 0$. The $a^m_{\omega LM}$ and $a^{m+}_{\omega LM}$ are annihilation and creation operators for photons of angular momentum L, projection M, having parity $(-1)^L$ and energy $\hbar\omega$, respectively.

We can now calculate the probability of emission or absorption of a photon of angular momentum L, projection M and energy $\hbar\omega$. It will, in first order perturbation theory, be given by

$$A^{LM}_{fi} = \frac{2}{\hbar^2 c} \sum |M^{LM}_{fi}|^2 \tag{3.59}$$

where M^{LM}_{fi} is obtained by using only the field describing an L,M photon from either (3.52) or (3.57) in (3.50). The electric type field is used if the parity difference between the initial and final atomic states (ψ_i and ψ_f, respectively) is $(-1)^L$; the magnetic type field, if it is $(-1)^{L+1}$.

To evaluate the matrix element M^{LM}_{fi}, we take advantage of the tensorial properties of the field amplitudes $\underline{A}$ and φ to write [3.19]

$$M^{LM}_{fi} = (-1)^{J_f - M_f} (n)^{1/2} \begin{pmatrix} J_f & L & J_i \\ -M_f & M & M_i \end{pmatrix} \left(J_f || \sum_j h_{\omega LM\lambda}(j) || J_i \right) \tag{3.60}$$

where n is the number of photons present in the field before an absorption or after an emission, ($|| \quad ||$) is a reduced matrix element, and the operator $h_{\omega LM\lambda}$ is simply

$$h_{\omega LM\lambda} = \underline{\alpha} \cdot \underline{A}^\lambda_{LM} - e\varphi^\lambda_{LM} \qquad \text{(absorption)}$$

$$= \underline{\alpha} \cdot (\underline{A}^\lambda_{LM})^* - e(\varphi^\lambda_{LM})^* \qquad \text{(emission)} \tag{3.61}$$

for λ either e (electric) or m (magnetic). This many-body reduced matrix element of (3.60) can be further broken down to give M^{LM}_{fi} in terms of single-particle reduced matrix elements [3.19]:

$$M^{LM}_{fi} = (-1)^{M_f + L} \sqrt{n} \begin{pmatrix} J_f & L & J_i \\ -M_f & M & M_i \end{pmatrix} [J_f, J_i]^{1/2} \sum (-1)^{J_c + j_i} \begin{Bmatrix} J_f & L & J_i \\ j_i & J_c & j_f \end{Bmatrix} (J_f\{|J_c, j_f)$$

$$\times (J_i\{|J_c, j_i)(j_f ||h_{\omega L\lambda}||j_c) \quad . \tag{3.62}$$

The quantities ($\{|$) are coefficients of fractional parantage [3.19,20].

The reduced matrix elements appearing in (3.62) can be evaluated in a straightforward manner by noting that [3.10]

$$\underline{\alpha} \cdot \underline{Y}_{J,K,M} = [(2K+1)/4\pi]^{1/2} (\underline{\alpha} \cdot \underline{C}^{(K)})^J {}_M (-1)^{J+K+1} \tag{3.63}$$

where $C^K_q = [(2K+1)/4\pi]^{1/2} Y_{Kq}$, and using the usual rules of angular momentum algebra [3.19]. One finds, for electric transitions [3.6,18]

$$(1/2\ell j||h_{\omega L e}||1/2\ell'j') = ae(\hbar\omega)^{1/2} \ (-1)^{L+j'+1/2}$$

$$\times \begin{pmatrix} j & L & j' \\ 1/2 & 0 & -1/2 \end{pmatrix} \left(\frac{[j,j']}{[L]}\right)^{1/2} \left\{ \left[\left(\frac{L}{L+1}\right)^{1/2} - G\right]\right.$$

$$\times \left[(\kappa-\kappa') \ I^+_{L+1} + (L+1) \ I^-_{L+1}\right] - \left[\left(\frac{L+1}{L}\right)^{1/2} + G\right]$$

$$\times \left[(\kappa-\kappa') \ I^+_{L-1} - L \ I^-_{L-1}\right] + G[L] \ J_L \frac{(E_f - E_i)}{|E_f - E_i|}\right\} \tag{3.64}$$

where $a=i^L$ for absorption, $a=(-1)^{L+1}(i)^L$ for emission. The quantities I and J are radial integrals defined by [3.18,21]

$$I^{\pm}_L = \int \Delta(\ell,\bar{\ell}',L)(F_j G'_j \pm G_j F'_j) \ j_L \ dr$$

$$J_L = \int \Delta(\ell,\ell',L)(F_j F'_j + G_j G'_j) \ j_L \ dr \tag{3.65}$$

where $\Delta(\ell,\ell',L)$ is equal to 1 if ℓ, ℓ', and L satisfy the triangular condition and their sum is even; it is zero otherwise. In a like manner, one obtains for magnetic transitions [3.6,18]

$$(1/2\ell j||h_{\omega L m}||1/2\ell'j') = -aei(\hbar\omega)^{1/2}I^+_L$$

$$\times (-1)^{L+j'+1/2}\begin{pmatrix} j & L & j' \\ 1/2 & 0 & -1/2 \end{pmatrix} \frac{(\kappa+\kappa')}{[L(L+1)]^{1/2}} \ [j,j',L]^{1/2} \quad . \tag{3.66}$$

As can easily be shown, in the case that both initial and final states are exact solutions to the same Dirac equation, all choices of G will lead to identical results [3.12,18]. However, when initial and final states are not exact solutions, or, perhaps, satisfy different Dirac equations (as will happen when both are obtained from Dirac-Hartree-Fock wave functions), the value of the matrix element will depend on G [3.18].

In order to understand the meaning of this gauge constant G, let us consider the nonrelativistic Pauli limit of the reduced matrix element of the electric transitions. By expanding (3.30) and keeping only the first nonvanishing term, one finds that

$$F_{n\ell j} \to R_{n\ell} \quad , \quad G_{n\ell j} \to \frac{-\hbar}{2mc}\left(\frac{d}{dr} + \frac{\kappa}{r}\right) R_{n\ell}$$

where $R_{n\ell}$ is a solution to the nonrelativistic Schrödinger equation. Keeping only the first term in the expansion of the wave functions and the first term in the expansion of the Bessel functions, one obtains immediately

$$I_L^+ \simeq -\frac{\hbar}{2mc}\frac{1}{(2L+1)!!}\left(\frac{\omega}{c}\right)^L (\kappa+\kappa'-L) \int R_{n\ell} R_{n'\ell'} r^{L-1} dr \ \Delta(\ell,\bar{\ell}',L)$$

$$I_L^- \simeq -\frac{\hbar}{2mc}\frac{1}{(2L+1)!!}\left(\frac{\omega}{c}\right)^L \int R_{n\ell}\left[(\kappa'-\kappa+L)r^{L-1} + 2r^L \frac{d}{dr}\right]R_{n'\ell'} dr \ \Delta(\ell,\bar{\ell}',L)$$

$$J_L \simeq \left(\frac{\omega}{c}\right)^L \frac{1}{(2L+1)!!} \int R_{n\ell} R_{n'\ell'} r^L dr \ \Delta(\ell,\bar{\ell}',L) \qquad . \tag{3.67}$$

The corresponding limit of the reduced matrix element of the electric transition operator of order L, (3.64), is

$$ae(\hbar)^{1/2} (-1)^{L+j'+1/2} \begin{pmatrix} j & L & j' \\ 1/2 & 0 & -1/2 \end{pmatrix} \left(\frac{[j,j']}{[L]}\right)^{1/2}$$

$$\times \left(\frac{\hbar}{2mc}\frac{1}{(2L-1)!!}\left(\frac{\omega}{c}\right)^{L-1}\left[\left(\frac{L+1}{L}\right)^{1/2} + G\right] - \left\{[\ell(\ell+1)\right.\right.$$

$$- \ell'(\ell'+1) - L^2 + L]\int R_{n\ell} R_{n'\ell'} r^{L-2} dr - 2L \int R_{n\ell} r^{L-1} \frac{d}{dr} R_{n'\ell'} dr\Big\}$$

$$- G \frac{(E_f-E_i)}{|E_f-E_i|}\left(\frac{\omega}{c}\right)^L \frac{1}{(2L-1)!!} \int R_{n\ell} R_{n'\ell'} r^L dr\Big) \qquad . \tag{3.68}$$

Let us consider for the moment only the L=1 term above. This is, of course, the electric dipole term. There are two values of G which lead to well-known operators [3.18]: if G=0, one obtains the usual velocity form of the dipole matrix element; if $G=-[(L+1)/L]^{1/2}=-(2)^{1/2}$, one obtains the usual length form of the dipole matrix element. We can therefore define two gauges which are particularly useful: the "velocity" gauge, which is in fact the Coulomb gauge (all terms identically equal to zero), for which

$$(1/2\ell j||h_{\omega Le}||1/2\ell'j')_V = ae(\hbar\omega)^{1/2} (-1)^{L+j'+1/2}$$

$$\times \left(\frac{[j,j']}{[L]}\right)^{1/2} \begin{pmatrix} j & L & j' \\ 1/2 & 0 & -1/2 \end{pmatrix}\Big\{\left(\frac{L}{L+1}\right)^{1/2}\left[(\kappa-\kappa') I_{L+1}^+ + (L+1) I_{L+1}^-\right]$$

$$- \left(\frac{L+1}{L}\right)^{1/2}\left[(\kappa-\kappa') I_{L-1}^+ - L I_{L-1}^-\right]\Big\} \tag{3.69}$$

and the "length" gauge, for which

$$(1/2\ell j||h_{\omega Le}||1/2\ell'j')_L = ae(\hbar\omega)^{1/2} [j,j',L]^{1/2} \left[\frac{(L+1)}{L}\right]^{1/2}$$

$$\times \begin{pmatrix} j & L & j' \\ 1/2 & 0 & -1/2 \end{pmatrix} (-1)^{L+j'+1/2} \Big\{\left(\frac{1}{L+1}\right)\left[(\kappa-\kappa') I_{L+1}^+\right.$$

$$+ (L+1) I_{L+1}^-\right] - \frac{(E_f-E_i)}{|E_f-E_i|} J_L\Big\} \qquad . \tag{3.70}$$

There is also a third value of G which has been used fairly often, $G=[L/(L+1)]^{1/2}$ $=(2)^{-1/2}$. This leads to the form of the operator suggested by BABUSHKIN [3.22]. In this case, the terms in (3.64) containing I_{L+1} vanish. However, as can be seen by reference to (3.68), this gauge does not lead to any familiar nonrelativistic limit, but rather to a combination of length and velocity forms. In fact

$$(1/2\ell j||h_{\omega Le}||1/2\ell'j')_{Babushkin} =$$

$$- (1/2\ell j||h_{\omega Le}||1/2\ell'j)_L + 2(1/2\ell j||h_{\omega Le}||1/2\ell'j')_V \quad . \tag{3.71}$$

As such, this gauge should probably be avoided when using approximate wave functions.

3.3.2 Effective Transition Operators

It is often both convenient and revealing to treat relativistic interactions using effective operators. There are many different, but equivalent viewpoint which can be used in defining these operators [3.23,23]. The simplest involves the state $|\psi>$ which is the nonrelativistic limit of the relativistic state $|\psi)$. Then one defines the effective operator H^e corresponding to the relativistic operator H by

$$(\psi|H|\psi') = <\psi|H^e|\psi'> \quad . \tag{3.72}$$

Since the transition operator is a one-electron operator, its equivalent operator can be expanded in the form [3.10]

$$\left[\sum_j h_{\omega LM}(r_j)\right]_{eff} = \sum_{\kappa,k,\ell_1,\ell_2} A^L_{\kappa k}(\ell_1,\ell_2) \, W^{(\kappa k)L} \, M(\ell_1,\ell_2) \tag{3.73}$$

where $\underline{W}^{(\kappa k)L}$ is a nonrelativistic double tensor operator [3.25] having rank κ in spin space, k in orbital space, and L in the space of $\underline{J}=\underline{L}+\underline{S}$. Its magnitude is defined by the equations

$$\underline{W}^{(\kappa k)}(\ell_1\ell_2) = \sum_j \underline{w}_j^{\kappa k} (\ell_1\ell_2)$$

$$(1/2\ell_\alpha||w^{(\kappa k)}(\ell_1\ell_2)||1/2\ell_\beta) = [\kappa,k]^{1/2} \delta(\ell_\alpha,\ell_1) \delta(\ell_\beta,\ell_2) \quad . \tag{3.74}$$

The coefficient $A^L_{\kappa k}$ is given by [3.10]

$$A^L_{\kappa k} (\ell_1,\ell_2) = \sum_{j_1,j_2} [j_1,j_2,\kappa,k]^{1/2} [L]^{-1/2} \begin{Bmatrix} 1/2 & 1/2 & \kappa \\ \ell_1 & \ell_2 & k \\ j_1 & j_2 & L \end{Bmatrix} (1/2\ell_1 j_1 ||h_{\omega L\lambda}||1/2\ell_2 j_2) \tag{3.75}$$

This description is particularly useful in making comparisons between relativistic operators and their nonrelativistic limits, since the relativistic operators are put into a form such that they have the transformation properties of nonrelativistic operators. In the case of the electric dipole transition operator, for example, the nonrelativistic operator (either length or velocity) transforms like a vector in the orbital space, and a scalar in the spin space, that is like $\underline{W}^{(01)1}$. The relativistic effective operator, on the other hand, can be expanded in terms of the three operators $\underline{W}^{(01)1}$, $\underline{W}^{(11)1}$ and $\underline{W}^{(12)1}$. The magnitudes of the coefficients of the last two of these tensors give an indication of the importance of relativistic effects, since these coefficients vanish identically in the nonrelativistic limit. In addition, these last two tensor operators clearly change the ordinary dipole selection rules. Where the nonrelativistic operator ($\underline{W}^{(01)1}$) leads to the usual selection rules $|\Delta S|$, $|\Delta L|$, $|\Delta J| \leq 1$, the two new operators $\underline{W}^{(11)1}$ and $\underline{W}^{(12)1}$ lead to the selection rules $|\Delta S|$, $|\Delta L|$, $|\Delta J| \leq 1$ and $|\Delta S|$, $|\Delta J| \leq 1$, $|\Delta L| \leq 2$, respectively.

The coefficients A of (3.73) can in the general case be evaluated using (3.75). For convenience, we give the explicit form of the relativistic equivalent operator for the electric dipole case:

$$\sum_j h_{\omega 1M}(r_j)\Big|_{eff} = \sum_{\ell_1,\ell_2} A^1_{01}(\ell_1,\ell_2)\, W^{(01)1}_M(\ell_1,\ell_2) + A^1_{11}(\ell_1,\ell_2)\, W^{(11)1}_M(\ell_1,\ell_2)$$

$$+ A^1_{12}(\ell_1,\ell_2)\, W^{(12)1}_M(\ell_1,\ell_2) \tag{3.76}$$

where

$$A^1_{01}(\ell_1,\ell_2) = [\ell_1,\ell_2,1,1/2]^{-1/2} \left\{ \left[(\ell_1+\ell_2+3)(\ell_1+\ell_2) \right]^{1/2} \right.$$

$$\times F_{++} - (\ell_2-1-\ell_1)^{1/2} F_{+-} + (\ell_1+1-\ell_2)^{1/2} F_{-+}$$

$$\left. + \left[(\ell_1+\ell_2+2)(\ell_1+\ell_2-1) \right]^{1/2} F_{--} \right\}$$

$$A^1_{11}(\ell_1,\ell_2) = 1/2\,[\ell_1,\ell_2,1]^{-1/2} \left\{ \left[(\ell_1+\ell_2)(\ell_1+\ell_2+3) \right]^{1/2} \right.$$

$$\times (\ell_2-\ell_1)F_{++} + (\ell_2+1-\ell_1)^{1/2} (\ell_2+\ell_2+1)F_{+-}$$

$$+ (\ell_1+1-\ell_2)^{1/2} (\ell_1+\ell_2+1)F_{-+}$$

$$\left. + \left[(\ell_1+\ell_2+2)(\ell_1+\ell_2-1) \right]^{1/2} (\ell_1-\ell_2)F_{--} \right\}$$

$$A_{12}^1(\ell_1,\ell_2) = 1/2[\ell_1,\ell_2,1]^{-1/2}\left\{-\left[(\ell_1+\ell_2)(\ell_1+\ell_2-1)\right]^{1/2}F_{++}\right.$$

$$+\left[(\ell_1+\ell_2+3)(\ell_2+1-\ell_1)(\ell_1+\ell_2-1)\right]^{-1/2}F_{+-}$$

$$+\left[(\ell_1+\ell_2-1)(\ell_1+1-\ell_2)(\ell_1+\ell_2+3)\right]^{1/2}F_{-+}$$

$$\left.+\left[(\ell_1+\ell_2+2)(\ell_1+\ell_2+3)\right]^{1/2}F_{--}\right\}\quad.$$

Here

$$F_{++} = (\ell_1 j_1=\ell_1+1/2||h_{\omega 1e}||\ell_2 j_2=\ell_1+1/2), \text{ etc.}$$

3.3.3 Dipole Oscillator, Line, and Momentum Strengths

Familiar nonrelativistic quantities associated with transition probabilities also have relativistic counterparts. For example, the relativistic dipole line strength $S(J_i,J_f)$ can be defined by

$$S(J_i,J_f) = \frac{3c^2}{2\hbar\omega^3}\left|(J_f||\sum_j h_{\omega 1e}(j)||J_i)_L\right|^2\quad.\tag{3.77}$$

In the nonrelativistic limit, this goes over to

$$S(J_i,J_f) \rightarrow \left|<J_f||\sum_j er_j||J_i>\right|^2\quad.$$

Likewise, one can define a relativistic dipole momentum strength $P(J_i,J_f)$ by

$$P(J_i,J_f) = \frac{3m^2c^2}{2e^2\hbar\omega}\left|(J_f||\sum_j h_{\omega 1e}(j)||J_i)_V\right|^2\tag{3.78}$$

which in the nonrelativistic limit goes to

$$P(J_i,J_f) \rightarrow \left|<J_f||\sum_j p_j||J_i>\right|^2\quad.$$

The relativistic length (f_L) and velocity (f_V) forms of the dipole oscillator strength can be defined in terms of these two quantities by

$$f_L = \frac{1}{[J_i]}\frac{2m\omega}{3\hbar^2e^2}\ S(J_i,J_f)\tag{3.79}$$

and

$$f_V = \frac{1}{[J_i]}\frac{2}{3\hbar\omega m}\ P(J_i,J_f)\quad.$$

3.4 Calculation of Wave Functions and Energies

As we have discussed above, there are two basic approaches to the calculation of relativistic effects: the "relativistic" approach, in which the starting point is the Dirac equation, with the Breit interaction being treated as a perturbation; and the "nonrelativistic" approach in which the starting point is the Schrödinger equation, with the Pauli limit interactions being treated as perturbations. Both of these approaches have in common a difficulty in definition of the zeroth order Hamiltonian, since a Hamiltonian containing the interelectron interaction $\sum e^2/r_{ij}$ leads to an equation which cannot be solved exactly.

There are three general approaches to defining the zeroth order Hamiltonian which are quite commonly used. They can be categorized roughly as "1/Z expansion", "parametric potential", and "Hartree-Fock". The techniques used for the relativistic and nonrelativistic calculations are so similar, it will be convenient to discuss both together.

3.4.1 1/Z Expansion

The simple arguments given in Section 3.1 are at the basis of this approach [3.2, 26-30]. The zeroth order Hamiltonian for an N electron atom is taken to be simply the sum of N hydrogenic Hamiltonians, either relativistic or nonrelativistic as one is doing a relativistic or a nonrelativistic 1/Z expansion, respectively. First order energy corrections are obtained by diagonalizing the matrix of the operator $\sum e^2/r_{ij}$ (nonrelativistic) or of the operators $\sum e^2/r_{ij}+H_B$ (relativistic) taken between all states of a common principal quantum number and parity. This set of states is called a complex [3.26]. In the nonrelativistic case, this leads to an expansion of the type [3.2,26-28]

$$E = Z^2 \sum Z^{-n} E_n \quad .$$
(3.80)

Numerous tables of the E_n are available in the literature [3.31-38]. The expansion in the relativistic case is more complicated since the zeroth order hydrogenic part itself has an expansion of the type

$$E^0 = \sum E_m (Z\alpha)^{2m} + Nmc^2 \quad .$$
(3.81)

Combining this with the expansion produced by the diagonalization of $\sum e^2/r_{ij}+H_B$ leads to an energy expansion of the form [3.29,32]

$$E = \sum E_{nm} Z^{-n} (Z\alpha)^{2m} + Nmc^2$$
(3.82)

when the complex contains only one state of a given J. In the case in which there
are several states of a given J in a complex, one is unable to expand the energy
obtained from the diagonalization in the form given by (3.82). The elements of the
matrix which must be diagonalized can, however, be given in terms of an expansion
of the form of (3.82). DOYLE [3.30] has, for the K- and L-shells, given the coefficients E_{00}, E_{01}, E_{10}, and E_{11} for such an expansion for all N=2 complexes and
most $3 \leq N < 10$ complexes. Expansions for which the first few terms are equivalent to
those of (3.82) can also be obtained in a nonrelativistic 1/Z expansion in which the
first order perturbation operator is taken to be not simply $\sum e^2/r_{ij}$, but $\sum e^2/r_{ij} + H_{BP}$,
where H_{BP} is the Breit-Pauli operator of (3.49) [3.30,39].

   Both relativistic and nonrelativistic expansions can be improved somewhat by
allowing screening of the nuclear field [3.26,30,40,41]. This screening can be introduced in a rather complicated fashion through variationally obtained screening
parameters for each orbital wave function, or more simply by replacing Z by Z-σ in
the energy expressions.

   Transition probabilities can, of course, be calculated in this scheme. The wave
functions in the nonrelativistic expansion can, as indicated in the first section,
be expanded in the form

$$\psi_{NR} = \sum_{n=0} Z^{-n} \psi_n \tag{3.83}$$

where ψ_0 is proportional to a nonrelativistic hydrogenic wave function. Matrix elements of an operator O_{NR} will then have an expansion

$$(\psi_{NR}|O_{NR}|\psi'_{NR}) = \sum Z^{-n} O_n \tag{3.84}$$

where the lowest nonvanishing term in (3.84) is simply the hydrogenic matrix element
of the operator O_{NR}. For the relativistic expansion, the equivalent expression for
the wave function is

$$\psi_R = \sum \psi_{nm} Z^{-n} (Z\alpha)^{2m} \tag{3.85}$$

where ψ_{00} is proportional to a relativistic hydrogenic wave function, which itself
can be expanded in powers of $(Z\alpha)^2$. Then, a matrix element of any operator O_R has
the expansion

$$(\psi_R|O_R|\psi'_R) = \sum O_{nm} Z^{-n} (Z\alpha)^{2m} \tag{3.86}$$

where the lowest nonvanishing O_{nm} is obtained by evaluating the operator to which
O_R reduces in the nonrelativistic limit between the states obtained by keeping only
the first term in a $(Z\alpha)^2$ expansion of ψ_{00} above. Thus, if in the nonrelativistic

limit O_R reduces to O_{NR}, and ψ_R reduces to ψ_{NR}, etc., the lowest nonvanishing term in (3.84) will be equal to the lowest nonvanishing term in (3.86). Once again, an expansion in which the first few terms are equivalent to the first few terms of (3.86) can be obtained by a nonrelativistic expansion in which the Breit-Pauli operators are included in the first order perturbation.

The coefficients for oscillator strengths calculated using an expansion such as given by (3.84) for the dipole matrix element have been tabulated for numerous transitions [3.36-38,42-47]. Comparison of the predicted values with either experimental values or values predicted using more sophisticated techniques shows that the nonrelativistic $1/Z$ expansion is best for $\Delta n=0$ transitions, being reasonably accurate for $\Delta n \neq 0$ transitions only at higher stages of ionization [3.48]. At higher stages of ionization, however, the expansion of (3.83) ceases to be valid, and one must use an expansion such as given by (3.86). It appears that expansions of the type (3.86) have not yet been used in calculations of transition probabilities. This is apparently because the effects of relativity on the transition energy are often much larger than the effects of relativity on the transition matrix element itself especially for the $\Delta n=0$ transitions. Thus the expansion of (3.82) is often of more practical importance than the expansion of (3.86).

3.4.2 Parametric Potentials

This approach is based either on the nonrelativistic Hamiltonian

$$H_0 = \sum \left(\frac{p_i^2}{2m} - eU(r_i) \right) \tag{3.87}$$

or on the relativistic Hamiltonian $H_D(i)$ [(3.40)] where $U(r_i)$ is a non-Coulombic central potential whose exact form is an analytic function of some set of parameters $\{\alpha\}$. Obviously, $U(r_i)$ is generally chosen such that $U(r) \rightarrow (Z+1-N)e/r$ as $r \rightarrow \infty$, and $U(r) \rightarrow Ze/r$ as $r \rightarrow 0$. Equally obviously, within these limitations, there are a large number of forms possible for $U(r)$.

Even if a specific form of $U(r)$ has been chosen, there are many ways in which to determine the parameters $\{\alpha\}$. For example, if there are several experimentally determined energy levels known for the atom of interest, $\{\alpha\}$ can be chosen so as to most nearly reproduce these levels. More generally, $\{\alpha\}$ can be chosen so as to reproduce any experimentally determined parameter. For highly ionized atoms, of course, one may have no experimental data at hand, and other means of determining $\{\alpha\}$ must be found. One obvious method in this case is to choose $\{\alpha\}$ so as to minimize the energy of the ground state of the ion.

One very simple form of parametric potential is the scaled Thomas-Fermi, introduced by STEWART and ROTENBERG [3.49]. This is an ordinary Thomas-Fermi potential in which the radius r is replaced by a scaled radius, αr; STEWART and ROTENBERG chose the single parameter α so as to reproduce measured energy levels. The drawbacks to such a potential include the well known [3.8] overestimation of the potential at small r by the Thomas-Fermi, and the lack of any shell structure in the potential.

KLAPISH [3.50] has introduced a considerably more complicated potential which has the benefit of explicitly taking into account the shell structure of the atom. This potential includes a parameter $\alpha_{n\ell}$ for each occupied subshell in the atom:

$$U(r) = \frac{e}{e} \sum_{n\ell} [q_{n\ell} \, f_{n\ell} \, (\alpha_{n\ell} r) + Z + 1 - N] \tag{3.88}$$

where $q_{n\ell}$ is the occupation number of the $n\ell$ subshell, and

$$f_{n\ell} = \exp(-\alpha_{n\ell} r) \sum_{j=0}^{2\ell+1} \frac{1}{j!} \left(1 - \frac{j}{2\ell+2}\right) (\alpha_{n\ell} r)^j \quad . \tag{3.89}$$

The form of this potential results from assuming that the charge distribution in a subshell is essentially given by the square of a simple Slater function

$$P \sim \left(r^{\ell+1} e^{-1/2\alpha_{n\ell} r}\right)^2 \quad . \tag{3.90}$$

This form of parametric potential may lead to such a large number of independent parameters for a heavy atom that one wishes to impose a further empirical constraint [3.50]

$$\alpha_{n\ell} = \alpha_n \left[\frac{\ell+1}{1-0.03\ell(\ell+1)}\right] \tag{3.91}$$

thereby reducing the number of parameters to the number of n shells which are fully or partially occupied. This type of potential has been fairly extensively used in both relativistic [3.51-54] and nonrelativisitc [3.55] calculations. We shall discuss some of these in the next chapter.

Another, and considerably less complicated parametric potential has been discussed by GREEN et al. [3.56]. It involves only two parameters, H and d:

$$U(r) = - \frac{e}{r} [(Z-1) \, \Omega(r) + 1] \tag{3.92}$$

where

$$\Omega(r) = [H(e^{r/d}-1) + 1]^{-1} \quad . \tag{3.93}$$

The values of the parameters H and d can, again, be determined by fitting of experimental data or by minimization of the ground state energy. The quantities d and K=H/d are found to vary considerably with atomic number and number of electrons; d is particularly sensitive to such variations. However, SZYDLIK and GREEN [3.57] have shown that both d and K vary smoothly along an isoelectronic sequence, at least for moderate degrees of ionization.

One difficulty associated with the use of these parametric potentials arises when the $\{\alpha\}$ are determined by reproducing measured energy values: it is often very difficult to determine what experimental energy the eigenvalue of the Schrödinger or Dirac equation should be compared to. For atoms having a single electron outside of closed sheels, the proper experimental energy is simply the ionization energy. When discussing an atom in which the outermost orbital contains several electrons, however, the ionization energy contains additional contributions such as term energy which are not contained in H_0. Use of parametric potentials in such cases is correspondingly less unambiguous and more complicated.

3.4.3 Hartree-Fock

In the Hartree-Fock procedure, one determines the potential and the electronic wave functions simultaneously in a self-consistent fashion [3.58-60]. That is, the wave functions are used to evaluate the interelectronic Coulomb term $\sum e^2/r_{ij}$, which, along with the nuclear potential, defines the potential in which the wave functions are evaluated. One difficulty is that, except in the closed shell case, such a procedure does not lead to a central potential. Since, as discussed in Section 3.3, a central potential is necessary if the wave functions are to be separable into angular and radial parts, this could lead to serious difficulties. These difficulties can be partially overcome by the expedient of defining the wave functions to be separable; however, the result of such an arbitrary separation is that the solutions obtained will not be exact solutions to the Hamiltonian equation. This in turn implies that perturbation theory may be required to obtain solutions of the desired accuracy.

The Hartree-Fock equations are obtained by a variational calculation based on the expression for the total atomic energy,

$$E = (\psi|H|\psi) \tag{3.94}$$

where E can be viewed as a function of the one electron radial wave functions. The Hamiltonian to be used in (3.94) will be, in the nonrelativistic case [3.58,59]

$$H = \sum_i \left(\frac{p_i^2}{2m} - \frac{Ze^2}{r_i} \right) + \sum_{i<j} \frac{e^2}{r_{ij}} \tag{3.95}$$

and, in the relativistic case [3.60],

$$H = \sum_i \left(c\underline{\alpha}_i \cdot \underline{p}_i - \beta_i mc^2 - \frac{Ze^2}{r_i} \right) + \sum_{i<j} \Lambda_{++}^{ij} \frac{e^2}{r_{ij}} \Lambda_{++}^{ij} \quad . \tag{3.96}$$

Note that if all of the particles described by $|\psi)$ of (3.94) are of positive energy, the projection operators in (3.96) can be dropped. The real complicating factor in this procedure is the Pauli principle, that is to say, the antisymmetry of the wave function ψ with respect to interchange of the one-electron wave functions. As a result of this, the integrodifferential equations which are obtained from the variational procedure are not quite in the form of a Hamiltonian equation, but rather are of the form

$$H_i \psi_i = E \psi_i + \sum_j C_{ij} \, \psi_j \tag{3.97}$$

where ψ_i is the wave function of the i-th electron, etc. It is this inhomogeneous term (the exchange term) which makes both the ordinary and relativistic Hartree-Fock equations so difficult to solve.

In the nonrelativistic case, there are, of course, N coupled equations of the type given by (3.97) which are to be solved simultaneously; in the relativistic case, since each wave function has both a large and a small component, there are 2N coupled equations to be solved. Consideration of the exact form of these equations is beyond the scope of this work. Detailed discussions of the nonrelativistic Hartree-Fock can be found, for instance, in the books by SLATER [3.58] or HARTREE [3.59]. Excellent recent reviews of the relativistic Hartree-Fock have been given by GRANT [3.60] and by LINDGREN and ROSEN [3.61]. The number of relativistic Hartree-Fock calculations carried out thus far is relatively small, and mainly concerned with neutral or only a few times ionized atoms [3.61-69]. However, recently some extensive studies [3.6,70] of highly ionized atoms have been carried out using the multiconfiguration Dirac-Hartree-Fock program of DESCLAUX [3.71].

Numerous schemes exist for simplifying the Hartree-Fock equations by approximating the exchange term. The simplest and most common of these was introduced by SLATER [3.72], and involves replacing the exchange by a local statistical exchange term

$$V_{ex}^S(r) = -\frac{3}{2} \left[\frac{3\rho(r)}{4\pi^2 r^2} \right]^{1/3} \tag{3.98}$$

where

$$\rho = \sum_i R_i^2 \qquad \text{(nonrelativistic)}$$

$$= \sum_i \left(F_i^2 + G_i^2 \right) \qquad \text{(relativistic)} \quad .$$

Use of this approximation leads to the Hartree-Fock-Slater equations. If the atom is approximated by an electron gas before the variational procedure, one obtains the Kohn-Sham exchange potential V_{ex}^{KS}, which is just 2/3 of V_{ex}^S [3.73-75]. One can also parameterize the exchange potential. Two well-known parameterizations are those introduced by LINDGREN and ROSEN [3.61]

$$V_{ex} = \frac{-3C}{2r} \left[\frac{3r^n \rho(r)^m}{4\pi^2} \right]^{1/3} \tag{3.99}$$

where C, n, and m are variable parameters; and the Xα potential of WILSON et al. [3.76]

$$V_{er} = \alpha V_{ex}^S \tag{3.100}$$

where α is a variable parameter. Naturally these modified Hartree-Fock equations cannot be expected to lead to wave functions which are as good as those obtained with the complete Hartree-Fock, although the relative ease with which they can be solved may make them more desirable for many purposes [3.61,77-82].

3.5 Results

In this section, we shall very briefly consider some of the calculations which indicate the importance of relativistic effects in atomic structure. Unfortunately, at this time, there have been very few relativistic calculations (in the sense of the discussion of Sec.3.2.1) carried out, particularly in highly ionized atoms; thus, the statements that one can make concerning the onset and importance of relativistic effects must all be preceded with a warning concerning our state of relative ignorance. In addition, I would like to reemphasize my introductory statement that this work is not meant to be a review of the literature, and therefore many excellent works may not be mentioned here.

In order to begin, some understanding as to the definition of relativistic effects must be reached. If we define the spin orbit interaction to be a relativistic effect (which, as can be seen from the discussion of Sec.3.1, we certainly should do), then relativistic effects are important in almost all atoms and ions. If we mean by relativistic effects those interactions which are "beyond the spin orbit", then we are immediately in relatively unknown territory. In addition, in a many-electron atom or ion, our discussion of Section 3.1, in which operators were described by their dependence on Z and α, must not be taken as gospel. Thus, for example, fifteen times ionized Zn still has fifteen electrons; what Z does one associate with the Zα description of any of the relativistic operators of Section 3.1-15, 30, or some other number? The answer is not unique, but must be a function of the specific operator involved. (Does it, for example, have its largest contribution near to the nucleus or at the outer edges of the electron cloud?) As a result, the operators can no longer be easily and unambiguously assigned a position in the hierarchy of strength and importance.

The spin orbit interaction is, however, almost always the strongest of the relativistic interactions; and we can, therefore, safely begin our discussion with it. The spin orbit interaction is composed of two parts - that due to the electron-nucleus interaction, which varies roughly as $Z^4\alpha^2$, and that produced by the electron--electron interaction, which varies roughly as $Z^3\alpha^2$. The importance of the spin orbit interaction in relatively highly ionized atoms is clearly demonstrated by the many calculations of COWAN and co-workers [3.83-88], who have used a nonrelativistic central field model to which they have added the spin orbit interaction and the Darwin and mass terms (which also vary as $Z^4\alpha^2$). Their extensive calculations of transition energies and intensities in 20 to 30 times ionized atoms have generally been in very close agreement with experiment. One obvious and important effect of the spin orbit interaction is to break down LS coupling, and, of course, this can be seen even in neutrals where the spin orbit interaction is generally weaker than it is in ions. Thus the dominant relativistic effect at intermediate levels of ionization (up to 30 to 40 times ionized) would seem to be the Darwin and mass terms and the spin orbit interaction with its resulting intermediate coupling. In addition, as noted in Section 3.1, the relative importance of these $Z^4\alpha^2$ terms on transition energies is greater for $\Delta n=0$ transitions than for $\Delta n \neq 0$ transitions. This is clearly demonstrated by relativistic Dirac-Hartree-Fock calculations in the Li sequence [3.6,70].

The remaining Pauli limit operators may, of course, also be important at low and intermediate levels of ionization [3.89,90]. The relativistic 1/Z expansions of DOYLE [3.30] and SAFRANOVA [3.39] indicate that the coefficients of the $Z^3\alpha^2$ terms are usually of the same magnitude as those of the $Z^4\alpha^2$ terms. However, as mentioned above, some parts of the $Z^3\alpha^2$ term comes from the electron-electron part of the spin orbit interaction; so it is very difficult to make a general statement based on these results which describes the importance of these remaining Pauli terms. It is to be expected that, in general, these terms are rather small, but many have importance due to the fact that some of them satisfy selection rules which are quite different from those satisfied by the spin orbit interaction. Thus, for example, they may break down LS coupling in ways which are different from those of the spin orbit interaction.

As one increases the stage of ionization beyond roughly 30, higher order relativistic terms - "beyond the spin orbit" - will begin to become important. For example, the next term in the electron-nucleus interaction is of order $Z^6\alpha^4$; $Z^6\alpha^4$ is equal to $Z^3\alpha^2$ at a Z of about 27. Thus the higher order relativistic terms arising from the electron-nucleus interaction will become more important than the Breit-Pauli terms somewhere beyond Z=30. In this highly ionized region, one is almost obligated to use relativistic calculations based on the Dirac equation, where the electron-nucleus interaction is included to all orders. That this is the case has been demonstrated by several recent calculations [3.6,70,91].

The discussion thus far has essentially been centered on energy contributions from the various operators. One is also interested in matrix elements of operators other than the Hamiltonian. These matrix elements will, of course, be affected by such things as intermediate coupling, which was discussed above; there are also relativistic effects in the one-electron matrix elements themselves. These one-electron effects can be produced in a number of ways, all of which are usually effects of relative order $Z^2\alpha^2$: changes in the nonrelativistic radial wave function ($R_{n\ell} \to R_{n\ell} + Z^2\alpha^2 R'$); introduction of the effects of the small components of the wave function ($G_{n\ell j} \approx Z R_{n1}$); and changes in the form of the operator itself (Sec.3.3.2). These relativistic changes in dipole transition matrix elements have been very clearly seen in the Li and Be isoelectronic sequences [3.6,70]. Since this change in the transition element is an effect of relative order $Z^2\alpha^2$ for allowed transitions, it is generally not of importance until atoms are roughly 30 times ionized.

"Beyond the spin orbit" effects can also be seen in the oscillator and line strengths of intercombination lines in the Be sequence [3.6,70]. Eq.(3.21), which is obtained by considering the spin orbit effect, would indicate that the oscillator strength of intercombination lines is always an increasing function of Z. Results indicate, however, that in Be the oscillator strength rises until about Z=50, remains constant until about Z=70, and then begins to drop slightly. This drop can be seen to be produced by the relativistic change in the one-electron matrix elements rather than by changes in coupling due to H_{so}.

Of course, these relativistic changes in matrix element can, in lowest order, be handled using perturbation theory in the nonrelativistic picture. For example, one can make the expansion

$$(\psi|0|\psi') + \sum \left[\frac{1}{\Delta E} (\psi|0|\psi'')(\psi''|H_{so}|\psi') + CC\right]$$

which will give $Z^2\alpha^2$ corrections to the original matrix element. This has been done, for example, by FOLEY and STERNHEIMER [3.92] in explaining inverted doublets in excited states of Cs (0 is, in this case, the Hamiltonian). However, a much more straightforward, less difficult, and more elegant method of studying such corrections (and higher order ones also) would seem to be to simply do a relativistic calculation; LUC-KEONIG [3.53] has used this relativistic approach to study the Cs question, and has obtained results which provide considerable insight into where such effects might occur.

Relativistic effects become important at very low Z when for some reason the nonrelativistic contribution is exactly zero. A very nice example of this is the magnetic dipole decay of the metastable $^3S_1$ state in He-like ions. Here, the matrix element of the nonrelativistic magnetic dipole operator vanishes identically, and the transition occurs through a combination of relativistic changes in the form of the magnetic dipole operator and of second order effects involving the Breit operator [3.14,54,93-95].

Thus far, we have not mentioned "indirect relativistic effects" which appear in relativistic self-consistent field (SCF) calculations [3.60,69,77]. These effects are very important, but very difficult to categorize and predict. They are produced by the relativistic contraction of the inner orbitals of lower angular momentum [3.96]. As a result, the outer electrons see a potential which is quite different from the nonrelativistic potential, with the nucleus being more effectively screened. Thus, although the outer electrons may themselves be very nonrelativistic, they move in a potential which clearly shows the effects of relativity. The consequence of this is that the wave functions of the outer (and perhaps very nonrelativistic) electrons obtained from a relativistic SCF calculation may be considerably different from those obtained from an equivalent nonrelativistic calculation. Where the core is large, it is possible that this will be the largest relativistic effect. Unfortunately, the complexity of a SCF calculation makes it very difficult to define where this important effect will be significant, and to predict its magnitude.

The study of relativistic effects in highly ionized atoms is obviously in its infancy. The present concurrence of interest in the subject and of the availability of the sophisticated computing equipment and programs necessary for such studies virtually guarantees that the next few years will see a greatly improved understanding of this subject. Hopefully, these studies will clearly point out where relativistic effects are important, and how they manifest themselves, for such knowledge is indispensible in carrying out accurate, efficient calculations in highly ionized atoms.

Acknowledgements. I would like to thank Dr. Dong L. LIN for reading this manuscript and for his very useful comments. The work described herein which was performed at the Johns Hopkins University was carried out with the support of the United States Energy Research and Development Administration.

References

3.1 P.A.M. Dirac: *The Principles of Quantum Mechanics*, 4th ed. (Oxford University Press, Oxford 1958)
3.2 E. Hylleraas: Z. Physik. 65, 209 (1930)
3.3 G. Breit: Phys. Rev. 36, 383 (1930)
3.4 H.A. Bethe: Phys. Rev. 72, 339 (1947)
3.5 S.M. Younger, A.W. Weiss: J. Research of NBS 79A, 629 (1975)

3.6 L. Armstrong, Jr., W.R. Fielder, D.L. Lin: Phys. Rev. A14, 1114 (1976)
3.7 P.A.M. Dirac: Proc. Roy. Soc. Ser. A 117, 610 (1928); 118, 351 (1928)
3.8 A. Messiah: *Quantum Mechanics*, Vol. II (John Wiley and Sons, New York 1962)
3.9 H.A. Bethe, E.E. Salpeter: *Quantum Mechanics of One- and Two-Electron Atoms*
     (Berlin, Göttingen, Heidelberg: Springer 1957)
3.10 L. Armstrong, Jr., S. Feneuille: In *Advances in Atomic and Molecular Physics*,
     Vol. 10, ed. by D.R. Bates and B. Bederson (Academic Press, New York 1974)
3.11 J.D. Bjorken, S.D. Drell: *Relativistic Quantum Fields* (McGraw-Hill, New York
     1965)
3.12 A.I. Akhiezer, V.B. Berestetskii: *Quantum Electrodynamics* (Interscience, New
     York 1965)
3.13 G. Feldman, T. Fulton, J. Townsend: Phys. Rev. A 8, 1149 (1973)
3.14 G.W.F. Drake: Phys. Rev. A 3, 908 (1971); 5, 1979 (1972)
3.15 A. Abragam, J.H. Van Vleck: Phys. Rev. 92, 1448 (1953); B.R. Judd, I. Lindgren:
     Phys. Rev. 122, 1802 (1961)
3.16 S. Feneuille, L. Armstrong, Jr.: Phys. Rev. A 8, 3 (1973)
3.17 J.B. Mann, W.R. Johnson: Phys. Rev. A 4, 41 (1971)
3.18 I.P. Grant: J. Phys. B 7, 1458 (1974)
3.19 A.R. Edmonds: *Angular Momentum in Quantum Mechanics* (Princeton University Press,
     Princeton 1974); B.R. Judd: *Operator Techniques in Atomic Spectroscopy* (McGraw-
     Hill, New York 1963)
3.20 L. Armstrong, Jr.: Phys. Rev. 172, 18 (1968)
3.21 H.R. Rosner, C.P. Bhalla: Z. Physik 231, 347 (1970)
3.22 F.A. Babushkin: Opt. Spectr. 13, 77 (1962); 19, 1 (1965); Acta Phys. Polon.
     25, 749 (1964)
3.23 P.G.H. Sandars, J. Beck: Proc. Roy. Soc. Ser. A 189, 97 (1965)
3.24 L. Armstrong, Jr.: J. Math. Phys. 7, 1891 (1966); 9, 1083 (1968)
3.25 S. Feneuille: J. Physique 28, 61 (1967)
3.26 D. Layzer: Ann. Phys. 8, 291 (1959)
3.27 J. Lindberg, H. Shull: J. Mol. Spect. 5, 1 (1960)
3.28 R.J.S. Crossley, C.A. Coulson: Proc. Roy. Soc. 81, 211 (1963)
3.29 D. Layzer, J. Bahcall: Ann. Phys. 17, 177 (1962)
3.30 H.T. Doyle: In *Adv. in Atomic and Molecular Physics*, Vol. 5, ed. by D.R. Bates
     and I. Estermann (Academic Press, New York 1969)
3.31 E.A. Godfredsen: Astrophys. J. 145, 308 (1966)
3.32 C.S. Sharma: Proc. Roy. Soc. A 304, 513 (1968)
3.33 D. Layzer, Z. Horak, M.N. Lewis, D.P. Thompson: Ann. Phys. 29, 101 (1964); 48,
     592 (1968)
3.34 M. Cohen, A. Dalgarno: Proc. Phys. Soc. 77, 165 (1961)
3.35 S. Seung, E.B. Wilson: J. Chem. Phys. 47, 5343 (1967)
3.36 J.S. Onello: Phys. Rev. A 11, 743 (1975)
3.37 J.S. Onello, L. Ford, A. Dalgarno: Phys. Rev. A 10, 9 (1974)
3.38 D.K. Watson, S.V. Oneil: Phys. Rev. A 12, 729 (1975)
3.39 U.I. Safranova: J. Quant. Spectr. Radiative Transfer 14, 251 (1974); 15, 231
     (1975)
3.40 A. Dalgarno, A.L. Stewart: Proc. Roy. Soc. A 257, 534 (1960)
3.41 M. Cohen, A. Dalgarno: Proc. Roy. Soc. A 261, 565 (1961); A 280 (1964); A 293,
     359 (1966)
3.42 A. Dalgarno, E.M. Parkinson: Phys. Rev. 176, 73 (1968)
3.43 M.A. Ali, R.J.S. Crossley: Mol. Phys. 15, 397 (1968); Intern. J. Quantum Chem.
     3, 17 (1969)
3.44 R.J.S. Crossley, A. Dalgarno: Proc. Roy. Soc. A 286, 510 (1965)
3.45 C.A. Laughlin, A. Dalgarno: Phys. Rev. A 8, 39 (1973)
3.46 G.W.F. Drake, A. Dalgarno: Phys. Rev. A 1, 1325 (1970)
3.47 G.W.F. Drake: Phys. Rev. A 5, 614 (1972)
3.48 A. Dalgarno: Nucl. Inst. Methods 110, 183 (1973)
3.49 J.C. Stewart, M. Rotenberg: Phys. Rev. 140, A 1508 (1965); 156, 230 (1967)
3.50 M. Klapish: Comput. Phys. Comm. 2, 239 (1971)
3.51 E. Luc-Koenig: J. Phys. (Paris) 33, 847 (1972)
3.52 E. Koenig: Physica 62, 393 (1972)
3.53 E. Luc-Koenig: Phys. Rev. A13, 2114 (1976)

3.54 S. Feneuille, E. Koenig: C. R. Acad. Sci. Ser. B 274, 46 (1972)
3.55 M. Aymar: Nucl. Inst. Methods 110, 211 (1973)
3.56 A.E.S. Green, D.L. Sellin, A.S. Zachor: Phys. Rev. 184, 1 (1969)
3.57 P.P. Szydlik, A.E.S. Green: Phys. Rev. A 9, 1885 (1974)
3.58 J.C. Slater: *Quantum Theory of Atomic Structure*, Vols. I and II (McGraw-Hill, New York 1960)
3.59 D.R. Hartree: *The Calculation of Atomic Structures* (Wiley, New York 1957)
3.60 I.P. Grant: Adv. Phys. 19, 747 (1970)
3.61 I. Lindgren, A. Rosen: Case Studies in At. Phys. 4, 93 (1974)
3.62 M.A. Coulthard: Proc. Roy. Soc. 91, 44 (1967)
3.63 F.C. Smith, W.R. Johnson: Phys. Rev. 160, 136 (1967)
3.64 Y.K. Kim: Phys. Rev. 154, 17 (1967); 159, 190 (1967)
3.65 J.B. Mann: J. Chem. Phys. 51, 841 (1969)
3.66 J.B. Mann, J.T. Waber: J. Chem. Phys. 53, 2397 (1970); At. Data 5
3.67 J.P. Desclaux, C.M. Moser, G. Verhaegen: Proc. Phys. Soc. B 4, 296 (1971)
3.68 J.P. Desclaux, D.F. Mayers, F. O'Brien: Proc. Phys. Soc. B 4, 631 (1971)
3.69 J.P. Desclaux, Y.K. Kim: J. Phys. B 8, 1177 (1975)
3.70 Y.K. Kim, J.P. Desclaux: Phys. Rev. Lett. 36, 139 (1976)
3.71 J.P. Desclaux: Comput. Phys. Comm. 9, 31 (1975)
3.72 J.C. Slater: Phys. Rev. 81, 385 (1951)
3.73 R. Gaspar: Acta Phys. Hung. 3, 263 (1954)
3.74 W. Kohn, L.J. Sham: Phys. Rev. A 140, 1133 (1965)
3.75 R.D. Cowan, A.C. Larson, D. Liberman, J.T. Waber: Phys. Rev. 144, 5 (1966)
3.76 T.M. Wilson, J.M. Wood, J.C. Slater: Phys. Rev. A 2, 620 (1970)
3.77 D.A. Liberman, D.T. Cromer, J.T. Waber: Comput. Phys. Commun. 2, 107 (1971); Phys. Rev. A 137, 27 (1965)
3.78 R.D. Cowan, J.B. Mann: In *Atomic Physics 2*, ed. by G.K. Woodgate and P.G.H. Sandars (Plenum, New York 1971) p. 215
3.79 A. Rosen, I. Lindgren: Phys. Rev. 176, 114 (1968)
3.80 D.T. Cromer, D.A. Liberman: J. Chem. Phys. 53, 1891 (1970)
3.81 C.P. Bhalla: Nucl. Instr. Methods 90, 149 (1970)
3.82 T.E.H. Walker, J.T. Waber: J. Phys. B 7, 674 (1974)
3.83 R.D. Cowan: Phys. Rev. 163, 54 (1967); J. Opt. Soc. Am. 56, 1416 (1966)
3.84 G.A. Doschek, U. Feldman, J. Davis, R.D. Cowan: Phys. Rev. A 12, 980 (1975)
3.85 P.G. Burkhalter, D.J. Nagel, R.D. Cowan: Phys. Rev. A 11, 782 (1975)
3.86 P.G. Burkhalter, U. Feldman, R.D. Cowan: J. Opt. Soc. Am. 64, 1058 (1974)
3.87 B.C. Fawcett, R.D. Cowan, R.W. Hayes: Astroph. J. 187, 377 (1974)
3.88 U. Feldman, G.A. Doschek, R.D. Cowan, L. Cohen: J. Opt. Soc. Am. 63, 1445 (1973)
3.89 B.R. Judd, H.M. Crosswhite, H. Crosswhite: Phys. Rev. 169, 130 (1968)
3.90 D.R. Beck, H. Odabasi: Ann. Phys. 67, 274 (1971)
3.91 Y.K. Kim: private communication
3.92 H.M. Foley, R.M. Sternheimer: Phys. Lett. 55A, 276 (1975)
3.93 G. Feinberg, J. Sucher: Phys. Rev. Lett. 26, 681 (1971)
3.94 W.R. Johnson, C.P. Lin: Phys. Rev. A 9, 1486 (1974)
3.95 D.L. Lin, G. Feinberg: Phys. Rev. A 10, 1425 (1974)
3.96 V.M. Burke, I.P. Grant: Proc. Phys. Soc. 90, 297 (1967)

4. Theory of Inelastic Atom-Atom Collisions

J. S. Briggs and K. Taulbjerg

With 4 Figures

The understanding of the inelastic scattering of atoms and ions requires a reliable theory of the motion of nuclei and electrons interacting via Coulomb forces under conditions where the center-of-mass energy far exceeds the binding energy of electrons around one or other nucleus. This many-body, Coulomb scattering problem is one of the more difficult problems in atomic physics, largely because of the long range of the Coulomb force. Despite much effort, even the simplest such problem— that of the scattering of protons and hydrogen atoms—is not completely understood, particularly when rearrangement or unbound final states are involved. For this reason the theory of ion-atom collisions has concentrated on developing methods and approximations by which various types of ion-atom collisions involving at most a few electrons may be described.

Two limiting situations where simplifications are readily made may be isolated. The first is where perturbation theory is applicable either by virtue of a large difference in the relative strength of the two nuclear fields or because of an interaction time which is short compared with electronic relaxation times. The second is where the collision is so slow that the motion itself may be regarded as a perturbation. Then an understanding of the collision may be based upon a knowledge of the *stationary* states of the electrons and nuclei, a problem which is far simpler to solve than the collision problem. Theoretical methods based upon these limiting physical situations have been tested by comparison with experimental results on excitation, ionization, and charge transfer in simple ion-atom collisions. Where data on collisions involving heavier atomic species required interpretation, the theory developed for one-electron systems has often been adapted in a simple way, for example by reducing the many-electron problem to an independent-electron problem or by taking advantage of scaling relationships.

Undeterred by the lack of reliable theory, experimenters have concentrated upon ion-atom collisions of increasing complexity over the last decade. These studies have been facilitated by the employment of ion sources and detectors of increasing sophistication and by the use of a great variety of ion accelerators, giving beams with energies up to tens of MeV and higher. Several aspects of these studies, in which phenomena not evident in the simpler collisions have been isolated, are worthy of mention.

The first is the study of specifically inner-shell, as distinct from outer-shell, processes. In heavy ion-atom collisions such processes are characterized by the dominance of the nuclear field over that provided by other electrons. As a result of this simplification, some progress has been made in the theoretical treatment. Experimentally both total and differential (in scattering angle) measurements have been made. Although the excitation mechanism for inner-shell electrons at low velocity is well understood, the precise role played by the outer-shell electrons has not been fully appreciated.

The atomic states prepared by heavy-ion collisions show rich structure of multiple excitation and ionization. Evidence for this structure has emerged from studies of charge states after a collision and by observations of the satellite structure of x-ray and Auger transitions. It is clear that the states resulting from heavy-ion collisions are far more complex than those produced by gentler means, e.g., electron or photon excitation. At the moment, little is known of the relative importance of multiple nuclear-electron scattering and direct or exchange electron-electron scattering in the preparation of these final states. In addition, the role played by electron correlations in deciding the response of an atom or ion to purely impulsive forces during an ion-atom collision has not been assessed.

Finally, improved experimental techniques have allowed atomic excitation by ion beams of increasing primary charge to be studied. Of particular interest here is the variation of excitation cross sections with increasing primary charge, the influence of electrons carried by target and projectile both in screening of the nuclear charges and in primary excitation, and the relative importance of "direct" and charge transfer processes. The latter is particularly relevant to the correct description of unbound final electronic states in the fields of nuclei of high charge, i.e., in understanding the processes of ionization and electron-loss.

The theoretical progress that has been made towards the resolution of some of the above questions posed by the results of collision experiments is fragmentary. The aim of this work is to examine the status of the theory of ion-atom collisions as applied to simple collision systems and hopefully to point out areas where the methods are relevant to the treatment of more complicated collisions. In this respect, the material and the reference to published work are selective rather than comprehensive. The theory established to date divides naturally into two parts: namely, the quantum-mechanical and the classical treatment of nuclear motion. In the first part, the plane-wave Born approximation is given prominence and, in particular, the treatment of electron-electron interactions is emphasized. In the second part, the approximate solution of the time-dependent electronic Schrödinger equation for an ion-atom collision is considered, mainly in terms of an expansion in known basis functions. However, alternative methods of solution that have been used for simpler one-electron problems are discussed and the possibility of their generalization to more complicated situations is assessed.

4.1 Quantum-Mechanical Treatment of Nuclear Motion

4.1.1 Expansion in a Finite Basis

The collision of two nuclei, A and B, each of which carries electrons into the collision, will be considered. The coordinate system shown in Fig.4.1 will be used.
For simplicity and clarity, initially only one electron on each nucleus will be
considered, although some of the modifications necessary to describe a many-electron
system will be indicated subsequently.

For a fixed total energy E of the collision, a solution of the Schrödinger equation

$$(H-E) \ \Psi(\underline{r},\underline{s},\underline{R}) = 0 \tag{4.1}$$

is sought. If the ratio of electron mass to atomic mass m/M is neglected, then the
total Hamiltonian H, which is a sum of kinetic energy, nucleus-nucleus, electron-
nucleus, and electron-electron interaction terms, may be written, in atomic units,

$$H = -\nabla_R^2/2M - \nabla_r^2/2 - \nabla_s^2/2 + Z_A Z_B/R - Z_A/s_A - Z_B/r_B$$

$$- Z_A/|\underline{r}_B - \underline{R}| - Z_B/|\underline{s}_A + \underline{R}| + 1/|\underline{R} - \underline{r}_B + \underline{s}_A| \quad . \tag{4.2}$$

In general, the exact solutions of (4.1) will be expanded in terms of some known
basis functions defined by

$$H_e \psi_n(\underline{r},\underline{s},\underline{R}) = E_n(R)\psi_n(\underline{r},\underline{s},\underline{R}) \quad , \tag{4.3}$$

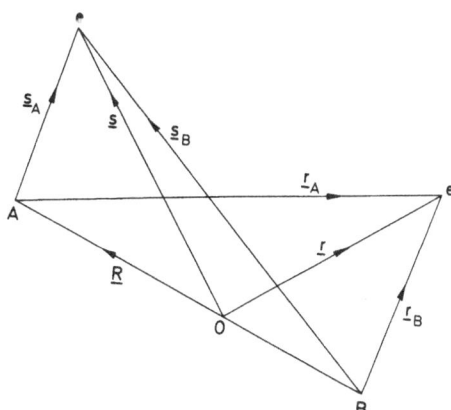

Fig. 4.1. Coordinate systems for two electrons e in the field of two colliding
nuclei A and B

where

$$H = -\nabla_R^2/2M + H_e + V \quad . \tag{4.4}$$

For the moment it is convenient not to specify the exact partition of the total Hamiltonian given by (4.4).

Using the expansion

$$\Psi(\underline{r},\underline{s},\underline{R}) = \sum_n F_n(\underline{R}) \psi_n(\underline{r},\underline{s},\underline{R}) \tag{4.5}$$

in (4.1) and projecting onto any basis function ψ_n of the set gives the coupled equations

$$[\nabla_R^2 + k_n^2(R)] F_n(\underline{R}) = 2M \sum_m <\psi_n|V|\psi_m> F_m(\underline{R}) - 2 \sum_{m \neq n} <\psi_n|\nabla_R|\psi_m> \cdot \nabla_R F_m(\underline{R})$$

$$- \sum_{m \neq n} <\psi_n|\nabla_R^2 + k_n^2(R)|\psi_m> F_m(\underline{R}) \quad , \tag{4.6}$$

where $k_n^2 = 2M[E-E_n(R)]$.

If the coupling terms on the right-hand side of (4.6) are neglected, then only elastic scattering is possible according to the equation

$$[\nabla_R^2 + k_n^2(R) - 2MV_{nn}] F_n(R) = 0 \tag{4.7}$$

for each state n. If it is considered that the coupling terms go to zero asymptotically and the energies $E_n(R)$ go to fixed values E_n independent of R, then the nuclear wave function for the incident channel may be written for large R

$$F_0(\underline{R}) \sim \exp(i\underline{k}_0 \cdot \underline{R}) + R^{-1} \exp(ik_0 R) f_0(\theta,\Phi) \quad , \tag{4.8}$$

where $k_0^2 = 2M(E-E_0)$ and θ, Φ are polar coordinates of $\underline{R}$. Similarly, for states other than the initial state

$$F_n(\underline{R}) \sim R^{-1} \exp(ik_n R) f_n(\theta,\Phi) \quad , \tag{4.9}$$

where $k_n^2 = 2M(E-E_n)$.

The cross section for excitation of a particular state n is then given by

$$\sigma_n = (k_n/k_0) \iint |f_n(\theta,\Phi)|^2 \sin\theta \, d\theta \, d\Phi \quad . \tag{4.10}$$

In practice, the scattering amplitudes f_n are obtained by examining the asymptotic behavior, according to (4.8) and (4.9), of the nuclear wave functions obtained as solutions of the coupled equations (4.6). Two particular situations where the expansion (4.5) has proved useful in ion-atom collisions will be considered explicitly. These situations are the limits of low and high impact velocity where basis states ψ_n which are a good representation of the states occupied by electrons during the collision may be chosen. Then the expansion (4.5) may be truncated to a few basis states with good accuracy.

4.1.2 The First Born Approximation for Excitation and Ionization

In a fast collision of two atoms, where "fast" means that the collision velocity is greater than characteristic orbital velocities of electrons in the bound states in question or when the influence of one atom on the other is a small perturbation, it is clear that an expansion basis of products of undistorted atomic functions based on nuclei A and B is appropriate. Furthermore, under these conditions the exchange part of the electron-electron interaction may be neglected, and electrons can be assigned to either nucleus A or B. Then it can be shown that the last two terms on the right-hand side of (4.6) become zero and the coupled equations are

$$(\nabla_R^2 + k_n^2) \ F_n(\underline{R}) = 2M \sum_m \ <\psi_n|V|\psi_m> \ F_m(\underline{R}) \tag{4.11}$$

where $k_n^2 = 2M(E-E_n)$ and E_n is the sum of the internal energies of electrons on A and B. Since the Green function satisfying

$$(\nabla_R^2 + k_n^2) \ G_n(\underline{R}) = \delta(\underline{R}-\underline{R}')$$

has the asymptotic form

$$G_n(\underline{R}) \sim -(4\pi R)^{-1} \ \exp[i(k_n R - \underline{k}_n \cdot \underline{R}')] \quad \text{as} \quad R \to \infty \quad ,$$

the scattering amplitude in (4.9) may be written

$$f_n(\theta,\Phi) = -(M/2\pi) \int \sum_m \ <\psi_n|V|\psi_m> \ F_m(\underline{R}') \ \exp(-i\underline{k}_n \cdot \underline{R}') \ d\underline{R}' \quad . \tag{4.12}$$

Equations (4.11) and (4.12) will now be approximated in the plane-wave Born approximation (PWBA). This involves replacing the sum of matrix elements on the right-hand side by the single first order interaction matrix element $<\psi_n|V|\psi_0>$ connecting initial and final states. In addition, the nuclear wave function $F_0(\underline{R}')$ is replaced by the plane-wave form $\exp(i\underline{k}_0 \cdot \underline{R}')$ to give

$$f_n(\theta,\Phi) = -(M/2\pi) \int \ <\psi_n|V|\psi_0> \ \exp[i(\underline{k}_0-\underline{k}_n) \cdot \underline{R}'] \ d\underline{R}' \quad . \tag{4.13}$$

Introducing the momentum transfer during the collision

$$\underline{K} = \underline{k}_0 - \underline{k}_n$$

such that $k_0 {\sim} k_n$ and

$$K^2 = k_0^2 + k_n^2 - 2k_0k_n \cos\theta \quad , \tag{4.14}$$

we find $KdK = k_0 k_n \sin\theta \, d\theta$. Assuming azimuthal symmetry in Φ and using the transformation (4.14) in (4.13), the angular integration in (4.10) may be converted to an integral over K, i.e.,

$$\sigma_n = (2\pi/k_0^2) \int |f_n(K)|^2 \, KdK \quad , \tag{4.15}$$

with

$$f_n(K) = -(M/2\pi) \int <\psi_n|V|\psi_0> \exp(i\underline{K}\cdot\underline{R}) \, d\underline{R} \quad . \tag{4.16}$$

The integration over K in (4.15) is restricted to the range implied by momentum and energy conservation. From (4.14)

$$K_{min} = k_0 - k_n = (k_0^2 - k_n^2)/(k_0 + k_n) \approx (k_0^2 - k_n^2)/2k_0 = (E_n - E_0)/v_I \tag{4.17}$$

where v_I, the incident relative velocity of the heavy-particle motion, is given by $k_0 = Mv_I$. The upper limit of the integration is given by

$$K_{max} = k_0 + k_n \approx 2k_0 = 2Mv_I \quad . \tag{4.18}$$

Since $|f_n(K)|^2$ is generally a rapidly decreasing function for large K and because of the large reduced mass M of the nuclei, K_{max} may be put equal to infinity without serious loss of accuracy.

If excitation of either target or projectile is considered, and charge transfer and electron exchange processes are neglected, then the amplitude $f_n(K)$ in the PWBA becomes

$$f_n(K) = -(M/2\pi) \iiint \exp(i\underline{K}\cdot\underline{R}) \, \psi_{nA}^*(\underline{s}_A) \, \psi_{nB}^*(\underline{r}_B)$$

$$[Z_A Z_B/R - Z_A/|\underline{r}_B - \underline{R}| - Z_B/|\underline{s}_A + \underline{R}| + 1/|\underline{R} - \underline{r}_B + \underline{s}_A|]$$

$$\psi_{0A}(\underline{s}_A) \, \psi_{0B}(\underline{r}_B) \, d\underline{s}_A \, d\underline{r}_B \, d\underline{R} \quad . \tag{4.19}$$

Three special cases of this general expression will be considered and the application of each of them in the calculation of excitation and ionization in heavy
ion-atom collisions will be described.

4.1.3 Incident Bare Nucleus

The case where the incident "atom" A is simply a bare nucleus of charge Z_A has received most attention, particularly in the treatment of ionization of the inner
shells of heavy atoms by incident light nuclei.

For a bare nucleus only the first two terms in the brackets [......] of (4.19)
survive and the wave functions on nucleus A are, of course, omitted. Making use of
the integral [4.1]

$$\int \exp(i\underline{K}\cdot\underline{R})/|\underline{r} - \underline{R}| \, d\underline{R} = (4\pi/K^2) \exp(i\underline{K}\cdot\underline{r}) \quad , \tag{4.20}$$

and considering only final states $n\neq0$, the transition amplitude becomes

$$f_n(K) = (2M/K^2) \, Z_A \, \varepsilon_{n0}^B(K) \quad , \tag{4.21}$$

where the inelastic form factor $\varepsilon_{n0}^B(K)$ is defined as

$$\varepsilon_{n0}^B(K) = \int d\underline{r}_B \, \psi_{nB}^*(\underline{r}_B) \, \exp(i\underline{K}\cdot\underline{r}_B) \, \psi_{0B}(\underline{r}_B) \quad . \tag{4.22}$$

The squared modulus of $\varepsilon_{n0}^B(K)$ is proportional to the generalized oscillator strength
(GOS). The GOS is a *target* property only and describes the effect of transferring
momentum $\underline{K}$ to the target atom in state 0 in exciting the atom to the state n. The
properties of the GOS have been discussed in detail by INOKUTI [4.2]. The total
cross section is obtained by integrating over all permissible momentum transfers
according to (4.15)

$$\sigma_n = (8\pi Z_A^2/v_I^2) \int_{K_{min}}^{K_{max}} |\varepsilon_{n0}^B|^2 \, dK/K^3 \quad . \tag{4.23}$$

The formula (4.23) has found wide application in the calculation of cross sections
for ionization of *outer* shells by swift heavy nuclei—in particular, protons and α-
particles (McDOWELL and COLEMAN [Ref.4.3, Chap.7]). Since, from (4.17) and (4.18)
and the properties of the GOS, it can be shown that the Born cross section for heavy-
particle collisions can be related to that for electron collisions, much of the
work on fast electron excitation [4.2] is of relevance to heavy-particle excitation.
In the case of outer shells, experimental and theoretical study has extended to the
separate cross sections for excitation of individual substates and the details of
the angular distributions of electrons ejected into the continuum [4.4-17]. Many
calculations have been reported using wave functions for many-electron atoms of near

Hartree-Fock accuracy [4.8,9] and even sophisticated correlated wave functions for few-electron atoms [4.10-12]. However, the emphasis in this chapter is on excitation processes in heavy atom-atom collisions where the understanding of the excitation processes is less advanced than that for electron impact. The attention has been concentrated mostly on the excitation of *inner-shell* electrons by heavy-particle impact. Here the electron-electron interactions become less important and simpler hydrogenic wave functions may be employed to give good results for total cross sections. In addition, as will be discussed further in Sec.4.1.5, when the charge of the incident bare ion is much less than the effective charge experienced by an inner-shell electron, the validity of the Born approximation may be extended to collision velocities which are slow compared with the inner-shell orbital velocity.

Using hydrogenic one-electron functions, the PWBA cross section (4.23) has been evaluated both for excitation to bound states [4.1,13,14] and for ionization [4.1, 15-20]. The use of the PWBA with hydrogenic wave functions in the particular case of inner-shell excitation has been discussed comprehensively by MERZBACHER and LEWIS [4.21] and by MADISON and MERZBACHER [4.22].

4.1.4 Incident Neutral Atom or Ion Carrying Electrons

When the incident particle is a neutral atom or ion carrying electrons into the collision, then additional interactions arise in the PWBA according to the general expression (4.19). In the following it will be assumed that one is interested in excitation to a particular state of the target atom B, but the considerations apply equally to excitation of the incident particle. The simplest case arises when the cross section for excitation of the target to state n when the incident particle *remains in its ground state* is required. Use of the Bethe integral (4.20) in (4.19) gives two nonvanishing terms,

$$
\begin{aligned}
f_n(K) &= (2M/K^2) \iint dr_B \, ds_A \, \psi_{0A}^*(s_A) \, \psi_{nB}^*(r_B) \left\{ - Z_A \exp(iK \cdot r_B) \right. \\
&\quad \left. + \exp[iK \cdot (r_B - s_A)] \right\} \psi_{0A}(s_A) \, \psi_{0B}(r_B) \\
&= (2M/K^2)[-Z_A \, \varepsilon_{n0}^B(K) + \varepsilon_{n0}^B(K) \, \varepsilon_{00}^A(K)] \quad ,
\end{aligned}
$$

so that the cross section may be written

$$
\sigma_{n0} = (8\pi/v_I^2) \int_{K_{min}}^{K_{max}} |\varepsilon_{n0}^B|^2 \, |Z_A - \varepsilon_{00}^A|^2 \, dK/K^3 \quad . \tag{4.24}
$$

Since the form factor ε_{00}^A involves only the squared modulus of the projectile wave function and can be written

$$
\varepsilon_{00}^A = \int ds_A \, |\psi_0(s_A)|^2 \, \cos(K \cdot s_A) \quad ,
$$

it is a real quantity and is bounded by a magnitude of unity. Hence, the presence
of the bound projectile electron acts only to screen the incident nucleus. In the
most general case, the screening effect may be positive or negative depending upon
the particular value of $\underline{K}$. In the simple example discussed below, ϵ_{00}^{A} is always
positive and the effective nuclear charge varies between Z_A and (Z_A-1).

When there is simultaneous excitation of electrons on both target and projectile,
the orthogonality of initial and final wave functions results in a nonzero contribution only from the electron-electron interaction term in (4.19). Hence, in the
PWBA, only this term can cause simultaneous excitation of target and projectile.
When the target is excited to state n and the projectile to state m, the cross section (4.15) with (4.19) and (4.20) becomes

$$\sigma_{nm} = (8\pi/v_I^2) \int_{K_{min}}^{K_{max}} |\epsilon_{n0}^{B}(K)|^2 \, |\epsilon_{m0}^{A}(K)|^2 \, dK/K^3 \quad . \tag{4.25}$$

Now K_{min} is decided by the sum of the excitation energy of both particles according
to (4.17).

When one is interested in the cross section for excitation of target state n,
irrespective of the final state of the projectile—i.e.,

$$\sigma_n = \sigma_{n0} + \sum_{m \neq 0} \sigma_{nm} \quad , \tag{4.26}$$

an approximate closure can be performed to give a simple closed form for the sum
(4.26). If K_{min} in (4.25) for different projectile states m is replaced by a constant
$\bar{K}$ independent of m, then

$$\sum_{m \neq 0} \sigma_{nm} = \sum_{m \neq 0} \int_{\bar{K}}^{K_{max}} |\epsilon_{n0}^{B}|^2 \, |\epsilon_{m0}^{A}| \, dK/K^3 = \int_{\bar{K}}^{K_{max}} |\epsilon_{n0}^{B}|^2 \, (1-|\epsilon_{00}^{A}|^2) \, dK/K^3 \quad , \tag{4.27}$$

since $\sum_m |\epsilon_{m0}^{A}|^2 = 1$ by closure, using (4.22). Note that the summation implies also an
integration over continuum states.

In the total cross section (4.26), simultaneous excitation of the projectile to
low-lying states is usually most important. Hence, the closure is usually performed
by setting $\bar{K}$ equal to the value of K_{min} appropriate to the lowest excited state.
Calculations of the PWBA cross section for atom-atom or "clothed" ion-atom collisions
have so far been made only for simple cases of one- and two-electron atoms either
by use of the closure relation or by explicit summation over excited target states
[4.23-30]. Nevertheless, certain systematic trends of these results should be applicable to more complex atomic collisions. The generalization of the results (4.24)
and (4.27) to the case of many-electron atoms is straightforward in principle, but
the expressions become cumbersome when open-shell atomic configurations are involved.

However, for the simple case of a closed-shell ground state formed from a single Slater determinant of orthonormal spin orbitals φ_i the generalization is relatively simple and yields

$$\sigma_{n0} = (8\pi/v_I^2) \int_{K_{min}}^{K_{max}} |\varepsilon_{n0}^B|^2 \left[|Z_A - \sum_j^{occ.} \omega_j <\varphi_j^A|\exp(i\underline{K}\cdot\underline{s}_A)|\varphi_j^A>|^2 \right] dK/K^3 \quad , \tag{4.28}$$

for the cross section when the projectile remains in its ground state. The summation in (4.28) extends over all occupied shells and ω_j is the statistical weight of shell j. As in (4.24), the diagonal matrix elements are real and their effect is to screen the projectile nucleus. As $\underline{K}\to 0$ the factor in brackets [......] tends to $(Z_A-N_A)^2$, where N_A is the number of electrons on nucleus A. As $\underline{K}\to\infty$ the matrix elements tend to zero and the nucleus is unscreened. This corresponds to what might be expected intuitively when one remembers that small momentum transfer collisions result from distant collisions and large momentum transfers from close collisions.

In the case of (4.27) the generalization to a closed-shell ground state reads

$$\sum_{m\neq 0} \sigma_{nm} = (8\pi/v_I^2) \int_{\underline{K}}^{K_{max}} |\varepsilon_{n0}^B|^2 \left[N_A - |\sum_j^{occ.} \omega_j <\varphi_j^A|\exp(-i\underline{K}\cdot\underline{s}_A)|\varphi_j^A>|^2 \right.$$

$$+ \sum_{i\neq j}^{occ.} \omega_i\, \omega_j <\varphi_i^A|\exp(i\underline{K}\cdot\underline{s}_A)|\varphi_i^A><\varphi_j^A|\exp(-i\underline{K}\cdot\underline{s}_A)|\varphi_j^A>$$

$$\left. + \sum_{i\neq j}^{occ.} \omega_i\, \omega_j <\varphi_i^A|\exp(i\underline{K}\cdot\underline{s}_A)|\varphi_j^A><\varphi_j^A|\exp(-i\underline{K}\cdot\underline{s}_A)|\varphi_i^A> \right] dK/K^3 \quad . \tag{4.29}$$

The extra two terms in this expression compared with (4.27) arise from interactions between the occupied subshells in the ground state. However, it should be emphasized that the assumption of a common K_{min} for shells differing widely in binding energy is a much more drastic approximation than for a one-electron atom. As $\underline{K}\to 0$ the factor in brackets [......] in (4.29) tends to $[N_A-N_A^2+N_A(N_A-1)]=0$. Again as $\underline{K}\to\infty$, only the first term N_A survives and the contribution to the cross section is the same as that from N_A independent free electrons of speed v_I incident upon the target atom.

These characteristics of the PWBA are illuminated by consideration of a particular simple example which may be used to interpret the results of the detailed calculations referred to above. Consider the excitation of a hydrogen atom in the 1s ground state to the 2p state by impact of a helium atom, also initially in the $1s^2$ ground state. For purposes of illustration, a simple hydrogenic wave function with effective $Z_A=1.345$ will be used for the helium ground state. Again it is instructive to consider separately the cross section for hydrogen excitation when the helium atom remains in its ground state and when it is excited.

In this example (4.28) becomes

$$\sigma_{n0} = (8\pi/v_I^2) \int_{K_{min}}^{K_{max}} |\epsilon_{n0}^B|^2 (2-2\epsilon_{00}^A)^2 \, dK/K^3 \quad , \tag{4.30}$$

where

$$\epsilon_{n0}^B = <\varphi_{2p}^B|\exp(i\underline{K}\cdot\underline{r}_B)|\varphi_{1s}^B>$$

and

$$\epsilon_{00}^A = <\varphi_{1s}^A|\exp(-i\underline{K}\cdot\underline{s}_A)|\varphi_{1s}^A> \quad .$$

Similarly in this case (4.29) reduces to (only one shell on each atom)

$$\sum_{m\neq0} \sigma_{nm} = (8\pi/v_I^2) \int_{K_{min}}^{K_{max}} |\epsilon_{n0}^B|^2 \left[2 - 2(\epsilon_{00}^A)^2\right] dK/K^3 \quad . \tag{4.31}$$

To compare the relative contributions of the two cross sections, the behavior of the two integrands as a function of momentum transfer will be examined. Aside from common constant factors, we can write for the helium atom

$$(2-2\epsilon_{00}^A)^2 \rightarrow 4[1 - 1/(1+t^2)^2]^2 \equiv A_0 \quad ,$$

$$2 - 2(\epsilon_{00}^A)^2 \rightarrow 2[1 - 1/(1+t^2)^4] \equiv A_{ex} \quad , \tag{4.32}$$

with $t=(K/2Z_A)$. In addition the hydrogen atom function for $1s\rightarrow2p$ excitation $|\epsilon_{n0}^B|^2/t^3$ constituting the remainder of the integral can be written, aside from constant factors

$$|\epsilon_{n0}^B|^2/t^3 \rightarrow \left\{t^2\left[(3Z_B/4Z_A)^2 + t^2\right]^3\right\}^{-1} \equiv B \quad . \tag{4.33}$$

The functions A_0, A_{ex} are plotted as a function of t in Fig.4.2a. Here it can be seen that, although A_0 is asymptotic to its unscreened value $Z_A^2=4$ for large t (close encounters), the effect of screening is quite dramatic for small t (distant encounters). Also, as might be expected, the screening effect changes most rapidly for $t\approx1$ corresponding to distances of the order of the projectile K-shell radius. The function A_{ex} is asymptotic to the value $N_A=2$ for large t, corresponding to the action of two fast free electrons. At small t the value of A_{ex} exceeds that of A_0.

In Fig.4.2b the integrands Z_A^2B, A_0B, and $A_{ex}B$ are plotted as a function of t. These three integrands give rise to cross sections, respectively, for bare nucleus, ground-state atom, and excitable atom impact according to (4.23), (4.30), and (4.31).

The range of integration is conveniently separated by the momentum transfer $t=1$, arising roughly from distances equal to the projectile K-shell radius. As $v_I \to \infty$ then the minimum momentum transfer $t_{min} \to 0$ and the region $t<1$ is most important. Then it is seen that the larger contribution to the total cross section is made by collisions in which the projectile is excited and that the total cross section for excitation by the neutral particle is lower than that for the bare nucleus alone. In addition both contributions from the neutral particle excitation (curves 2 and 3) are nonsingular at $t=0$. Then as $t_{min} \to 0$ the integral over t is bounded and from (4.30) and (4.31) the cross section varies as v_I^{-2} for large v_I. This is contrasted with the singular behavior of the integrand for bare nuclear excitation (curve 1) that gives rise to the well-known $v_I^{-2} \ln(v_I^2)$ variation of the excitation cross section. The drastic effect of screening on the contribution from distant encounters is evident from Fig.4.2b for $t<1$.

For close encounters ($t>1$) the integrand (curve 2) for ground state projectile impact exceeds that for impact in which the projectile is excited (i.e., $Z_A^2 > N_A$ as shown by the asymptotic behavior in Fig.4.2a). In addition the different values of t_{min} for the two processes weights strongly the contribution from ground state impacts. For example, for excitation of the target only $K_{min} = \Delta E(target)/v_I$, but for simultaneous excitation $K_{min} = [\Delta E(target) + \Delta E(projectile)]/v_I$. This means that for

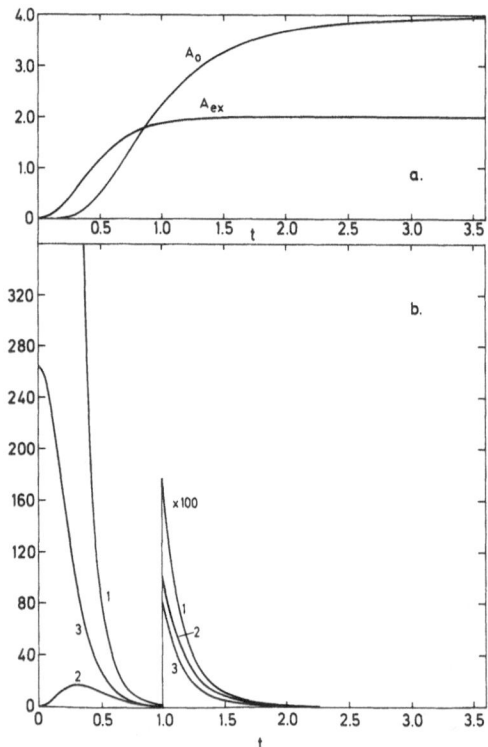

Fig. 4.2a and b. Plot of integrands discussed in the text: a) The functions A_0 and A_{ex} defined by (4.32). b) The functions $Z_A^2 B$ (curve 1), $A_0 B$ (curve 2) and $A_{ex} B$ (curve 3), where B is defined by (4.33)

low-velocity impacts the total cross section is dominated by ground-state scattering of the projectile. Since this is opposite to the situation at high velocity, the two contributions to (4.26) may give rise to a double peak in the total cross section for excitation of a particular target level. This is particularly pronounced when the excitation energy of the projectile is greater than that of the target [4.24,28]. Although at low velocity, where the cross section is decided mostly by close encounters (inside the projectile shell radius), the incident atom behaves much like a bare nucleus—the effect of screening still is not negligible. For example in the He-H encounter illustrated by Fig.4.2b, at $v_I=0.1$ a.u., t_{min} for ground-state projectile impact is ~1.4 and there the screening still acts to reduce the cross section by ~30% from that for bare nucleus impact. Finally, although the PWBA is plainly invalid for $v_I=0.1$ a.u. in the example He-H, the qualitative features illustrated by this example are expected to apply equally to situations where the Born approximation *is* valid at low velocity.

4.1.5 Comments on the Validity of the PWBA

The region of validity of the PWBA as a function of relative collision velocity in the case of heavy ion-atom or atom-atom impacts depends critically upon the relative magnitude of the nuclear charges involved and more particularly upon the *effective* nuclear charges experienced by particular electrons in their initial atomic orbits. Two separate conditions must be satisfied simultaneously. The first is that the relative heavy-particle motion may be described adequately by plane waves. The second is that the potential causing transitions between initial and final electronic states may be treated as a small perturbation.

For collisions involving particles of charge Z_A and Z_B interacting via the Coulomb field, the plane-wave approximation generally has been taken as the most restrictive since it requires the condition

$$Z_A Z_B/v_I \ll 1 \quad . \tag{4.34}$$

However, it is widely recognized [4.21,31] that for many ion-atom collisions, if one is interested in total cross sections, this condition is unnecessarily restrictive. In particular the very close collisions where the plane-wave approximation is least good are often not important for the total cross section. This is discussed more quantitatively below. In addition, for the more distant collisions, the screening effect of the "spectator" electrons may reduce considerably the effective internuclear interaction. Accordingly, for consideration of ionization of outer-shell electrons in heavy-ion collisions, some effective screened charges may be used in (4.24) to make this condition much less restrictive.

The most restrictive condition on the validity of the PWBA in a general ion-atom collision is then provided by the perturbation treatment of the electron-electron interaction which gives rise to the "screening" of the nuclei and to simultaneous excitation of projectile and target. In any given collision the PWBA may be valid for interaction of electrons in some subshells but not in others. It is clear that if perturbation of one electron by the other is to be small and exchange effects are to be neglected, then the collision velocity must be larger than the electron orbital velocities. Then, for the interaction between a particular shell on the target and a particular shell on the projectile,

$$v_I \gg v_{el} \quad , \quad \text{i.e.,} \quad \tilde{Z}/v_I \ll 1 \tag{4.35}$$

is a sufficient condition, where $\tilde{Z}$ and v_{el} are the effective nuclear charge and orbital velocity, respectively, of the orbit of the more strongly bound of the two electrons. However, for consideration of the excitation of a particular target shell, the contribution to the total cross section from excitation of projectile shells for which (4.35) is violated may be very small since on more general grounds the inverse of (4.35) expresses the condition that excitation of a particular shell is not probable.

The condition (4.35) also applies to the ground-state projectile impacts giving rise to screening of the projectile nucleus. In particular it applies when the relevant shells on target and projectile are rather close in binding energy (or effective nuclear charge $\tilde{Z}_A \sim \tilde{Z}_B$) provided that the electron-electron interaction is comparable to the effective nuclear charge $\tilde{Z}$ experienced by the electron in its orbit. However, when the effective nuclear charges are dominant, the Born approximation to the electron-electron interaction may be valid. Since simultaneous projectile excitation will be improbable under these conditions, then the main effect appears as screening of the nuclei.

Finally there remains the condition that the electron-other nucleus interaction be treated as a perturbation. That the effect of a nucleus of charge Z_A on an electron of velocity v_I relative to it should be small is expressed by the condition

$$Z_A/v_I \ll 1 \quad . \tag{4.36}$$

Then, even when (4.35) is satisfied so that the mean relative velocity of the electron is close to v_I, (4.36) will form a more restrictive condition. However, (4.36) applies, strictly speaking, to free electrons and so is likely to be a valid condition for outer electrons having $\tilde{Z} \sim 1$. For inner electrons, where $\tilde{Z} \sim Z_B$, there exists a region where both (4.36) and (4.35) may be violated, and yet the Born approximation for the electron-other nucleus interaction may be valid. This occurs when

$$Z_A/\tilde{Z} \sim Z_A/Z_B \ll 1 \quad , \tag{4.37}$$

so that the effect of the incident nucleus is always a small perturbation even in slow encounters. Then the main limitation on the use of PWBA arises from the plane-wave treatment of the *nuclear* motion. Even here, however, where close encounters are important, it can be shown that the restrictive condition (4.34) does not apply when one is interested in total cross sections. This is because the Massey criterion [4.32] suggests that the maximum contribution to the total cross section comes from impacts close to the adiabatic distance

$$d \sim v_I/\Delta E \tag{4.38}$$

where ΔE is the excitation energy which for inner-shells can be put equal to $(1/2)Z_B^2$ a.u. However, rectilinear nuclear motion is likely to be a good approximation provided that the internuclear potential energy at this distance is small compared with the initial kinetic energy of nuclear motion, i.e.,

$$Z_A Z_B/d \ll M v_I^2/2 \tag{4.39}$$

where M is the reduced mass of the nuclei. With (4.38) the condition (4.39) may be written

$$Z_A Z_B^3/M v_I^3 \ll 1 \quad . \tag{4.40}$$

In the case of inner-shell excitation by the impact of light nuclei on heavy atoms, for which (4.37) and, hence, the Born treatment of the electron-other nucleus interaction is valid, $M \sim M_A \sim 2Z_A$ so that (4.40) becomes

$$(Z_B/v_I)^3/(2 \times 1836) \ll 1 \quad . \tag{4.41}$$

This is plainly a far less restrictive condition than (4.34) and allows the validity of PWBA to be extended to slow collisions so long as (4.37) is fulfilled. In fact, since (4.37) is independent of v_I, in the region where (4.41) is no longer valid a Born treatment of the electron-other nucleus interaction is still permissible. Suitable modifications to the treatment of the nuclear motion can then be made, e.g., by using distorted waves instead of plane waves.

In summary, it is clear that no single prescription can be given for the validity of PWBA. In particular it is important to isolate the validity of the plane-wave treatment of nuclear motion from the first order treatment of the electron-electron and electron-other nucleus interactions. For total cross sections the plane-wave treatment may have extended validity, depending upon the type of excitation of interest according to criterion (4.40). Details of differential cross sections, however, may require the more restrictive condition (4.34). The validity of the use of perturbation theory at a given collision velocity depends critically upon the magnitude of the relevant effective nuclear fields to which an electron is subjected. Each case must be examined separately according to the initial states and binding energies of the electrons, the particular perturbing field and the criteria (4.35-37).

4.1.6 The Binary Encounter Approximation

In this subsection we consider only the excitation or ionization of an atom by an incident bare nucleus. Furthermore, discussion will be restricted to the case $Z_A << Z_B$ so that the Born approximation is expected to remain valid below the peak in the excitation cross section, i.e., for slow collisions. This is the case that has been most widely studied in PWBA and applies to the ionization of inner-shells by incident light nuclei, e.g., protons and α-particles.

GARCIA [4.33] has examined a purely classical approach to the interaction of an incident light nucleus with an inner-shell electron. In this binary encounter approximation (BEA) the potential of the target nucleus is neglected in the actual scattering process but its influence on the initial and final electron states enters in two ways. First, the electron acquires a momentum distribution by virtue of being initially bound in the target potential. Secondly, the target potential acts to require the energy transferred to the inner-shell electron by the incident nucleus to exceed the minimum energy necessary for ionization of the electron in the field of the target atom, i.e., the ionization potential. This field is due both to the target nucleus and to the other spectator target electrons. Within these limitations, the BEA treats the interaction of the incident nucleus and the inner-shell electron as a binary encounter between two *free* classical particles. The ionization cross section is then obtained by integration over all possible energy transfers greater than the ionization potential, due attention being paid to the satisfaction of energy and momentum conservation in the free-particle binary collision.

Although the BEA assumptions would appear to be of doubtful validity and are known to lead to the wrong behavior at high velocity [4.34], this procedure is remarkably successful in the low-velocity region when $Z_A << Z_B$ [4.35]. It is widely appreciated that this is mainly because of the fortunate circumstance that the classical differential cross section for Coulomb scattering of free particles is identical to the quantum-mechanical result. Furthermore, BATES and McDONOUGH [4.36,37] and MADISON and MERZBACHER [4.22] have emphasized that the exact form of the expression for the total ionization cross section in BEA can be obtained from the PWBA. This involves merely the replacement of the final state of the ionized electron in the field of the target nucleus, i.e., a Coulomb wave, by that of a free electron of the same momentum, i.e. a plane wave. From (4.23) the differential cross section for ejection of an electron into the continuum with momentum $\underline{k}$ is given in PWBA as

$$d\sigma/d\underline{k} = (8\pi/v_I^2)\ Z_A^2 \int_{K_{min}}^{K_{max}} |\varepsilon_{\underline{k}0}|^2\ dK/K^3 \tag{4.42}$$

where

$$\varepsilon_{\underline{k}0} = \int d\underline{r}\ \psi_{\underline{k}}^* \exp(i\underline{K}\cdot\underline{r})\ \psi_0$$

and $\psi_{\underline{k}}(\underline{r})$ is a Coulomb wave in the field of nucleus B. For values of k such that Z_B/k is small, the Coulomb wave may be replaced by a plane wave $(2\pi)^{-3/2} \exp(i\underline{k}\cdot\underline{r})$. Then (4.42) becomes

$$d\sigma/d\underline{k} = (8\pi/v_I^2)\ Z_A^2 \int_{K_{min}}^{K_{max}} |\psi_0(\underline{k}-\underline{K})|^2\ dK/K^3 \tag{4.43}$$

where $\psi_0(\underline{k}-\underline{K})$ is the Fourier component of $\psi_0(\underline{r})$ of momentum $\underline{k}-\underline{K}$. The total cross section for ionization is obtained by integration over all final momenta $\underline{k}$

$$\sigma = (8\pi/v_I^2)\ Z_A^2 \int d\underline{k} \int_{K_{min}}^{K_{max}} |\psi_0(\underline{k}-\underline{K})|^2\ dK/K^3 \quad. \tag{4.44}$$

By making the transformation $\underline{q}=\underline{k}-\underline{K}$, the angular integration over $d\hat{\underline{k}}$ may be transformed to an integral over q to give

$$\sigma = (16\pi^2/v_I^2)\ Z_A^2 \int_0^\infty kdk \int_{K_{min}}^{K_{max}} dK/K^4 \int_{q_{min}}^{q_{max}} |\psi_0(q)|^2\ qdq \tag{4.45}$$

where it has been assumed that the momentum distribution $|\psi_0(q)|^2$ is isotropic. The limits on the integrations are obtained from (4.17) and (4.18) with $\underline{q}=\underline{k}-\underline{K}$, i.e.,

$$|k - K| \le q \le k + K$$

and

$$K_{min} = (k^2/2+I)/v_I \quad, \quad K_{max} \sim \infty \tag{4.46}$$

where I is the ionization potential of the electron in its initial state on Z_B. The expression (4.45), subject to the limits (4.46) is exactly the cross section for scattering of a free electron from an initial state of momentum q to a final state of momentum $\underline{k}$ in the Coulomb field of a nucleus Z_A. As BATES and McDONOUGH [4.36] have noted, rearrangement of the integration limits in (4.45) leads exactly to the procedure employed in the BEA [4.33]. One important difference is that the lower limit K_{min}, which appears naturally in (4.45) and avoids the singularity in the free-electron Coulomb differential cross section for K=0, is injected artificially in the BEA by limiting the *energy transfer* to have the lower limit I.

Although the BEA can be obtained in this way from the PWBA and, hence, the use of classical mechanics may be justified, this derivation still does not explain the reasons for the close agreement of the BEA with the experiments in the low-velocity region [4.35] and with the PWBA [4.38]. However, this agreement suggests that the PWBA calculated using only plane waves for all final electron-ionized states should not differ appreciably from that obtained using Coulomb waves. At first sight this would appear not to hold. The condition that an electron of velocity v moving in the Coulomb field of a nucleus of charge Z_B can be considered to behave like a plane wave is usually expressed as $Z_B/v \ll 1$. However, MADISON and MERZBACHER [4.22] have shown that the mean $\bar{v}$ of the velocity of all electrons ejected into the continuum from inner-shells is such that $Z_B/\bar{v} \geq 1$. Hence the plane-wave approximation would appear to be invalid. Then the argument of GARCIA [4.33] that the BEA is valid because of the dominance of large energy transfers with respect to the binding energy would also appear not to hold. Hence at this time the precise reason for the close agreement [4.38] of BEA and PWBA total cross sections at low velocity remains obscure.

4.1.7 Charge Exchange in Fast Collisions

Thus far only transitions of electrons from initial bound states in one or other nuclear potential to final bound or continuum states in the *same* nuclear potential have been considered. Transitions of electrons from the field of one nucleus to the field of the other during an ion-atom collision may also contribute to the ionization of a particular partner. Although such processes have been discussed in detail for outer-shell electrons under the impact of ions with small effective charge, the influence of charge transfer on the ionization of inner-shells, particularly in the case of highly charged incident ions, is only beginning to be studied [4.39-41].

Unfortunately the problem of charge transfer, unlike that of direct excitation, is not adequately treated by the PWBA. In particular, the nonorthogonality of initial and final atomic states means that there is not a unique form of the PWBA for charge transfer and the different forms give different results. These difficulties can be removed by proceeding to the second Born approximation or by explicitly extending the first Born approximation to cancel the terms arising from the overlap of initial and final atomic states. This latter procedure is discussed in more detail in the impact parameter treatment of Section 4.2.3.

In the case of charge transfer it is not permissible at the outset to ignore terms of order m/M where m is the electron mass (=1 a.u.) as was done in Section 4.1.1. For simplicity only one electron in the field of two nuclei will be discussed; the coordinate system for the rearrangement problem is shown in Fig.4.3. The functions $\psi_{nA}(r_A)$ and $\psi_{mB}(r_B)$ are eigenfunctions in the field of nuclei A and B, respec-

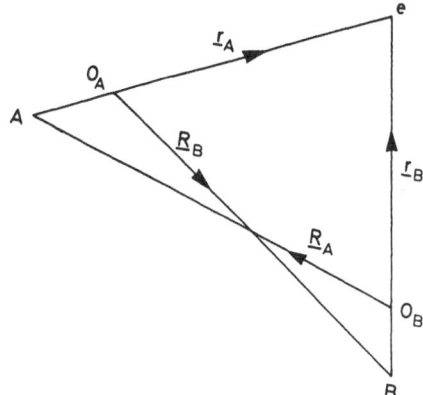

Fig. 4.3. Coordinate systems for an electron e in the field of two colliding nuclei A and B

tively. The complete sets of such functions are not orthogonal and diagonalize different electronic Hamiltonians H_e^A and H_e^B. The total wave function can be expanded in one such complete set, e.g.,

$$\Psi(\underline{r}_A, \underline{R}_B) = \sum_m \psi_{mA}(\underline{r}_A) \, F_m(\underline{R}_B) \tag{4.47}$$

where ψ_{mA} describes the state of the electron around nucleus A and $F_m(\underline{R}_B)$ describes the motion of nucleus B relative to the center of mass of nucleus A and the electron. As in Section 4.1.1, substitution of (4.47) in the Schrödinger equation leads to a set of coupled equations

$$\left(\nabla_{\underline{R}_A}^2 + k_m^2\right) F_m(\underline{R}_B) = 2M \int \psi_{mA}^*(\underline{r}_A) \, (Z_A Z_B / |\underline{r}_A - \underline{r}_B| - Z_B / r_B) \, \Psi(\underline{r}_A, \underline{R}_B) \, d\underline{r}_A \tag{4.48}$$

where $k_m^2 = 2M(E - E_m)$ and M is the reduced mass

$$M = M_B(M_A + m)/(M_A + M_B + m) \quad . \tag{4.49}$$

(In this subsection for clarity m refers to the electron mass, although in a.u. it has the value unity and is omitted elsewhere).

A first Born approximation for charge exchange may be made by replacing the exact scattering wave function on the rhs of (4.48) by its initial atomic form, i.e.,

$$\Psi = \psi_{nB}(\underline{r}_B) \, \exp(i\underline{k}_n \cdot \underline{R}_A) \quad , \tag{4.50}$$

where now ψ_{nB} is the bound state on nucleus B and $\exp(i\underline{k} \cdot \underline{R}_A)$ describes the plane-wave motion of A relative to the center of mass of atom B. Then the first Born approximation cross section may be written

$$\sigma = 2\pi(k_m/k_n) \int |f_m(\theta)|^2 \sin\theta \ d\theta \quad , \tag{4.51}$$

where θ is the angle between $\underline{k}_n$ and $\underline{k}_m$ and

$$f_m = -(M/2\pi) \int \psi^*_{mA}(\underline{r}_A) \ [(Z_A Z_B/|\underline{r}_B-\underline{r}_A|) - Z_B/r_B]$$

$$\times \ \psi_{nB}(\underline{r}_B) \ \exp(i\underline{k}_n \cdot \underline{R}_A - i\underline{k}_m \cdot \underline{R}_B) \ d\underline{r}_A \ d\underline{R}_B \quad . \tag{4.52}$$

Now, unlike the situation for *excitation* of an electron, this Born approximation for charge exchange involves the internuclear potential. This undesirable feature arises from the nonorthogonality of initial and final states in the first Born approximation. This contribution does not occur in the exact expression for the transition amplitude (see Ref.4.48, p.144). This led OPPENHEIMER [4.42] and BRINKMAN and KRAMERS [4.43] to calculate the Born approximation omitting the internuclear term. On the other hand, BATES and DALGARNO [4.44] and JACKSON and SCHIFF [4.45] argued in favor of retaining the internuclear potential. This is not the only ambiguity with regard to the first Born approximation. The interaction in (4.52) is the interaction between the nucleus B and the atom A after the electron has transferred and for this reason is called the "post" interaction. An alternative form of the first Born approximation may be derived [Ref.4.3, p. 381] involving the "prior" interaction between nucleus A and the atom B. The two forms do not necessarily lead to equivalent results.

DRISKO [4.46] examined the asymptotic form of the *second* Born approximation and showed that these ambiguities may be removed; in particular, the contribution of the internuclear potential does not appear at high velocity when both first and second Born terms are evaluated. DRISKO [4.46], CHESHIRE [4.47] and DETTMANN [4.48] showed that the cross section for charge transfer including first and second Born approximations has the asymptotic form for large incident velocity

$$\sigma_2 \sim \left[0.295 + 5\pi v_I/2^{11}(Z_A+Z_B)\right] \sigma_1 \quad \text{a.u.} \quad , \tag{4.53}$$

where σ_1 is the cross section calculated by BRINKMANN and KRAMERS in the first Born approximation neglecting the internuclear potential contribution in (4.52). This cross section has the asymptotic form

$$\sigma_1 \sim \pi \ 2^{18} \ (Z_A Z_B)^5/(5v_I^{12}) \quad \text{a.u.} \tag{4.54}$$

for transfer from a 1s state on Z_B to a 1s state on Z_A.

Although strict conditions for the validity of the Born expansions are difficult to determine, the form (4.53) should certainly be valid when the condition

$$Z_A \ Z_B/v_I \ll 1 \tag{4.55}$$

is satisfied. This guarantees not only that the plane-wave treatment of nuclear motion is valid but that the collision is fast compared with electron orbital velocities in initial and final states.

At lower velocities, BATES and McCARROLL [4.49] have shown how the ambiguities arising from nonorthogonality of initial and final states may be removed by a more exact solution of the coupled equations (4.48). This more exact solution is most easily achieved in the impact parameter formulation that is considered in Section 4.2.3. At still lower velocities, where the impact velocity falls below the typical electron orbital velocities, a Born approximation based upon atomic eigenfunctions is inadequate unless $Z_A << Z_B$. Then the multiple interactions between both nuclei and the electrons must be taken into account. One way of achieving this is to expand the exact scattering states in *molecular* wave functions as described in the next section. Under these conditions, where both nuclei are treated on an equal footing, the distinction between excitation and charge transfer processes need not be made.

In Section 4.1.2-4, ionization of a target or projectile was considered in first Born approximation under the assumption that it is adequate to describe the final state of the ionized electron as being decided solely by the field of the parent nucleus to which it is initially bound. Such a description of final states should be adequate in the case of excitation or charge transfer. Then the bound states have finite extension in space and the exciting nucleus is far removed from the excited electron after the collision. However, for continuum states, extending over all space, this is not so. In this situation it is more meaningful to consider the *momentum* of the ejected electron. As described in Section 4.1.5 the effect of the Coulomb field of a nucleus of charge Z_B on an electron of velocity $\underline{v}$ relative to it is measured by the Sommerfeld parameter Z_B/v. The Sommerfeld parameter for the same electron in the field of a nucleus of charge Z_A moving with velocity $\underline{v}_I$ relative to nucleus B is given by Z_A/v' where $\underline{v}' = |\underline{v} - \underline{v}_I|$. Then the field of one or other nucleus is more important according to

$$Z_B/v \gtrless Z_A/|\underline{v} - \underline{v}_I| \quad . \tag{4.56}$$

In the simple case of ionization of atom B by a fast bare nucleus with $Z_A \lesssim Z_B$, the dominant contribution to the ionization cross section comes from electrons with final velocity $\underline{v} << \underline{v}_I$. Then since

$$Z_B/v > Z_A/v' \approx Z_A/v_I \quad , \tag{4.57}$$

the description of the final states as Coulomb waves solely in the field of the nucleus Z_B is adequate for computation of the total cross section. However, if one is interested in details of the differential cross section $d\sigma/d\underline{v}$ (e.g., RUDD and MACEK [4.50]), such a description may not be adequate. In particular, for final electron states with momenta $\underline{v} \sim \underline{v}_I$ so that $\underline{v}' = \underline{v} - \underline{v}_I$ is very small we have

$$Z_A/v' > Z_B/v \approx Z_B/v_I \quad . \tag{4.58}$$

Such states with momentum $\underline{v}'$ small relative to Z_A may be considered as occupied by electrons moving along with nucleus A and predominantly in the same direction. In a $Z_A{\to}Z_B$ collision such electrons may be described as moving solely in the field of nucleus A. This feature has been called "charge transfer to the continuum" by MACEK [4.51] who gave the first theoretical description of the effect.

Normally when perturbation theory holds, only small contributions to the ionization of a target atom result from charge transfer to the continuum (and indeed from charge transfer to bound states).

However, when $Z_A{>}Z_B$—as, for example, the collision of a highly stripped nucleus with a neutral target atom—this will not always be the case. In particular in this situation, $Z_A/|\underline{v}{-}\underline{v}_I|$ may be comparable with Z_B/v, even for electron velocities $\underline{v}$ rather small—i.e., close to the ionization threshold of the atom B. In the limiting situation $Z_A{\gg}Z_B$, it would appear that most of the free electrons could be described as moving in the field of Z_A alone—i.e., charge transfer to the continuum would appear as the dominant process of direct target ionization. Although in these situations the Born approximation is probably not valid either for ionization into the field of Z_B alone or for charge transfer into the continuum of Z_A (indeed the processes are not distinguishable in such a case), the Born cross sections may give some idea of the relative magnitude of the two processes as the ratio Z_A/Z_B is varied. In this connection, it is worth emphasizing the rather different Z_A and v_I dependence of the cross sections for ionization and charge transfer for a bare incident nucleus. Asymptotically that for direct ionization is proportional to $Z_A^2 \ln(v_I^2)/v_I^2$ [4.1,2] while that for charge transfer is proportional to Z_A^5/v_I^{12} [see (4.54)].

4.1.8 Slow Collisions of Atoms

In the limit of a very slow collision, i.e., where $v_I{\ll}v_{el}$ (the orbital velocity of the initially bound electron), it is clear that in the collision region the electron is subject equally to the fields of both nuclei. Then the best basis in which to expand the exact scattering wave function appears to be a basis composed of *molecular wave functions*. These functions are solutions of the full Hamiltonian (4.2) for fixed internuclear distance $\underline{R}$, i.e., for the stationary molecule. Now considering, for simplicity, only one electron in Fig.4.1 and choosing the center of mass 0 of the nuclei A and B as origin, the basis set $\chi(\underline{r},\underline{R})$ are solutions of

$$H_e \; \chi_n(\underline{r},\underline{R}) = E_n(R) \; \chi_n(\underline{r},\underline{R}) \tag{4.59}$$

with $H_e{=}{-}\nabla_r^2/2{-}Z_A/r_A{-}Z_B/r_B$. Expansion of the exact scattering wave function in terms of this basis, as in (4.5), leads to the set of coupled equations of the perturbed stationary state (PSS) method (MOTT and MASSEY [Ref.4.32, p. 428])

$$[\nabla_R^2 + k_n^2(R)] \; F_n(\underline{R}) = - \sum_{m \neq n} 2\langle x_n | \nabla_R | x_m \rangle \cdot \nabla_R \; F_m(\underline{R})$$

$$- \sum_{m \neq n} \langle x_n | \nabla_R^2 + k_n^2(R) | x_m \rangle \; F_m(\underline{R}) \quad , \qquad (4.60)$$

where $k_n^2 = 2M[E - E_n(R)]$.

Since the molecular basis diagonalizes the full electronic Hamiltonian, only the dynamic couplings involving the ∇_R operator appear and the potential couplings $\langle x_n | V | x_m \rangle$ of (4.6) are eliminated. The PSS method has been widely used to describe excitation and charge transfer in very slow atom-atom collisions. Indeed the distinction between the two processes is not made in the PSS method and depends only upon the asymptotic behavior of molecular wave functions. Since the excitation probability of a particular molecular level depends largely upon the dynamic couplings in the close or molecular region, the cross sections for excitation and charge transfer are generally of the same order of magnitude.

However, as the collision velocity is increased and more of the asymptotic region of the encounter becomes important, it is clear that the PSS method has serious limitations, first recognized by BATES and McCARROLL [4.49]. An electron which asymptotically moves with one or other nucleus, say A, is described by a function which assumes a constant asymptotic form in terms of $\underline{r}_A$, the distance from nucleus A. It is clear that a function in terms of the coordinate $\underline{r}$ does not assume a constant asymptotic form; hence, an expansion of the form (4.5) in terms of a *finite* basis of such functions cannot correctly satisfy the boundary conditions (4.8) and (4.9). BATES and McCARROLL (see also MOTT and MASSEY [Ref.4.32, p. 428]) showed how this difficulty can be overcome by a suitable modification of the zero-velocity molecular basis functions. The same problem arises when a molecular electronic basis set is used with the classical treatment of nuclear motion; and so a detailed discussion of the resolution of this problem is deferred until Section 4.2 which considers classical trajectory methods.

The quantum-mechanical treatment of nuclear motion is tractable in the high-velocity region where the plane-wave approximation is valid, i.e., where $k_n(R) \sim$ constant. At lower velocities, where it is desirable to solve the coupled equations of the PSS method (4.60), the effect of the potential $E_n(R)$ on the nuclear motion must be considered. Then the problem is only made tractable by expansion of the nuclear wave functions $F_n(\underline{R})$ in terms of separate radial and angular parts. i.e.,

$$F_n(\underline{R}) = R^{-1} \sum_{\ell} (2\ell+1) \; G_{n,\ell}(R) \; P_{\ell}(\cos\theta) \quad . \qquad (4.61)$$

Then a set of coupled equations for each ℓ-value separately can be derived from (4.60). However, the equations for the radial wave functions $G_{n,\ell}(R)$ now contain the repulsive centrifugal potential $\ell(\ell+1)/R^2$. For low-velocity collisions this potential acts to prevent the penetration of all but the lowest angular momentum component wave functions into the interaction region. In this situation the expansion (4.61) converges rapidly with increasing ℓ-value, i.e., angular momentum. However, as the initial kinetic energy is increased, more and more ℓ-values are required for convergence of the expansion. Then in the interaction region the variation with ℓ becomes quasi-continuous, and the quantum-mechanical angular momentum $\ell(\ell+1)$ may be associated with the classical angular momentum Mbv_I of a collision at impact parameter b according to

$$Mbv_I = [\ell(\ell+1)]^{1/2} \sim \ell \quad . \tag{4.62}$$

The classical treatment of nuclear motion generally required that $k_n^2 = 2M[E-E_n(R)]$ is always large, i.e., the wavelength associated with nuclear motion is small compared with the variation in the potential $E_n(R)$. For heavy particle motion this is true away from the classical turning points. However, the foregoing considerations apply to elastic scattering on the potential $E_n(R)$. If a transition is made to a higher potential surface $E_m(R)$, this will occur predominantly for impact parameters where

$$(E_n-E_m)\ b \sim v_I$$

which with (4.62) gives

$$\ell \sim Mv_I^2/(E_m-E_n) \quad . \tag{4.63}$$

Hence, if the incident kinetic energy is much greater than the inelastic energy of the transition in the interaction region, then many angular momenta contribute and it may be admissible to use a classical description of nuclear motion, for inelastic scattering also. If this condition is not met, i.e., close to the inelastic thresh-old, then a quantum-mechanical treatment of nuclear motion is necessary. Although a vast literature exists on the approximate solution of (4.60) (MOTT and MASSEY [4.32]) for slow collisions of atoms close to the inelastic threshold, in most col-lisions of atoms at keV or MeV impact energies a classical description of nuclear motion is adequate. Nevertheless, the collision may still be slow compared with typ-ical electronic orbital velocities, so that a molecular description of the electron motion is a good approximation. Throughout the rest of this chapter the nuclear mo-tion is treated classically. The main advantage of this approach, apart from its intuitive appeal, is that it gives rise to a *first order* set of coupled equations instead of the *second order* equations (4.60) or (4.6) of the fully quantum-mechan-ical approach.

4.2 Classical Treatment of Nuclear Motion

4.2.1 Coupled Equations in the Impact Parameter Method

In the classical-trajectory formulation of the theory of inelastic processes in
ion-atom collisions, the electronic motion is considered in the time-dependent field
determined by the motion of the projectile nucleus relative to the target nucleus.
Denoting collectively the electron coordinates with respect to the origin 0 of some
inertial frame by $\underline{x}=(\underline{r}_1,\ldots,\underline{r}_i,\ldots \underline{r}_N)$, the wave function for the electronic system
$\Psi(\underline{x},t)$ is determined as a solution of the time-dependent electronic Schrödinger
equation

$$\left[H_e(t) - i \frac{\partial}{\partial t}\Big|_{\underline{x}} \right] \Psi(\underline{x},t) = 0 \quad , \tag{4.64}$$

subject to appropriate boundary conditions. It is emphasized explicitly in (4.64)
that time differentiation is performed with fixed electron coordinates in the chosen
reference frame. The Hamiltonian

$$H_e(t) = \sum_{i=1}^{N} \left[-(1/2)\nabla_{r_i}^2 - Z_A/r_{Ai} - Z_B/r_{Bi} \right] + \sum_{i<j}^{N} 1/|\underline{r}_i - \underline{r}_j| \tag{4.65}$$

is time dependent through the electron coordinates with respect to the two nuclei,
$\underline{r}_{Ai}=\underline{r}_i-\underline{R}_A(t)$ and $\underline{r}_{Bi}=\underline{r}_i-\underline{R}_B(t)$, where $\underline{R}_A$ and $\underline{R}_B$ are the position vectors of the nuclei A and B.

It is an obvious requirement that the theory must be Galilei invariant, i.e.,
that the classical-trajectory formulation must be independent of the choice of inertial reference frame. Consider an arbitrary reference frame with origin $\tilde{0}$ trans-
lating uniformly with constant velocity $\underline{v}$ with respect to 0. In this frame the time-
dependent Schrödinger equation reads

$$\left[H_e(t) - i \frac{\partial}{\partial t}\Big|_{\underline{\tilde{x}}} \right] \tilde{\Psi}(\underline{\tilde{x}},t) = 0 \quad . \tag{4.66}$$

Galilei invariance requires that

$$\Psi(\underline{x},t) = U \ \tilde{\Psi}(\underline{\tilde{x}},t) \tag{4.67}$$

where U is some unitary operator. Hence,

$$U\left(H_e - i \frac{\partial}{\partial t}\Big|_{\underline{\tilde{x}}} \right) U^{\dagger} = H_e - i \frac{\partial}{\partial t}\Big|_{\underline{x}} \quad . \tag{4.68}$$

Utilizing the identity

$$\frac{\partial}{\partial t}\Big|_{\tilde{x}} \ g(\underline{x},t) = \frac{\partial}{\partial t}\Big|_{\underline{x}} \ g(\underline{x},t) + \sum_{k=1}^{N} \underline{v} \cdot \nabla_{\underline{r}_k} \ g(\underline{x},t) \qquad (4.69)$$

for an arbitrary function $g(\underline{x},t)$, it is seen that (4.68) is satisfied by

$$U = \exp\left\{ i\left[\sum_{k=1}^{N} \underline{v} \cdot \underline{r}_k - N(1/2)v^2 t\right]\right\} \qquad . \qquad (4.70)$$

This operator is conventionally referred to as a translation factor and is seen to account for the electronic energy and linear momentum associated with the translational motion of $\tilde{O}$ with respect to O.

The Galilei transformation becomes particularly important when dealing with the boundary conditions imposed on the time-dependent Schrödinger equation. These conditions are determined from the initial state of the electronic system, i.e., the state of the projectile ion and target atom before they interact. The separated-atom states belonging to, say, nucleus A,

$$\varphi_n^A(\underline{x}_A,t) = \varphi_n^A(\underline{x}_A) \ \exp\left(-i \ E_n^A \ t\right) \qquad , \qquad (4.71)$$

are defined as solutions of a time-dependent Schrödinger equation with reference to a frame with origin in nucleus A. To construct the boundary condition on (4.64), the atomic state (4.71) must be multiplied by the rhs of (4.70) with $\underline{v}=\underline{v}_A=d\underline{R}_A/dt$ and $N=N_A$, the number of electrons on atom A. Correspondingly, the initial atomic state of B must be transformed to the common reference frame by a translation factor with $\underline{v}=\underline{v}_B=d\underline{R}_B/dt$ and $N=N_B$, the number of electrons on atom B. The asymptotic form of the wave function Ψ is thus given in terms of traveling separated-atom states

$$\Psi(\underline{x},t) \xrightarrow[t\to-\infty]{} \varphi_n^A(\underline{x}_A) \ \exp\left(i \sum_{k=1}^{N_A} \underline{v}_A\cdot\underline{r}_k\right) \exp\left\{-i\left[E_n^A + (1/2)N_A \ v_A^2\right]t\right\}$$

$$\times \ \varphi_n^B(\underline{x}_B) \ \exp\left(i \sum_{k=N_A}^{N} \underline{v}_B\cdot\underline{r}_k\right) \exp\left\{-i\left[E_n^B + (1/2)N_B \ v_B^2\right]t\right\} \qquad . \qquad (4.72)$$

The solution of (4.64) which satisfies the initial condition (4.72) is denoted $\Psi^+(\underline{x},t)$. The probability P_n for finding the electronic system in a particular state $\Psi_n(\underline{x},t)$ after the collision is then obtained by projection of Ψ^+ onto Ψ_n

$$P_n = \lim_{t\to\infty} |<\Psi_n(\underline{x},t)|\Psi^+(\underline{x},t)>|^2 \qquad . \qquad (4.73)$$

In constructing the final states of interest it should be recalled that electronic states on the separating collision partners must be Galilei transformed to the chosen reference frame in accordance with (4.67) and (4.70) before these components of the final state are inserted in (4.73).

The time-dependent electronic wave function depends on the nuclear position and velocity vectors. However, the Galilei invariance ensures that the choice of reference frame is only a matter of convenience. Then, the electronic wave function depends only on the relative internuclear motion $\underline{R}=\underline{R}(\underline{b},\underline{v}_I,t)$ where $\underline{b}$ and $\underline{v}_I$ are the classical impact parameter and velocity, respectively. The probability P_n is a function of $\underline{b}$ and $\underline{v}_I$, too. The total cross section for finding the electronic system in the state Ψ_n after the collision is obtained by the integration over impact parameter

$$\sigma_n(v_I) = \int d^2\underline{b} \ P_n(\underline{b},v_I) \quad . \tag{4.74}$$

In practice, an exact solution of the time-dependent Schrödinger equation is not possible. Rather, a solution must be sought in some approximate form. This is usually achieved by constructing a trial function in terms of basis functions (x_k) which are considered to span the important part of Hilbert space that is occupied by the electronic system during the collision. At the present stage, the functions x_k in the basis set remain unspecified; in general they may depend upon the time-varying internuclear vector and the internuclear velocity. Upon substitution of the linear combination

$$\Psi(\underline{x},t) = \sum_k c_k(t) \ x_k(\underline{x},t) \tag{4.75}$$

into (4.64) the time-dependent Schrödinger equation appears in the form of a system of coupled equations

$$i \ \underline{S}(d/dt) \ \underline{C} = \underline{M} \ \underline{C} \tag{4.76}$$

where $\underline{C}$ is the column matrix of expansion coefficients (c_k), $\underline{S}$ is the overlap matrix with elements $S_{k\ell}=\langle x_k|x_\ell\rangle$, and $\underline{M}$ is the coupling matrix with the elements $M_{k\ell}=\langle x_k|H_e-i\partial/\partial t|_x|x_\ell\rangle$.

The Hermiticity of the electronic Hamiltonian implies that the exact solution of (4.64) is of constant norm, i.e., $(d/dt) \ \langle\Psi|\Psi\rangle=0$. Probability conservation is ensured also in the approximate case so long as (4.76) is solved exactly [4.52]: The norm of the approximate wave function (4.75) is given by

$$N^2 = \langle\Psi|\Psi\rangle = \underline{C}^\dagger \ \underline{S} \ \underline{C} \quad . \tag{4.77}$$

Utilizing (4.76) and the relation $\underline{S}=\underline{S}^\dagger$, it is found that

$$i(d/dt)N^2 = \underline{C}^\dagger(i d\underline{S}/dt+\underline{M}-\underline{M}^\dagger) \ \underline{C} \quad . \tag{4.78}$$

Since the Hermiticity of H_e implies the relation [4.52]

$$i \ d\underline{S}/dt = \underline{\underline{M}}^{\dagger} - \underline{\underline{M}} \qquad (4.79)$$

the rhs of (4.78) vanishes exactly, i.e., the approximate wave function (4.75) has constant norm. Eq.(4.79) implies that the coupling matrix is Hermitian, if the basis states are orthogonal.

In actual calculations, the character of the basis set to be used in the expansion of the electronic wave function and the number of states to be included depend upon the violence of the collision under consideration and upon the initial and final electronic states that are of interest. The *asymptotic* conditions require that the basis functions at large separations approach stationary separated-atom functions multiplied by appropriate translation factors according to (4.72). For finite internuclear separations, the form of the basis may be chosen on physical grounds with the object of achieving an effectively convergent expansion with respect to any particular inelastic scattering amplitude. However, it is customary to separate off the explicit $\underline{v}$ dependence of the basis as a translational factor which evolves into the plane-wave form (4.70) as R→∞. The remainder of the basis function then depends upon the coordinates $\underline{x},\underline{R}$ only and evolves into separated-atom functions as R→∞. These spatial functions are often taken to be those which diagonalize some part of the electronic Hamiltonian for fixed R, e.g., an atomic or molecular basis, although other choices will be discussed below. The translational factors for finite R are usually chosen by physical intuition. Consider for simplicity a one-electron system. Representing the translation factor for the n-th basis function in the flexible form $\exp[i \ f_n(\underline{r},\underline{R}) \ \underline{v} \cdot \underline{r}]$, the correct asymptotic behavior is ensured by requiring that $f_n \underline{v} \rightarrow \underline{v}_A$ when R→∞ with r_A finite and $f_n \underline{v} \rightarrow \underline{v}_B$ when R→∞ with r_B finite. An expansion in terms of "traveling" basis functions may then be expressed as

$$\Psi(\underline{r},t) = \sum_n c_n(t) \ \varphi_n(\underline{r},\underline{R}) \ \exp[i \ f_n(\underline{r},\underline{R}) \ \underline{v} \cdot \underline{r}] \ \exp(-i\int^t \varepsilon_n dt') \quad . \qquad (4.80)$$

The time-dependent phase integral is arbitrary and appears as a matter of convenience. The overlap matrix $\underline{S}$ in the system of coupled equations (4.76) then reads

$$S_{k\ell} = <\varphi_k| \exp[-i(f_k-f_\ell) \ \underline{v} \cdot \underline{r}]|\varphi_\ell> \ \exp[i\int^t (\varepsilon_k-\varepsilon_\ell) \ dt'] \qquad (4.81)$$

and the coupling matrix $\underline{\underline{M}}$ becomes

$$M_{k\ell} = <\varphi_k| \exp[-i(f_k-f_\ell) \ \underline{v} \cdot \underline{r}] \ (\hat{P}_\ell-i\hat{D}_\ell)|\varphi_\ell> \ \exp[i\int^t (\varepsilon_k-\varepsilon_\ell) \ dt'] \qquad (4.82)$$

in terms of the two coupling operators

$$\hat{P}_\ell = H_e + (1/2)[\nabla_r(f_\ell \underline{v} \cdot \underline{r})]^2 + [\frac{\partial}{\partial t}|_r \ (f_\ell \underline{v} \cdot \underline{r})] - \varepsilon_\ell \qquad (4.83)$$

and

$$\hat{D}_\ell = \frac{\partial}{\partial t}\Big|_{\underline{r}} + [\nabla_r(f_\ell\underline{v}\cdot\underline{r})] \cdot \nabla_r + (1/2)[\nabla_r^2(f_\ell\underline{v}\cdot\underline{r})] \quad . \tag{4.84}$$

This partition of the coupling operator is arbitrary but is made such that the real part contains all terms quadratic in $\underline{v}$, i.e., energy or "potential" couplings, and the imaginary part contains terms linear in $\underline{v}$, i.e., dynamic couplings. This general expression for the coupling matrix is rather complicated, but in particular cases the complexity can be reduced significantly.

In the case of an atomic expansion and, particularly, in the straight-line trajectory approximation, plane-wave translation factors become appropriate. Then, the factors f_n are assumed to be constants: $f_n\underline{v}=\underline{v}_n$, with $\underline{v}_n=\underline{v}_A$ or $\underline{v}_n=\underline{v}_B$ in accordance with the asymptotic condition on f_n. Then

$$\hat{P}_\ell = H_e + (1/2)\underline{v}_\ell^2 - \epsilon_\ell \tag{4.85}$$

and

$$\hat{D}_\ell = \frac{\partial}{\partial t}\Big|_{\underline{r}} + \underline{v}_\ell \cdot \nabla_r = \frac{\partial}{\partial t}\Big|_{\underline{r}_\ell} \tag{4.86}$$

where $\underline{r}_\ell=\underline{r}_A$ if $f_\ell\underline{v}=\underline{v}_A$ and $\underline{r}_\ell=\underline{r}_B$ if $f_\ell\underline{v}=\underline{v}_B$. The form of the rhs of (4.86), which is obtained using (4.69), demonstrates that the plane-wave translation factors have the effect of shifting the origin of coordinates to one or other of the two nuclear centers. Eq.(4.85) is strictly valid only in the straight-line trajectory approximation. In the case of a curved trajectory, it is seen from (4.83) that the term

$$\hat{A}_\ell = (d\underline{v}_\ell/dt) \cdot \underline{r} \quad , \tag{4.87}$$

which accounts for the perturbation exerted on the electron by the acceleration of the nuclei during the collision, in principle should be included on the rhs of (4.85). In actual evaluations $\hat{A}_\ell$ has been neglected although it may be expected to be of some significance in case of a strongly deflected trajectory.

In the case of a molecular expansion, the two nuclei appear on the same footing at small separations. Then less rigid translation factors may become relevant [4.53-55]. Coupled-state calculations are then significantly simplified if the same $f(\underline{r},\underline{R})$ is chosen for all basis functions since this choice results in a diagonal overlap matrix and a simple velocity dependence in the coupling matrix.

In the following sections specific cases will be discussed. First, we consider excitation or ionization of target atoms by impact of fast ions, where a description in terms of time-independent functions, centered at the colliding nuclei, furnishes a natural starting point. At lower impact velocities it is desirable to extend or modify the atomic basis in order to represent distortion of the asymptotic atomic functions during the collision. At still lower velocities the molecular functions, which account for distortion exactly in the limit of zero velocity, become an appropriate basis.

4.2.2 Expansion in Target Eigenfunctions

In this section excitation or ionization of the target atom is considered under conditions such that the target electrons remain rather undisturbed throughout the collision. This situation arises with incident projectiles of low charge relative to the effective nuclear charge experienced by the active electron in its initial state. It also arises even with highly charged projectiles when the effective collision time becomes sufficiently short at high impact velocities. An expansion in atomic eigenstates for the target is then appropriate.

The basis set of target-atom eigenfunctions $[\varphi_k^B(\underline{x}_B)]$ is orthonormal and time independent. Accordingly, the overlap matrix is diagonal and the dynamic part of the coupling matrix vanishes [see (4.82) and (4.86)]. Then, neglecting the acceleration term (4.87), the system of coupled equations (4.76) may be written

$$i(dc_n/dt) = \sum_k c_k(t) \, V_{nk}(t) \, \exp[i \int^t (\varepsilon_n - \varepsilon_k) \, dt'] \quad , \tag{4.88}$$

where

$$V_{nk} = <\varphi_n^B(\underline{x}_B) \,|\, H_e - \varepsilon_k \,|\, \varphi_k^B(\underline{x}_B)> \tag{4.89}$$

and (ε_k) are arbitrary time-dependent energies, which may be chosen for convenience since the full solution of (4.88-89) is independent of this choice. As the basis functions (φ_k^B) are eigenstates for the unperturbed target Hamiltonian, the coupling element may be written

$$V_{nk} = <\varphi_n^B(\underline{x}_B) \,|\, \left(-Z_A \sum_{i=1}^{N_B} |\underline{R} - \underline{r}_{Bi}|^{-1}\right) + (E_k^0 - \varepsilon_k) \, \delta_{nk} \,|\, \varphi_k^B(\underline{x}_B)> \tag{4.90}$$

where (E_k^0) are the unperturbed energies of the target states.

The boundary conditions on (4.88) are given by

$$c_n(-\infty) = \delta_{n0} \quad , \tag{4.91}$$

where $\varphi_0^B(\underline{x}_B)$ is the initial state of the target atom. The probability for transition to a particular atomic state n is, in accordance with (4.73), given by

$$P_n(\underline{b},v_I) = |c_n(\infty)|^2 \quad , \tag{4.92}$$

from which the total cross section (4.74) is obtained by integration over impact parameter.

An approximate solution of (4.88) may be based on the assumption that (4.91) holds to a good approximation throughout the collision, i.e., that the incident projectile acts as a weak perturbation. Then, ignoring small terms on the rhs of (4.88), the equations are decoupled and may be directly integrated resulting in the first order solution

$$c_n(t) = - i \int_{-\infty}^t dt' \, V_{n0}(t') \exp[i \int^{t'}(\varepsilon_n - \varepsilon_0) \, dt''] \quad . \tag{4.93}$$

It is seen that this amplitude, in contrast to the full solution of (4.88-90), depends on the choice of the energies (ε_n), which in (4.88) appear as arbitrary phases. Indeed, the familiar impact-parameter-Born approximation as well as the so-called distortion approximation [4.3] are included in (4.93) as special cases merely by specification of the phases (ε_n). The first Born approximation is obtained if ε_n is identified with the unperturbed energy E_n^0 of the basis function $\varphi_n^B(\underline{x}_B)$:

$$c_n^B(t) = - i \int_{-\infty}^t dt' \, V_{n0}(t') \exp(i\omega_{n0}^0 t') \quad , \tag{4.94}$$

where $\omega_{n0}^0 = E_n^0 - E_0^0$, and V_{nk} is given by (4.90) with $\varepsilon_k = E_k^0$. The distortion approximation is obtained by choosing

$$\varepsilon_n = E_n^1 = E_n^0 - Z_A \langle\varphi_n^B|\sum_{i=1}^{N_B} |\underline{R} - \underline{r}_{Bi}|^{-1}|\varphi_n^B\rangle \quad , \tag{4.95}$$

i.e., the perturbed energy of the n-th basis function, so that

$$c_n^D(t) = - i \int_{-\infty}^t dt \, V_{n0}(t') \exp[i \int^{t'} \omega_{n0}^1(t'') \, dt''] \quad , \tag{4.96}$$

where $\omega_{n0}^1 = E_n^1 - E_0^1$ and V_{nk} is given by (4.90) with $\varepsilon_k = E_k^1$.

It is apparent that the above first order solutions and the plane-wave-Born approximation (Sec.4.1.2) basically are founded on the same approximation scheme. Indeed, the *total* PWBA cross section for transition to an arbitrary atomic level has been shown by several authors to be identical to the result in the straight-line trajectory first Born approximation. A general proof of this equivalence has been given by BETHE and JACKIW [4.56].

The distortion approximation (4.96) is generally considered to be superior to the first Born approximation (4.94) since this particular choice of phase factors ensures that the diagonal terms in the coupling matrix (4.90) vanish, i.e., that the secular terms are removed from (4.88) before the first order approximation is applied. Various calculations in the first order approximation have been reported for excitation of outer-shells by projectiles of low charge [4.57-59]. These calculations indicate that the excitation cross section in the distortion approximation departs strongly from the Born-approximation result for impact velocities less than roughly twice the initial electron orbital velocity. This departure is most pronounced for optical-ly forbidden one-electron transitions which generally require closer distance of approach where the distortion effect is larger.

Although these first-order distortion calculations generally improve the agreement between theory and experiment, they are still basically invalid for $v_I \lesssim v_{el}$, since for outer-shells the projectile charge is of the same order as the effective charge seen by the electrons from their parent nucleus. For ionization of inner-shells, however, the conditions for validity of the first order approximation may be met even when $v_I < v_{el}$ (see Sec.4.1.5). For this reason, first order treatments of inner--shell *ionization* have been relatively successful in comparison with experiments. The impact parameter dependent probability $P(\underline{b})$ for ionization of an inner-shell electron is obtained by integration of the transition probability over final states in the continuum and is a natural intermediate result in the calculation of the total ionization cross section. The ionization probability is of particular interest since $P(\underline{b})$, or the related differential ionization cross section, is measurable by various methods, e.g., the coincidence or the inelastic-energy-loss technique. In the independent electron picture the first Born expression for the ionization probability becomes

$$P(\underline{b}) = Z_A^2 \int d^3 \underline{k} \ |\int_{-\infty}^{\infty} dt \ <\psi_{\underline{k}}(\underline{r})|\{|\underline{R} - \underline{r}|\}^{-1}|\psi_0(\underline{r})> \times \exp[i(k^2/2+U) \ t]|^2 \quad ,(4.97)$$

where ψ_0 and $\psi_{\underline{k}}$ are the initial and the final wave function for the active electron and U is the ionization energy of the shell under consideration.

The evaluation of the rhs of (4.97) is difficult even when simple wave functions are substituted for the initial and the final states of the electron. Assuming a straight-line trajectory $\underline{R}=\underline{b}+\underline{v}_I t$ and introducing a partial wave expansion for the final wave function of the ejected electron, the multi-dimensional integral may be reduced to a complicated but manageable two-dimensional integral [4.60-62]. The final numerical evaluations have been performed by HANSTEEN and co-workers [4.63,64] using nonrelativistic hydrogenic wave functions for initial and final states. Con-ventionally these results in the semi-classical approximation are referred to as SCA calculations. Recent preliminary calculations [4.65] have shown that the use of re-lativistic hydrogenic wave functions becomes essential for heavy atoms with $Z \gtrsim 40$.

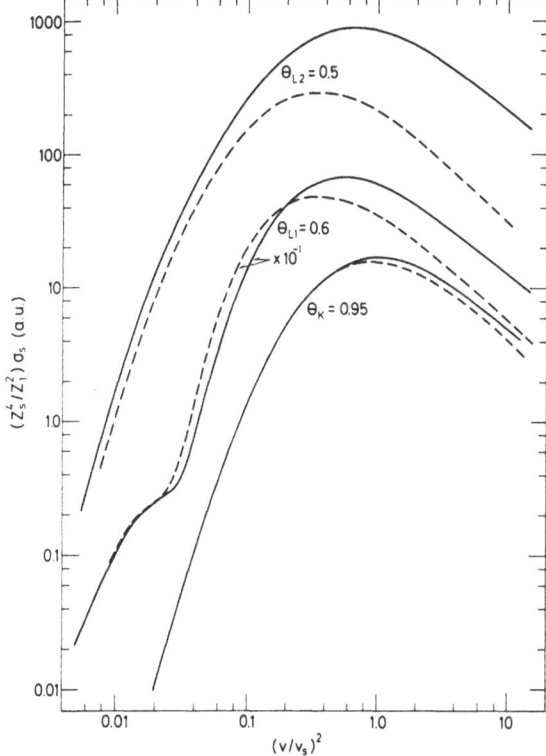

Fig. 4.4. Comparison of K, L1, and L2 ionization cross sections calculated in the plane-wave Born approximation (fully drawn curves) and in the semi-classical approximation (dashed curves). Results are shown for typical values of the screening parameter θ_S, which is defined as the ratio of the experimental and the ideal hydrogenic ionization potential for the shell s under consideration. Further, Z_1 is the charge of the projectile, Z_S is the effective charge of the target nucleus, and $v_S = Z_S/n_S$ (au) is the initial orbital velocity of the electron

Since the impact-parameter Born and the plane-wave Born approximation are equivalent insofar as total cross sections are concerned, provided that identical representations are used for the initial and final states of the ejected electron, it is of interest to compare available SCA and PWBA results for the total ionization cross section. Typical SCA [4.64] and PWBA [4.19,18] cross sections are shown in Fig.4.4. These results are numerically accurate to within a few percent. Accordingly, the large differences which are found must be due to differences in the representation of initial and final states of the ejected electron. Indeed, as discussed by ARTHURS [4.66], incorporation of screening due to outer electrons introduces difficulties in the assessment of the lower limit on the integration over final electron states.

Independently of the value of the screening parameter

$$\theta_S = U_S/I_S \quad , \tag{4.98}$$

where U_s and I_s are the experimental and the ideal hydrogenic ionization potential, respectively, of the shell under consideration, HANSTEEN and co-workers [4.63,64] have identified the wave number k_c in the Coulomb functions for the ejected electron with the wave number k of the true continuum function; i.e., in the integration over final states $(k_c^2)_{min}=0$ is taken as lower limit. It should be realized, however, that the hydrogenic approximation to the wave function is only valid close to the nucleus where the effective single-electron potential may be written

$$V_{eff}(r) = -Z_s/r + (I_s-U_s) \quad , \tag{4.99}$$

such that the energy of the Coulomb wave functions is given by $T_c=k_c^2/2$ with reference to an energy level placed (I_s-U_s) *above* the ionization level.

MERZBACHER and co-workers [4.21], on the other hand, introduced in their PWBA calculations an integration over energy transfer

$$W = k_c^2/2 + I_s \tag{4.100}$$

instead of the integration over wave number and took the lower limit to be $W_{min}=U_s$, the experimental ionization energy. Although the corresponding minimum wave number for the hydrogenic wave functions is imaginary $(k_c^2)_{min}=-(I_s-U_s)$, this limit is equivalent to $(k^2)_{min}=0$ for the wave number of the true continuum wave function. It should be recalled that the integration over Coulomb functions with imaginary k_c values is not restricted to states in the discrete spectrum since these functions, rather than fulfilling the usual conditions for bound states at larger radial distances, must match the true continuum wave function for the ejected electron.

In view of the different treatments of the final states of the ejected electron it is not surprising that the SCA and the PWBA results diverge for Θ_s smaller than unity as shown in Fig.4.4. Although there remain some uncertainties in both approaches, particularly in respect to the normalization of the continuum wave functions, it is obvious that the approach used in the PWBA calculations is more consistent with the screened hydrogenic model than is the method used by HANSTEEN and co-workers in their SCA calculations. Fortunately, improved calculations using single-particle wave functions based upon Hartree-Fock-Slater potentials [4.68] are now becoming available and may test the calculations within the screened hydrogenic model. For example CHOI [4.69] has calculated the L-shell ionization cross section of argon in collision with heavy charged particles and found excellent agreement with the PWBA calculations of MERZBACHER and co-workers. This is in line with results for copper K-shell ionization [4.67].

Due to the time-dependent phase factors, the first order distortion approximation (4.96) is more complicated than the Born approximation (4.94) and has not been evaluated explicitly for ionization. If the collision velocity is much smaller than the velocity of the active electron, ionization is likely only in collisions with impact parameter much smaller than the orbital radius of the initial state [4.60] because ionization is then effectively limited to the high-momentum Fourier components of the initial wave function. Then, the energy-phase factors in (4.94) and (4.96) will differ significantly over the important part of the time (trajectory) integrals, and the result of a distortion calculation is expected to depart appreciably from the first Born approximation. At higher projectile velocities the more distant parts of the trajectory integrals contribute such that it may be expected that the distortion calculations approach the first Born result.

These qualitative arguments are in line with the estimates of the effect of "binding" performed by BRANDT and co-workers [4.70,71]. These authors evaluated the perturbed binding energy of the initial state at the distance of closest approach of the projectile and used this constant binding energy in an approximate distortion calculation. In this approximation the distortion amplitude (4.96) appears as a modified first-Born amplitude and may be evaluated accordingly.

Starting from a molecular description of the initial state of the electron, thus accounting for the relaxation of the initial state under the influence of the incident projectile, BASBAS et al. [4.72] have discussed the effect of distortion of the initial state in a more formal manner. This approach was further developed by ANDERSEN et al. [4.73] in conjunction with measurements of the impact-parameter dependence of K-shell ionization. These authors suggest that the SCA formalism may be appropriately modified to account for the adiabatic relaxation of the initial state by determining the important parameters, namely the binding energy and the effective orbital radius of the initial wave function, variationally at the distance of closest approach, b.

Consider for simplicity ionization of a K-shell electron initially described by the hydrogenic wave function $\varphi_0 = \pi^{-1/2} r_K^{-3/2} \times \exp(-r/r_K)$ where r_K is the K-shell radius. The first order correction to the binding energy due to the presence of the projectile charge at the distance b is given by

$$\Delta U - (Z_A/b)[1 - (1+b/r_K) \exp(-2b/r_K)] \quad . \tag{4.101}$$

The perturbed binding energy may then be expressed as [4.73]

$$U = -1/(2r_K^2) + (Z_B-0.3)/r_K + \Delta U(b,r_K) - U_{sc} \tag{4.102}$$

where the first term represents the kinetic energy, the second term represents the potential energy due to attraction to the target nucleus appropriately screened by the other K-shell electron, the third term is given by (4.101), and the last term accounts for the screening from the outer electrons. The binding energy U and the radius r_K are then determined by allowing the wave function to relax by a variation of r_K to maximize U. The outer screening is considered constant under this variation. As a result U and r_K appear as functions of impact parameter. It was shown by ANDER-SEN et al. [4.73] that the application of the modified parameters $U(b)$ and $r_K(b)$ may change the results of Born calculations very significantly, e.g., in the case of oxygen impact on copper by a factor of $\sim 1/3$, in good accord with the experimental results.

When a more accurate solution of (4.76) is sought without recourse to actual numerical integration of the coupled equations, the expressions rapidly become rather cumbersome. For example a second order approximation is obtained by writing (4.88) as

$$c_n(t) = - i \int_{-\infty}^{t} dt' \, c_0 \, V_{n0} \, \exp[i \int^{t'} dt''(\varepsilon_n - \varepsilon_0)]$$

$$- i \int_{-\infty}^{t} dt' \sum_{m \neq 0} c_m \, V_{nm} \, \exp[i \int^{t'} dt''(\varepsilon_n - \varepsilon_m)] \qquad (4.103)$$

for the final excited state of interest and then approximating $c_0 \sim 1$ and

$$c_m(t) \sim - i \int_{-\infty}^{t} dt' \, V_{m0} \, \exp[i \int^{t'} dt''(\varepsilon_m - \varepsilon_0)] \quad . \qquad (4.104)$$

This approximation made in (4.103) results in

$$c_n \simeq - i \int_{-\infty}^{t} dt' \, V_{n0} \, \exp\left[i \int_{-\infty}^{t'} dt''(\varepsilon_n - \varepsilon_0)\right] - \int_{-\infty}^{t} dt' \sum_{m \neq 0} V_{nm} \, \exp[i \int^{t'} dt''(\varepsilon_n - \varepsilon_m)]$$

$$\times \int_{-\infty}^{t'} dt'' \, V_{m0} \, \exp[i \int^{t''} dt'''(\varepsilon_m - \varepsilon_0)] \quad . \qquad (4.105)$$

The second term on the rhs of (4.105) allows for $0 \to n$ transitions via the two-step $0 \to m \to n$ transition. Inclusion of these virtual transitions may be thought of as a distortion or polarization of the initial atomic wave function. A simplified version of this second approximation has been considered, but only for the simplest case of proton-hydrogen collisions [4.74]. Since the numerical labor involved in the integration of (4.105) is probably little different from that of solving coupled equations for a small basis set, the main advantage of the second Born approach is that it allows an approximate closure to be made over all intermediate states m including continuum states. If this is done, then the asymptotic high-velocity excitation cross section is obtained exact to second order since the target wave functions form a complete set. This is equivalent to an expansion of the exact scattering wave function Ψ of (4.64) to second order in the projectile potential V_A. CHESHIRE and SULLIVAN [4.75] recognized that inclusion of continuum states is a desirable feature in the description of the excitation process. They adopted a different procedure from the eigenfunction expansion of the form (4.80) by a direct expansion of the scattering wave function in spherical harmonics of order ℓ, i.e.,

$$\Psi(\underline{r},t) = r^{-1} \sum_{\ell,m} R_{\ell m}(r,t) \, Y_{\ell m}(\theta,\varphi) \quad . \tag{4.106}$$

A similar expansion of the perturbing potential

$$Z_A/|\underline{R} - \underline{r}_B| = 4\pi \, Z_A \sum_{\ell,m} [\gamma_\ell(r,R)/(2\ell+1)] \, Y^*_{\ell m}(\theta,\varphi) \, Y_{\ell m}(\Theta,\Phi) \quad , \tag{4.107}$$

where θ,φ and Θ,Φ are angular coordinates of $\underline{r}$ and $\underline{R}$, respectively, and substitution in the Schrödinger equation

$$\left(-\nabla_r^2/2 - Z_B/r_B - Z_A/|\underline{R}-\underline{r}_B| - i\frac{\partial}{\partial t}\right) \Psi(\underline{r}_B,t) = 0 \tag{4.108}$$

leads to a set of coupled *second order* differential equations for the radial wave functions $R_{\ell m}$. These equations are solved subject to appropriate initial boundary conditions and the final atomic amplitude is projected from the scattering wave function $\Psi(\underline{r},\infty)$ constructed according to (4.106). Clearly for excitation from or to an atomic state of given ℓ value, the coupled equations up to at least that value must be solved. This involves $(2\ell+1)$ coupled second order differential equations. However, the solution of this problem is equivalent to the solution by an eigenfunction expansion of $R_{\ell m}(r,t)$ in terms of SA wave functions $\psi_{n\ell m}$ *for all n values including the continuum*. This attraction of the method is largely offset by computational difficulties but it was applied by CHESHIRE and SULLIVAN to the simple case of $H^+ + H(1s)$ $\rightarrow H^+ + H(2s,2p)$ at high velocity where only $\ell=0$ and $\ell=1$ partial waves contribute appreciably to the electronic scattering wave function. In addition the method is not feasible for charge transfer processes since it is not possible to describe bound states around nucleus A in terms of expansion in a few ℓ-values around nucleus B.

The main attraction of the first and second order approximation (4.94,96,105) is that the coupled equations (4.76) are decoupled to a single equation for each transition amplitude. Ideally, of course, one would seek an exact solution of the scattering problem by a full solution of the infinite set of coupled equations (4.76). In practice two possibilities are open. One is to solve the coupled equations approximately for an infinite set of basis states, i.e., to consider a limited number of transitions among an infinity of states. The first and second order approximations described above fall into this category. The other is to consider an infinite number of transitions between a limited number of states, i.e., to solve the coupled equations exactly in a truncated basis. In this situation the accuracy of the approximate solution of the scattering problem that is obtained depends upon the initial choice of basis functions. This choice is made by physical intuition with a view to represent as closely as possible the dynamical behavior of electrons during the collision.

4.2.3 Eigenfunction Expansions on More Than One Center

When charge transfer processes are considered explicitly, it is desirable to include in the basis set functions that can represent adequately the bound states of an electron around the incident nucleus. This can be achieved most readily by simply including in the basis set traveling atomic functions on the incident nucleus Z_A as well as on the target nucleus Z_B, i.e., by expanding

$$\Psi(\underline{r},t) = \sum_n a_n(t)\ \varphi_n^A(\underline{r}_A)\ \exp(i\underline{v}_A\cdot\underline{r})\ \exp(-i\int^t \varepsilon_n^A dt')$$

$$+ \sum_n b_n(t)\ \varphi_n^B(\underline{r}_B)\ \exp(i\underline{v}_B\cdot\underline{r})\ \exp(-i\int^t \varepsilon_n^B dt') \quad . \tag{4.109}$$

Now the coupled equations (4.76) can be written explicitly in terms of amplitudes on A and amplitudes on B, i.e.,

$$i\ da_n/dt + i \sum_m S_{nm}\ db_m/dt = \sum_m H_{nm}\ a_m + \sum_m K_{nm}\ b_m$$

$$i\ db_n/dt + i \sum_m S_{nm}^\dagger\ da_n/dt = \sum_m \bar{K}_{nm}\ a_m + \sum_m \bar{H}_{nm}\ b_m \quad . \tag{4.110}$$

With the choice of phase factors in (4.109) such that

$$\varepsilon_n^A = E_n^A + (1/2)v_A^2$$

$$\varepsilon_n^B = E_n^B + (1/2)v_B^2 \quad ,$$

the matrix elements in (4.110) become [see (4.81-86)]

$$H_{nm} = <\varphi_n^A|V_B|\varphi_m^A> \exp\left[i \int^t (E_n^A - E_m^A)\ dt'\right]$$

$$\bar{H}_{nm} = <\varphi_n^B|V_A|\varphi_m^B> \exp\left[i \int^t (E_n^B - E_m^B)\ dt'\right]$$

$$K_{nm} = <\varphi_n^A|\exp(-i\underline{v}\cdot\underline{r})V_A|\varphi_m^B> \exp\left\{i \int^t \left[(E_n^A - E_m^B) + (1/2)(v_A^2 - v_B^2)\right]dt'\right\}$$

$$\bar{K}_{nm} = <\varphi_n^B|\exp(i\underline{v}\cdot\underline{r})V_B|\varphi_m^A> \exp\left\{i \int^t \left[(E_n^B - E_m^A) + (1/2)(v_B^2 - v_A^2)\right]dt'\right\} \tag{4.111}$$

and

$$S_{nm} = <\varphi_n^A|\exp(-i\underline{v}\cdot\underline{r})|\varphi_m^B> \exp\left\{i \int^t \left[(E_n^A - E_m^B) + (1/2)(v_A^2 - v_B^2)\right]dt'\right\} \quad . \tag{4.112}$$

The solution of these coupled equations removes some of the undesirable features of the PWBA approximation for charge transfer referred to in Section 4.1.7. In particular it is readily seen that inclusion of the internuclear potential $Z_A Z_B/R$ in the electronic Hamiltonian gives no contribution to the transition cross section so long as the overlap matrix $\underline{S}$ is retained in the equations. Any function F(R) added to the Hamiltonian H_e gives rise to a new set of coupled equations

$$i \underline{S} \, d\underline{C}/dt = \underline{\underline{M}} \, \underline{C} + F \underline{\underline{S}} \, \underline{C} \tag{4.113}$$

replacing (4.76). The term in F may be removed by the phase transformation

$$\underline{C}' = \underline{C} \, \exp(i\int^t F dt') \quad . \tag{4.114}$$

This transformation does not affect the transition probabilities at infinity in an exact solution of the coupled equations (4.110). However, this may not be the case for an approximate solution.

The coupled equations (4.110) for charge transfer have been solved with basis sets of different sizes for various simple systems (see [4.76] and [Ref.4.3, Chap.4]). The simplest approximation is to consider only the initial state on B and the final state on A in the expansion. At the highest velocities the charge transfer amplitude tends to that obtained from a first order solution of (4.110) when overlap is neglected, i.e.,

$$a_n(t) = \int_{-\infty}^{t} dt' \, <\varphi_n^A | \exp(i\underline{v} \cdot \underline{r}) V_A | \varphi_m^B>$$

$$\times \exp\left\{ i \int^{t'} dt'' \left[(E_n^A - E_m^B) + 1/2(v_A^2 - v_B^2) \right] \right\} \quad . \tag{4.115}$$

This provides the same cross section as the Brinkman-Kramers approximation in the wave treatment of the first Born approximation. Hence the two-state expansion does not provide the asymptotic limit (4.53) of the second Born approximation. For this reason the truncated atomic basis expansions of the type (4.109) are of most use at low velocities where indeed they have been most successful.

The most elaborate calculations using solely *SA eigenfunctions* have been those of WILETS and GALLAHER [4.77] who included all 1s, 2s, and 2p wave functions on each center in the description of $H^+ + H(1s)$ collision. A direct numerical solution of the coupled equations (4.110) then provides the cross section for both excitation and charge transfer to each of these states. As a general rule, it has been found that these calculations agree best with experiment in an intermediate velocity range $1/2 \lesssim v_I/v_{el} \lesssim 2$ which corresponds for example to 5-100 KeV in $H^+ + H(1s)$ collisions. At higher velocities the limited atomic basis suffers the same deficiencies as the one-center expansion bases of Section 4.2.2. In this situation the excitation and chargee transfer processes may be effectively separated since the latter are so much less probable than the former. However, for both, the deficiency of the limited basis can be ascribed to the truncation of the basis set on the nucleus around which the final state is bound.

To improve the basis further, it is necessary in some way to include the effect of *continuum* states. In the case of excitation the inclusion of virtual continuum states has been discussed already in Section 4.2.2. In the case of charge transfer, CHESHIRE [4.78] has developed a second order method in which the effect of continuum states is included exactly by closure. However, this formal development leads to a set of second order coupled differential equations in time. For this reason the computational difficulties are formidable, and too limited use of this method has been reported [4.78,79] for its value to be assessed. CHESHIRE [4.47] has also developed a continuum distorted-wave method which may be approximated to yield the correct asymptotic form of the charge exchange cross section, i.e., the form (4.53) of the second Born approximation. However, like the second Born approximation, although an infinite number of states are included, only at most two-step transitions are allowed between them. In particular, at lower velocities, back-coupling to the initial states becomes important and the approximation method used by CHESHIRE [4.47] fails.

At velocities below $v_I/v_{el} \sim 1$ it appears that improvement of a truncated SA eigenfunction expansion is best achieved by modification of the limited basis set while still requiring exact solution of (4.110) rather than by extension of the basis set with approximate solution of the coupled equations as is necessary at high velocity. This is because in slow collisions the oscillating phase factors in the off-diagonal terms K_{nm}, $\bar{K}_{nm}$, H_{nm}, and $\bar{H}_{nm}$ effectively damp out transitions to high-lying states for which the energy gap is largest. Hence the effect of truncation of the basis to low-lying states is not serious. By contrast the diagonal terms H_{nn} and $\bar{H}_{nn}$ do not contain oscillatory phase factors. Rather they represent the distortion of the energies of the SA basis during the collision. However, in a strict SA basis, this distortion is only included as a first order perturbation in the other nuclear potential, i.e., $<\varphi_n^A|V_B|\varphi_n^A>$ or $<\varphi_n^B|V_A|\varphi_n^B>$. For this reason attention has been directed at low velocity towards a more accurate representation of this energy distortion of low-lying states including the initial state. This is simply an attempt to allow for the near-adiabatic motion of electrons when $v_I < v_{el}$ and leads naturally to a *molecular basis*.

The first attempt to account adequately for this distorting effect at low velocity again was made in an ingenious development of CHESHIRE [4.80]. He suggested that rather than a fixed SA basis, whose eigenfunctions are decided by the Coulomb field of a *fixed* nuclear charge Z_A or Z_B, expansion be made in terms of functions of atomic hydrogenic form but with variable charge $Z_n(t)$ depending upon the particular basis function n. CHESHIRE considered the simplest case of a symmetric $H^+ + H(1s)$ collision although the method is easily generalized to asymmetric cases. The scattering wave function is expanded

$$\Psi(\underline{r},t) = \sum_n a_n \, \varphi_n^A(\underline{r}_A, Z_n) \, \exp[-i \, \underline{v}_A \cdot \underline{r} - i(1/2)v_A^2 t - i \, E_n t]$$

$$+ \sum_n b_n \, \varphi_n^B(\underline{r}_B, Z_n) \, \exp[-i \, \underline{v}_B \cdot \underline{r} - i(1/2)v_B^2 - i \, E_n t] \quad . \tag{4.116}$$

Note that for convenience the SA diagonal energies E_n are retained in the phase factor. From symmetry these energies and the variable charges Z_n are the same for given
n on nucleus A and B. Since the coupled equations (4.110) can be derived from a variational principle by considering linear variations in the coefficients a_n, b_n
(see, e.g., [4.76]), CHESHIRE recognized that the optimum values of the $Z_n(t)$ can
similarly be obtained from a variational condition which he showed to be of the form

$$(\partial/\partial Z_n) \, <\Psi|H - i \, d/dt|\Psi> + i(d/dt) \, <\Psi|(\partial/\partial Z_n)|\Psi> = 0 \quad . \tag{4.117}$$

This leads to a set of coupled equations for the $Z_n(t)$ which must be solved simultaneously with those of (4.110) for the amplitudes a_n, b_n. However, for economy in
computation, CHESHIRE chose to decouple the equations for the Z_n from each other,
and more importantly, from the equations for the a_n, b_n.

This last step removes much of the attraction of the method since it means that
at low velocity the variable-charge basis set merely becomes an approximation to
the molecular orbitals of the H_2^+ molecular ion. Nevertheless CHESHIRE showed that
the results obtained using this method give close agreement in the case of H^+-H(1s)
both with calculations using a molecular basis at low velocity [4.81] and using an
atomic basis at higher velocity [4.82].

The variable-charge method would be difficult to generalize to more complicated
collisions involving many electrons, since the electron-electron interaction can
play a large role in the effective charge seen by an electron. However, the method
may be useful for inner-shell electrons. In particular, in the united atom (UA)
limit, the effective charge becomes that of the UA nucleus; and in this sense the
method is superior to the first order distortion approximation where only the first
order perturbation approximation to the UA energy is obtained. In addition, the
power of the variational approach should be noted since this method can be used to
optimize the variation of *any* parameters upon which the initial trial wave function
is chosen to depend.

A simpler method of including the effect of formation of a molecular complex at
low collision velocities was chosen by CHESHIRE et al. [4.83]. Again they considered
excitation and charge transfer between 1s, 2s, and 2p levels in H^+-H(1s) collisions
and retained these particular eigenfunctions in order to represent well the states
of interest at infinite separation. Recognizing that the form of basis functions is
arbitrary, they elected to include "pseudo-states" which for computational convenience were still centered on nuclei A and B. These states were designed to have a
high overlap with the low-lying state of the united atom He^+. They argued that the

inclusion of such pseudo-states should give a more rapidly convergent expansion of $\Psi(\underline{r},t)$ at low velocity than the inclusion of higher bound SA eigenfunctions, since the entire set of the latter have only a 76% overlap with the $He^+(1s)$ state. The pseudo-state wave functions were chosen to have the simple analytic form

$$\varphi_{n\ell m}(\underline{r}) = \exp(-\lambda_n r)\ g_n(r)\ Y_{\ell m}(\theta,\varphi) \qquad (4.118)$$

where $\underline{r}$ is $\underline{r}_A$ or $\underline{r}_B$ and $g(r)$ is a polynomial of maximum order 2. In practice the λ_n were fixed and the coefficients of the polynomial g_n were determined by requiring that the pseudo-states be orthogonal to all other functions in the set. The expanded basis set gives almost 100% overlap with the low-lying states of He^+, in contrast to the expansion in SA eigenfunctions alone. The results of the calculation of CHESHIRE et al. [4.83] are in good agreement with experiments of MORGAN et al. [4.84] on excitation and charge transfer in H^+-H(1s) collisions above 1 KeV. However, it is difficult to see how the pseudo-state method could be easily generalized in the case of more complex collisions, particularly where many electrons are involved.

More recently, ANDERSON et al. [4.85] have suggested that the simplest way to guarantee high overlap with He^+ states is to incorporate these states into the basis set, i.e., to use a three-center expansion of the scattering wave function $\Psi(\underline{r},t)$. To date this suggestion has been tested only on the charge exchange H^+-H(1s)$\rightarrow$H(1s)-H^+ using a five-state H(1s,2s) and $He^+(1s)$ expansion. The results are in surprisingly close agreement with the pseudo-state calculation of CHESHIRE et al. [4.83]. This method appears to be simpler and more direct than the pseudo-state approach.

The foregoing methods, utilizing basis functions of one-center or atomic form, appear to be capable of giving a good representation of charge transfer and excitation among low-lying states at intermediate velocity, e.g., $1/2 \lesssim v_I/v_{el} \lesssim 2$. Their attraction lies in their simple analytic form and the relative ease with which the translational factors $\exp(i\underline{v}\cdot\underline{r})$ can be incorporated exactly. These factors are essential to describe the change of electron momentum when electrons transfer from one nucleus to the other. However, nowadays molecular wave functions are more readily available in the form of multicenter expansions in terms of functions of simple analytic form. Such molecular functions would appear to form a more flexible basis for low and intermediate velocities since they not only guarantee the correct asymptotic atomic form but take account of adiabatic distortion exactly. In addition, experiment [4.84,86] indicates that when $v_I << v_{el}$, the cross sections for charge transfer and excitation to a particular state become equal in a symmetric collision. This constitutes an experimental proof that the translational factors are not important. This greatly facilitates the use of a molecular basis since expansion of the effect of translational factors to lowest order in $\underline{v}$ renders the coupling matrix elements (4.82) independent of velocity so that they need not be calculated afresh for solution of the coupled equations at different impact velocities.

4.2.4 Molecular Eigenfunction Expansion

In spite of the pronounced adiabatic character of low-velocity collisions, large
cross sections may occur for particular inelastic processes. An expansion in terms
of stationary functions is inappropriate in such situations because a large excitation cross section is intimately related to the near degeneracy of adiabatically
relaxing electronic states of the collision complex some time during the encounter.
A proper account of these phenomena therefore is feasible only if the time-dependent
electronic wave function is expanded in a basis of molecular states.

In many situations it is appropriate to describe a collision between heavy atoms
in the independent electron picture and to consider excitation processes as single-
particle transitions in which the spectator electrons remain independent of the
state of the active electron. In particular this approximation is useful in the description of inner-shell excitation processes. For simplicity the following discussion is limited to such cases. For a comprehensive review see [4.87].

The formalism developed in Section 4.2.1 may be adopted also in the case of a
molecular expansion of the time-dependent electron wave function

$$\Psi(\underline{r},t) = \sum_n c_n(t)\ \varphi_n(\underline{r},\underline{R})\ \exp[i\ f_n(\underline{r},\underline{R})\ \underline{v}\ \cdot\ \underline{r}]\ \exp(-i\int^t \epsilon_n dt') \qquad (4.119)$$

where (φ_n) is a basis of eigenstates for the effective one-electron Hamiltonian:
$\hat{h}_e \varphi_n(\underline{r},\underline{R}) = E_n(R)\varphi_n(\underline{r},\underline{R})$. Then, the time evolution of the system is governed by the
Schrödinger equation, in the form (4.76) with the overlap matrix given by (4.81) and
the coupling matrix given by (4.82-84). The choice of translation factors $\exp(if_n \underline{v}\cdot\underline{r})$
still remains to be discussed. Two distinct cases will be considered.

The translation factors were introduced in Section 4.2.1 to ensure Galilei invariance and proper asymptotic boundary conditions. These conditions require that $f_n \underline{v}$
at large separations approach the velocity of the nucleus to which the electron in
the n-th basis state is attached at infinite separations. As in case of an atomic
expansion, it is suggestive to ascribe the same constant value to f_n also at finite
separations, i.e., to use plane-wave translation factors. As shown by BATES and
WILLIAMS [4.88] this choice ensures that spurious long-range couplings do not arise.
It is convenient in this case to distinguish between "direct" and "exchange" elements
in the overlap matrix $S_{k\ell}$ and in the coupling matrix $M_{k\ell}$ according to whether φ_k and
φ_ℓ asymptotically are attached to the same nucleus or to either of the two nuclei,
respectively. The direct parts of the overlap matrix are diagonal

$$S_{k\ell}^D = \delta_{k\ell} \qquad (4.120)\cdot$$

whereas the exchange elements become

$$S_{k\ell}^E = <\varphi_k|\exp(\pm i\underline{v}\cdot\underline{r})|\varphi_\ell> \exp\left\{i\ \int^t \left[(E_k-E_\ell) + (1/2)(v_k^2-v_\ell^2)\right]dt'\right\} \qquad (4.121)$$

For convenience the arbitrary phases (ε_k) have been determined by (4.87) since this choice ensures that the elements of the potential operator (4.85) vanish so that the coupling matrix becomes

$$M_{k\ell}^D = - i<\varphi_k|\frac{\partial}{\partial t}|_{\underline{r}_\ell}|\varphi_\ell> \exp[i \int^t (E_k - E_\ell) \, dt'] \tag{4.122}$$

for direct terms, and

$$M_{k\ell}^E = -i<\varphi_k|\exp(\pm i\underline{v}\cdot\underline{r})|\frac{\partial}{\partial t}|_{\underline{r}_\ell}|\varphi_\ell> \exp\left\{i \int^t \left[(E_k - E_\ell) + (1/2)(v_k^2 - v_\ell^2)\right]dt'\right\} \tag{4.123}$$

for exchange terms.

The plane-wave translation factors constrain the electron to move along with one or other nucleus also at finite R. This is at variance with the adiabatic relaxation of the electron wave function under the influence of the combined field of the nuclei. Accordingly, these factors are not entirely compatible with the use of a molecular expansion. Alternative forms for the translation factors may be used so long as the strict asymptotic conditions are fulfilled. SCHNEIDERMAN and RUSSEK [4.53] have suggested the use of a common factor $f_n = f(\underline{r}, \underline{R})$ for all basis functions. This choice has the immediate advantage that the matrices appearing in the Schrödinger equation (4.76) are significantly simplified. More importantly, by choosing a suitable function $f(\underline{r}, \underline{R})$, it is possible to fulfill various conditions that may be considered to be appropriate in slow collisions, e.g., that $f(\underline{r}, \underline{R})$ vanishes as the nuclei approach each other closely. Various specific forms for the Schneiderman-Russek factor have been considered in actual calculations [4.54,55]. However, further studies are required to establish appropriate rules or optimization procedures [4.89-91] governing these factors.

The Schneiderman-Russek factors are attractive in particular because the internuclear velocity is factorized in the overlap matrix and in the coupling matrix, such that the calculation of these elements may be performed independently of the collision parameters. In fact the overlap matrix reduces to the unit matrix and the coupling matrix becomes

$$M_{k\ell} = <\varphi_k|\hat{P} - i \hat{D}|\varphi_\ell> \exp[i \int^t (\varepsilon_k - \varepsilon_\ell) \, dt'] \tag{4.124}$$

where, according to (4.83,84)

$$\hat{P} = (E_\ell - \varepsilon_\ell) \, \delta_{k\ell} + (1/2)[\nabla_r(f\underline{v}\cdot\underline{r})]^2 + [\frac{\partial}{\partial t}|_{\underline{r}}(f\underline{v}\cdot\underline{r})] \tag{4.125}$$

and

$$\hat{D} = \frac{\partial}{\partial t}|_{\underline{r}} + [\nabla_r(f\underline{v}\cdot\underline{r})] \cdot \nabla_r + (1/2)[\nabla_r^2(f\underline{v}\cdot\underline{r})] \quad . \tag{4.126}$$

In accord with (4.79) it is readily seen that $\underline{M}$ is Hermitian. The evaluation of the matrix elements of the dynamical coupling operator (4.126) may be significantly simplified by using the identity [4.55]

$$(1/2)\nabla_r^2 \, g(\underline{r}) + [\nabla_r \, g(\underline{r})] \cdot \nabla_r = -[\hat{h}_e, g(\underline{r})] \tag{4.127}$$

where $\hat{h}_e$ is the electronic Hamiltonian.

In the case of plane-wave translation factors, the matrix elements (4.121) and (4.123) appear to depend in a rather complicated manner upon the trajectory parameters b and v_I such that they must be evaluated explicitly for each trajectory. Since the molecular expansion is appropriate in particular for slow collisions, it is suggestive to expand the matrix elements in powers of v retaining only the lowest order terms. Then, the Schrödinger equation (4.76) appears with elements

$$S_{k\ell} = \delta_{k\ell} \quad , \tag{4.128}$$

and

$$M_{k\ell} = -i \langle \varphi_k | \frac{\partial}{\partial t} |_{\underline{r}_\ell} | \varphi_\ell \rangle \, \exp[i \int^t (E_k - E_\ell) \, dt'] \tag{4.129}$$

for direct as well as exchange elements. Note, however, that $M_{k\ell} \neq M_{\ell k}^*$ in the exchange elements, i.e., that the coupling matrix is non-Hermitian in disagreement with (4.76) when $\underline{S}$ is given by (4.128). Probability conservation is thus fulfilled only to lowest order in v. Since exact probability conservation is a useful property in coupled state calculations, several authors [4.88,92] have ensured this by forcing the hermiticity of $\underline{M}$. Usually this has involved some average procedure applied to the exchange coupling elements. However, this introduces first order errors in the calculated transition amplitudes [4.52] and must in most cases be abandoned.

These problems do not arise when the Schneiderman-Russek factors are used. Since the internuclear velocity only appears in factorized form in (4.125,126), the coupling elements may be evaluated numerically once and for all independently of the trajectory parameters. Thus, in principle there is no difficulty in retaining the nonlinear terms in (4.125), i.e., the term proportional to v^2 and the term proportional to $d\underline{v}/dt$. However, it is interesting to note that probability consevation in this case is exactly fulfilled in the low velocity approximation in contrast to the case of plane-wave translation factors. The elements of the potential coupling operator vanish in the low-velocity approximation if the eigenenergies (E_k) are substituted for the arbitrary phase factors (ε_k). Then, the coupling elements become

$$M_{k\ell} = -i[<\varphi_k|\frac{\partial}{\partial t}|_{\underline{r}}|\varphi_\ell> + (E_\ell - E_k)<\varphi_k|f(\underline{r},\underline{R})\underline{v} \cdot \underline{r}|\varphi_\ell>]$$

$$\times \exp[i \int^t (E_k - E_\ell) \, dt'] \tag{4.130}$$

where (4.127) has been used. Since $\underline{\underline{S}}$ is the unit matrix and $\underline{\underline{M}}$ is Hermitian in this approximation, the unitarity condition (4.79) is seen to be satisfied.

For simplicity, we have not, in the discussion in this section, accounted for the special complications that arise when dealing with homo-nuclear collision systems, due to the symmetry properties of the molecular eigenstates. These details have been discussed by BATES and co-workers [4.49,88].

The molecular expansion has been used in the calculation of excitation, charge transfer and ionization processes in hydrogen in collision with protons [4.88,91, 93] and in collision with α-particles [4.94]. It has also been used in the calculation of inner-shell excitation processes in ion-atom collisions, either simulated by appropriate one-electron model calculations [4.55,95] or within the framework of the molecular-orbital independent-particle picture [4.92]. In general, these calculations compare favorably with experimental results. It is an important feature of such calculations that the results depend sensitively upon the choice of internuclear trajectory [4.95]; this is so, not only in case of differential (impact-parameter dependent) cross sections, but often also in the case of total cross sections, particularly in the projectile-energy threshold region which is strongly governed by the internuclear repulsion.

The K-shell excitation process has been most thoroughly studied and is now so well understood that accurate theoretical predictions, based on scaling principles [4.96], are available for arbitrary collision systems [4.97]. Molecular coupled-state calculations [4.55] have also substantiated the semi-empirical formulation due to MEYERHOF [4.98] for K-shell vacancy sharing in asymmetric collisions.

The calculation of excitation processes for other atomic levels in slow atomic collisions is significantly more complicated, and only a few results are available [4.92,99]. Therefore, the interpretation of experimental data has presently to rely upon simplified models [4.100-102]. However, much attention is currently focused on a quantitative description of these processes. Multi-electron effects may become important in these situations, particularly in the calculation of outer-shell phenomena. Then, an expansion in molecular states rather than independent particle orbitals is required [4.103].

References

4.1 H.A. Bethe: Ann. Physik $\underline{5}$, 325 (1930)
4.2 M. Inokuti: Rev. Mod. Phys. $\underline{43}$, 297 (1971)
4.3 M.R.C. McDowell, J.P. Coleman:*Introduction to the Theory of Ion-Atom Collisions* (North-Holland, Amsterdam 1970)
4.4 W.J.B. Oldham: Phys. Rev. $\underline{161}$, 1 (1967)
4.5 S.T. Manson: Phys. Rev. A3, 1260 (1971)
4.6 D.H. Madison: Phys. Rev. A8, 2449 (1973)
4.7 S.T. Manson, L.H. Toburen, D.H. Madison, N. Stolterfoht: Phys. Rev. $\underline{A12}$, 60 (1975)
4.8 G. Peach: Proc. Phys. Soc. $\underline{85}$, 709 (1965); J. Phys. $\underline{B1}$, 1088 (1968); J. Phys. $\underline{B3}$, 328 (1971)
4.9 S.T. Manson: Phys. Rev. $\underline{182}$, 97 (1969); Phys. Rev. $\underline{A5}$, 668 (1972); Phys. Rev. $\underline{A6}$, 1013 (1972)
4.10 Y.K. Kim, M. Inokuti: Phys. Rev. $\underline{175}$, 176 (1968); Phys. Rev. $\underline{181}$, 205 (1968); Phys. Rev. $\underline{184}$, 38 (1969); Phys. Rev. A1, 1132 (1970); Phys. Rev. A3, 665 (1971)
4.11 K.L. Bell, D.J. Kennedy, A.E. Kingston: J. Phys. $\underline{B1}$, 204 and 1028 (1968); J. Phys. B2, 26 (1969)
4.12 J.S. Briggs, Y.K. Kim: Phys. Rev. A3, 1342 (1971)
4.13 H.S.W. Massey, C.B.O. Mohr: Proc. Roy. Soc. A132, 605 (1931)
4.14 M. Inokuti: Argonne National Laboratory Report No. ANL-6769 (1963)
4.15 M.C. Walske: Phys. Rev. $\underline{101}$, 904 (1956)
4.16 K. Omidvar: Phys. Rev. $\underline{140}$, A 26 and A 38 (1965)
4.17 K. Omidvar, H.L. Kyle, E.C. Sullivan: Phys. Rev. A5, 1174 (1972)
4.18 G.S. Khandelwal, B.H. Choi, E. Merzbacher: Atomic Data 1, 103 (1969)
4.19 B.H. Choi, E. Merzbacher, G.S. Khandelwal: Atomic Data $\underline{5}$, 291 (1973)
4.20 B.H. Choi: Phys. Rev. A7, 2056 (1973)
4.21 E. Merzbacher, H.W. Lewis: "X-ray Production by Heavy Charged Particles", in *Handbuch der Physik*, ed. by S. Flügge, Vol. 34 (Berlin-Göttingen-Heidelberg: Springer 1958) pp. 166-192
4.22 D.H. Madison, E. Merzbacher: Theory of Charged-Particle Excitation. In *Atomic Inner-Shell Processes*, ed. by B. Crasemann, Vol. 1 (Academic Press, New York San Francisco, London 1975) pp. 1-72
4.23 D.R. Bates, G.W. Griffing: Proc. Phys. Soc. A66, 961 (1953)
4.24 D.R. Bates, G.W. Griffing: Proc. Phys. Soc. A68, 90 (1955)
4.25 I.M. Cheshire, H.L. Kyle: Phys. Lett. $\underline{17}$, 115 (1965)
4.26 T.A. Green, J.M. Peek: Phys. Rev. $\underline{169}$, 37 (1968)
4.27 K.L. Bell, V. Dose, A.E. Kingston: J. Phys. B3, 129 (1970)
4.28 K.L. Bell, A.E. Kingston: J. Phys. B4, 162 (1971)
4.29 D. Belkic, R. Gayet: J. Phys. $\underline{B8}$, 442 (1975)
4.30 F. Drepper, J.S. Briggs: J. Phys. B9, 2063 (1976)
4.31 W. Henneberg: Z. Physik $\underline{86}$, 592 (1933)
4.32 N.F. Mott and H.S.W. Massey: *The Theory of Atomic Collisions*, 3rd ed. (Oxford University Press 1965)
4.33 J.D. Garcia: Phys. Rev. A1, 1402 (1970)
4.34 L. Vriens: Case Stud. At. Phys. 1, 337 (1969)
4.35 J.D. Garcia, R.J. Fortner, T.M. Kavanagh: Rev. Mod. Phys. $\underline{45}$, 111 (1973)
4.36 D.R. Bates, W.R. McDonough: J. Phys. B3, L 83 (1970)
4.37 D.R. Bates, W.R. McDonough: J. Phys. B5, L 107 (1972)
4.38 K. Taulbjerg: Scaling Properties of Inner-Shell Excitation Cross Sections, in *Invited Talks, 2nd Int. Conf. on Inner-Shell Ionization Phenomena*, ed. by W. Mehlhorn and R. Brenn (University of Freiburg, 1976) pp. 130-144
4.39 J.R. MacDonald, P. Richard, C.L. Cocke, M.D. Brown, I.A. Sellin: Phys. Rev. Lett. 31, 684 (1973)
4.40 J.H. McGuire: Phys. Rev. A8, 2760 (1973)
4.41 A.M. Halpern, J. Law: Phys. Rev. Lett. $\underline{31}$, 4 (1973)
4.42 J.R. Oppenheimer: Phys. Rev. 31, 349 (1928)
4.43 H.C. Brinkman, H.A. Kramers: Proc. Acad. Sci. (Amsterdam) 33, 973 (1930)
4.44 D.R. Bates, A. Dalgarno: Proc. Phys. Soc. $\underline{A65}$, 919 (1952)

152

4.45 J.D. Jackson, H. Schiff: Phys. Rev. 89, 359 (1953)
4.46 R.M. Drisko: Thesis. Carnegie Institute of Technology (1955)
4.47 I.M. Cheshire: Proc. Phys. Soc. 84, 89 (1964)
4.48 K. Dettmann: "High Energy Treatment of Atomic Collisions", in *Springer Tracts in Modern Physics*, ed. by G. Höhler, Vol. 58 (Springer, Berlin, Heidelberg, New York 1971) pp. 119-206
4.49 D.R. Bates, R. McCarroll: Proc. Roy. Soc. A245, 175 (1958)
4.50 M.E. Rudd, J.H. Macek: Case Stud. At. Phys. 3, 46 (1972)
4.51 J.H. Macek: Phys. Rev. A1, 235 (1970)
4.52 T.A. Green: Proc. Phys. Soc. 86, 1017 (1965)
4.53 S.B. Schneiderman, A. Russek: Phys. Rev. 181, 311 (1969)
4.54 H. Levy, W.R. Thorson: Phys. Rev. 181, 256 (1969)
4.55 K. Taulbjerg, J. Vaaben, B. Fastrup: Phys. Rev. A12, 2325 (1975)
4.56 H.A. Bethe, R.W. Jackiw: *Intermediate Quantum Mechanics*, 2nd. ed. (Benjamin, New York 1968) p. 326
4.57 D.R. Bates: Proc. Phys. Soc. 73, 227 (1959)
4.58 R.J. Bell: Proc. Phys. Soc. 78, 903 (1961)
4.59 J.S. Briggs, A.G. Roberts: J. Phys. B7, 1370 (1974)
4.60 J. Bang, J.M. Hansteen: Kgl. Dan. Vidensk. Selsk. Mat. Fys. Medd. 31, no. 13 (1959)
4.61 D.R. Bates, R.J. Tweed: J. Phys. B7, L 117 (1974)
4.62 L. Kocbach: J. Phys. B7, L 486 (1974)
4.63 J.M. Hansteen, O.P. Mosebekk: Z. Physik 234, 281 (1970); Nucl. Phys. A201, 541 (1973)
4.64 J.M. Hansteen, O.M. Johnsen, L. Kocbach: Atom. Data Nucl. Data Tables 15, 305 (1975)
4.65 P.A. Amundsen, L. Kocbach: J. Phys. B8, L122 (1975)
4.66 A.M. Arthurs: Proc. Phys. Soc. 73, 681 (1959)
4.67 L. Kocbach: Private communication
4.68 F. Herman, S. Skillmann: *Atomic Structure Calculations* (Prentice-Hall Englewood Cliffs, New Jersey 1963)
4.69 B.H. Choi: Phys. Rev. A11, 2004 (1975)
4.70 W. Brandt, R. Laubert, I. Sellin: Phys. Rev. 151, 56 (1966)
4.71 G. Basbas, W. Brandt, R. Laubert: Phys. Rev. A7, 983 (1973)
4.72 G. Basbas, W. Brandt, R.H. Ritchie: Phys. Rev. A7, 1971 (1973)
4.73 J.U. Andersen, E. Laegsgaard, M. Lund, C.D. Moak: Nucl. Instr. Methods 132, 507 (1976)
4.74 A.R. Holt, B.L. Moiseiwitsch: J. Phys. B1, 36 (1968); J. Phys. B2, 1253 (1969)
4.75 I.M. Cheshire, E.C. Sullivan: Phys. Rev. 160, 4 (1967)
4.76 B.H. Bransden: Rep. Prog. Phys. 35, 949 (1972)
4.77 L. Wilets, D.F. Gallaher: Phys. Rev. 147, 13 (1966)
4.78 I.M. Cheshire: Phys. Rev. 138, A992 (1965)
4.79 G.W. Catlow, M.R.C. McDowell, I.M. Cheshire: J. Phys. B3, 183 (1970)
4.80 I.M. Cheshire: J. Phys. B1, 428 (1968)
4.81 A.F. Ferguson: Proc. Roy Soc. A264, 540 (1961)
4.82 R. McCarroll: Proc. Roy. Soc. A264, 547 (1961)
4.83 I.M. Cheshire, D.F. Gallaher, A.J. Taylor: J. Phys. B3, 813 (1970)
4.84 T.J. Morgan, J. Geddes, H.B. Gilbody: J. Phys. B6, 2118 (1973)
4.85 D.G.M. Anderson, M.J. Antal, M.B. McElroy: J. Phys. B8, L 118 (1974)
4.86 R. Hippler, K.H. Schartner: J. Phys. B8, 2528 (1975)
4.87 J.S. Briggs: Rep. Prog. Phys. 39, 217 (1976)
4.88 D.R. Bates, D.A. Williams: Proc. Phys. Soc. 83, 425 (1964)
4.89 M.E. Riley, T.A. Green: Phys. Rev. A4, 619 (1971);
     T.A. Green, M.E. Riley: *Abstracts of Papers, IXth Int. Conf. on Physics of Electronic and Atomic Collisions*, ed. by J.S. Risley and R. Geballe (University of Washington Press 1975) p. 71
4.90 C.F. Lebeda, W.R. Thorson, H. Levy: Phys. Rev. A4, 900 (1971)
4.91 V. SethuRaman, W.R. Thorson, C.F. Lebeda: Phys. Rev. A8, 1316 (1973)
4.92 J.S. Briggs, K. Taulbjerg: J. Phys. B8, 1909 (1975); B9, 1641 (1976)
4.93 H. Rosenthal: Phys. Rev. Lett. 27, 635 (1971)
4.94 R.D. Piacentini, A. Salin: J. Phys. B7, 1666 (1974)

4.95 J.S. Briggs, J.H. Macek: J. Phys. B5, 579 (1972)
4.96 J.S. Briggs, J.H. Macek: J. Phys. B6, 982, 2484 (1973)
4.97 K. Taulbjerg, J.S. Briggs, J. Vaaben: J. Phys. B9, 1351 (1976)
4.98 W.E. Meyerhof: Phys. Rev. Lett. 31, 1341 (1973)
4.99 R. Albat, N. Gruen, B. Wirsam: J. Phys. B8, 2520 (1975)
4.100 L. Landau: Phys. Z. Sowjetunion 2, 46 (1932); C. Zener: Proc. Roy. Soc. A137, 696 (1932)
4.101 E.E. Nikitin: Izv. A.N.SSSR Ser. Fiz. 27, 996 (1963); "The Theory of Nonadiabatic Transitions: Recent Development with Exponential Models", in *Advances in Quantum Chemistry*, ed. by P.O. Löwdin, Vol. 5 (Academic Press, New York 1970) p. 135; Yu.N. Demkov: Sov. Phys. JETP 18, 138 (1964)
4.102 Q.C. Kessel: Bull. Amer. Phys. Soc. 14, 946 (1969)
4.103 J.C. Brenot, D. Dhuicq, J.P. Gauyacq, J. Pommier, V. Sidis, M. Barat, E. Pollack: Phys. Rev. A11, 1245 (1975); V. Sidis, N. Stolterfoht, M. Barat: *Abstracts of Cont. Papers, 2nd Int. Conf. on Inner-Shell Ionization Phenomena* (Freiburg 1976) p. 68

5. Excitation in Energetic Ion-Atom Collisions Accompanied by Electron Emission

N. Stolterfoht

With 28 Figures

When a fast charged particle is incident on an atom or a molecule, generally one or more electrons are ejected in the collision. The production of electrons by ion impact is of principal interest in many branches of physics and related sciences. Thus, electron spectroscopy in ion-atom collisions has received a great deal of attention in the past. Atomic structures as well as dynamic excitation mechanisms have been studied by electron detection. These efforts have revealed important information about the electron production mechanisms in collisions between ions and atoms.

   The purpose of the present chapter is to review recent studies of inner- and outer-shell excitation in ion-atom collisions accompanied by electron emission. Emphasis is given to the excitation mechanisms rather than to the atomic structure aspects. Projectile energies in the keV range as well as in the MeV range will be treated. The amount of work done in this field is tremendous so that no attempt is made to give a complete summary here. For instance, the present chapter will be restricted to projectile energies higher than about 5 keV, and electron measurements using the beam-foil method are not included. Also, instrumental details will not be given; the discussion of the apparatus is left to the original references. In this chapter, specific topics are selected which are instructive for the understanding of electron excitation mechanisms in ion-atom collisions.

   For reasons of convenient representation, this chapter is structured to describe excitation of outer-shells and inner-shells separately. This choice is somewhat arbitrary, as the related excitation mechanisms are rather similar in the two cases. However, differences occur in the experimental methods for studying outer- and inner--shell excitation by electron spectroscopy. Generally, in the latter case, Auger electrons of well-defined energies are detected, whereas in the first case continuous electron spectra are measured.

   Following the background Section 5.1, a brief survey about electron production mechanisms is given in Section 5.2. Then, more detailed aspects of the field will be discussed. Section 5.3 is devoted to outer-shell ionization of the projectile and the target atom. In Section 5.4 the inner-shell excitation of the collision particles is treated. A few examples of Auger spectra will be given and, then, inner vacancy production mechanisms will be discussed within the framework of excitation models, such as direct excitation and electron promotion. Literature till spring 1976 is taken into consideration.

5.1 Historical Background

Since the atomic structure of matter was recognized, considerable interest has been
devoted to the interaction of fast charged particles with atomic electrons. As early
as 1912 RUTHERFORD presented his famous formula to treat collisions between point
charged particles. This formula was based on classical theories of mechanics and
electrodynamics and it may be applied to a structureless charged particle incident
on a free electron. With the discovery of quantum mechanics, ion-electron collisions
could be treated more adequately. In 1933 BETHE [5.1] used the Born approximation
[5.2] to calculate ion impact excitation and ionization of electrons bound in atomic
orbitals. The theory yields useful expressions for the energy loss of the projectile
and the energy of the ejected electrons [5.3,4]. On the other hand, it should be
emphasized that classical (or semi-classical) methods for the description of ion-
atom collisions still have considerable usage to date. On the basis of Rutherford's
formula, GRYSZINSKY [5.5] developed the binary encounter theory which has been found
successful in a wide range of applications.

Although in the past there was a considerable interest in the analysis of electrons ejected by ion impact, only a few electron measurements had been carried out
before the early 1960s. BLAUTH [5.6], MOE and PETSCH [5.7], and BERRY [5.8] were the
first to measure electron production in ion-atom collisions using projectiles of
several keV. The experiments were made for a restricted number of electron observation angles, and the results were often hampered by problems concerning the detection
of low-energy electrons. In 1963, JORGENSEN et al. [5.9,10] reported on a series of
pioneering measurements of absolute cross sections for electron production in ion-
atom collisions. In those experiments nearly complete energy and angular distributions of ejected electrons were measured for the first time. Since then, similar
studies of electron production cross sections in ion-atom collisions were carried
out at laboratories at Richland [5.11], Berlin [5.12], and Seattle [5.13].

Most electrons produced in ion-atom collisions exhibit a continuous energy distribution. However, it was soon recognized that electron spectroscopy is well suited
to analyze structures superimposed on the continuous electron background. Such
structures are produced by Auger or autoionization processes following atomic excitation above the ionization threshold. This may occur through inner-shell vacancy
creation or double electron excitation. In the early experiments, BERRY [5.8] attributed certain features in the electron spectra to autoionization transitions in the
excited atoms. Definite identifications of electron peaks produced by autoionization
and Auger processes were made by RUDD et al. [5.14] using ion impact on He and Ar.
To date high-resolution Auger spectroscopy has a high level of development. Related
studies yield information about atomic structure as well as dynamics of the collision process.

In particular, the dynamics of inner-shell excitation mechanisms have been the subject of a great deal of attention in the past. For collisions in which the projectile velocity is much smaller than that of the relevant electron, inner-shell excitation usually takes place via coupling of molecular orbitals (MO) in the quasi-molecule formed during the collision. In 1965 FANO and LICHTEN [5.15] proposed the application of the MO concept to interpret successfully inner-shell vacancy production observed in energy-loss experiments [5.16]. The Fano-Lichten model had certainly a stimulating effect on studies of excitation mechanisms in slow collisions. Many experiments were performed applying the method of x-ray detection, and the field of inner-shell excitation was rapidly expanding during the late 1960s. At the same time the first Auger experiments were made by KESSEL et al. [5.17] and CACAK et al. [5.18] to measure inner-shell vacancy production in slow Ne^++Ne and Ar^++Ar collisions. To date, x-ray and Auger-electron spectroscopy may be considered as alternative techniques in the field of inner-shell excitation.

Most experiments of electron production in ion-atom collisions have been made at projectile energies lower than a few MeV. Only recently, higher projectile energies near 30 MeV were applied by BURCH et al. [5.19] using particles from a tandem van de Graaff accelerator. Subsequently, similar work was carried out at other laboratories [5.20,21]. Although no qualitatively new process for electron production was observed using high-energy projectiles, the corresponding electron spectra indicate pronounced structures which made possible the detailed study of individual electron production mechanisms [5.22]. Primarily heavy ions were used in the experiments. Heavy-ion impact often causes kinematic line broadening whose influence, however, decreases with increasing projectile velocity. Thus, energetic heavy ions may serve as a powerful tool in the field of high-resolution Auger spectroscopy.

It should be emphasized that comprehensive articles have appeared [5.23-25] reviewing studies of electron production in ion-atom collisions. More specific reports were given by BURCH [5.26] and STOLTERFOHT [5.27]; related work was summarized in [5.28-30]. For electron studies using incident energies below ~5 keV not treated in this chapter, the reader is referred to the work of GERBER et al. [5.31] and RISLEY [5.13]. Also, for beam foil electron studies not included here see, for example, the work by HASELTON et al. [5.32].

Finally, it should be noted that atomic units will be used throughout this chapter.

5.2 General Considerations

High-energy data are particularly suited for a survey of the field, as the corresponding electron spectra indicate pronounced structures, each of which can be attributed to a certain excitation or deexcitation process in the projectile or target

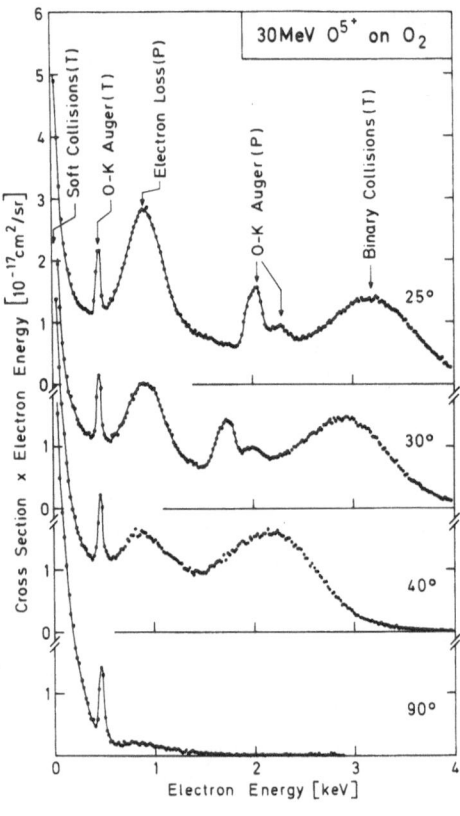

Fig. 5.1. Cross section times electron energy for electron production in 30-MeV $O^{5+}+O_2$ collisions. The observation angle is varied as indicated (from STOLTERFOHT et al. [5.22])

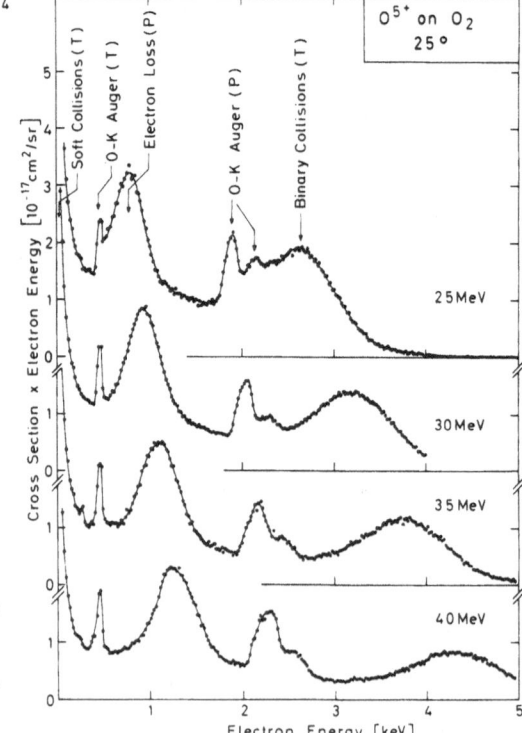

Fig. 5.2. Cross section times electron energy for electron production in $O^{5+}+O_2$ collisions at 25^0 emission angle. The projectile energy is varied as indicated (from STOLTERFOHT et al. [5.22])

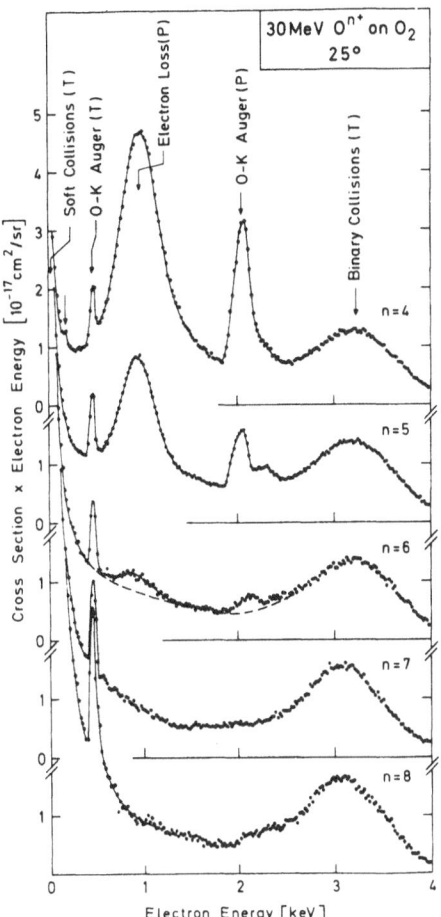

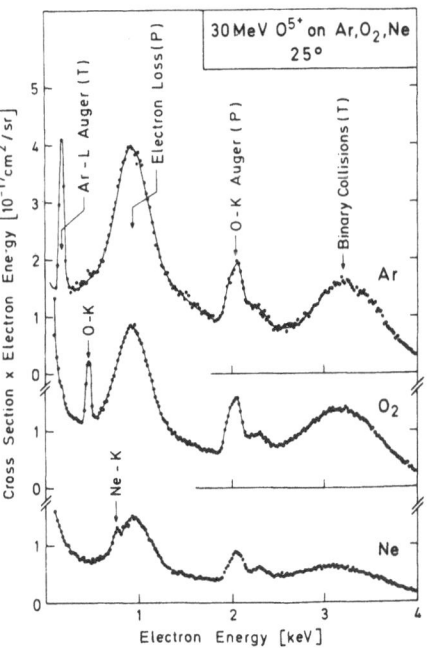

Fig. 5.3. Cross section times electron energy for electron production in 30-MeV $O^{n+}+O_2$ collision at 25^0 emission angle. The projectile charge state n is varied as indicated (from STOLTERFOHT et al. [5.22])

Fig.5.4. Cross section times electron energy for electron production at 25^0 by 30-MeV O^{5+} impact. Targets are Ar, O_2, and Ne as indicated (from STOLTERFOHT et al. [5.22])

atom. In Figs.5.1 to 5.4, double differential cross sections are plotted for electron production in 25- to 40-MeV O^{n+} impact on Ne, O_2, and Ar. The data result from a cooperative work between the University of Washington, Seattle, and the Hahn-Meitner-Institut, Berlin [5.22]. For the purpose of convenient graphical display, the cross sections are multiplied by the electron energy. In the spectra the peaks are labeled P and T, as the corresponding electrons originate from the projectile and target, respectively. In Figs.5.1 to 5.4, specific experimental parameters are varied, i.e., the electron observation angle, the projectile energy, the projectile charge state, and the target species, respectively. It is seen that the spectral structures generally change rapidly with the variation of one of these quantities.

The peaks labeled "Soft Collision (T)" and "Binary Collisions (T)" are produced by direct outer-shell ionization of the target particle. In the theoretical description based on the Born approximation, "soft collisions" and "binary collisions" correspond, respectively, to the minimum and maximum momentum transfer by the projectile. The peak labeled "electron loss" represents outer-shell electrons ejected from the projectile due to elastic scattering in the screened Coulomb field of the target. The centroid energy of the peak corresponds to an electron velocity equal to that of the projectile. The "O-K Auger (T)" peak is produced by Auger electrons following K-vacancy creation in the target atom, and "O-K Auger (P)" denotes the corresponding projectile peak shifted in energy by kinematic (Doppler) effects.

Fig.5.1 shows electron production cross sections in the angular range of 25° to 90°. It was found that the data do not substantially change at angles larger than 90° [5.22]. Soft collision electrons are ejected with nearly zero energy at all angles, whereas the binary collision peak is strongly shifted to lower energies as the observation angle increases. At angles larger than 90° the binary collision peak vanishes. A simple analysis based on energy and momentum conservation yields for the centroid energy E_b of the binary collision peak:

$$E_b = 4t\cos^2\theta \quad \text{for} \quad \theta \leq \pi/2 \tag{5.1}$$

with $t=E_i/\lambda$, where E_i is the projectile energy and λ is the projectile mass in units of the electron mass. It is found [5.22] that the centroid energy of the binary collision peak agrees well with E_b for a given observation angle θ.

As the centroid energy E_ℓ of the electron loss peak corresponds to a velocity equal to the projectile velocity, it follows that

$$E_\ell = t \quad . \tag{5.2}$$

The centroid energy E_ℓ is independent of the observation angle; however, the intensity of the electron loss peak is seen to decrease strongly with increasing θ. The angular dependence is well understood in the simple picture that quasi-free projectile electrons are elastically scattered in the Coulomb field of the target. Indeed, elastically scattered electrons are strongly forward directed.

Recently, DREPPER and BRIGGS [5.33] pointed out some important features of the electron loss process. With neglect of the ionic and molecular nature of the projectile and target, respectively, the collision system $O^{n+}+O_2$ is symmetric. Hence, the electron spectra which one would obtain in the projectile rest frame are equal to those in Fig.5.1, except that the labels T and P would be interchanged. After transformation into the laboratory frame, soft collision and binary collision electrons from the projectile are both found to contribute to the electron loss peak [5.33]. This can readily be verified for the angles 0° and 180°. Soft collision electrons ejected at (nearly) zero energy in the projectile rest frame are found at 0° in the laboratory frame and, there, they have a velocity equal to the projectile velocity v_i. Again, in the projectile frame, binary collision electrons are primarily ejected in the direction of the incident target atom and they have the velocity $2v_i$, see also (5.1). In the laboratory frame these electrons are ejected at 180° and they have the velocity $v_i=2v_i-v_i$. Hence, in the laboratory frame, soft collision and binary collision electrons from the projectile have the same energy; however they are ejected at opposite angles, i.e., at forward and backward angles, respectively. In this picture, the dominance of the electron loss peak at forward angles is also understood. Soft collision electrons are preferentially produced and these electrons, in turn, are ejected at forward angles.

The Auger electrons from the target produce a relatively narrow peak whose position does not vary with observation angle. This indicated that the recoil energy transferred to the target atom is not significant in the high-energy collisions studied here. Furthermore, the peak intensity is found to be isotropic which is typical for target Auger electrons following K-shell excitation [5.34]. However, Auger electrons form the projectile are seen to vary significantly in energy and intensity as the observation angle is changed. These variations are caused by kinematic effects, since the projectile Auger electrons originate from a moving emitter. A simple velocity vector diagram yields the following relation [5.23] between the Auger energy E' in the emitter rest frame and the corresponding laboratory energy E:

$$E' = E + t - 2(Et)^{1/2} \cos\theta \cos\beta \tag{5.3}$$

where β is the scattering angle of the projectile. A similar analysis can be made for the intensities I' and I of the Auger electrons in the emitter and laboratory frame, respectively [5.34]. It follows that

$$\frac{I}{I'} = \frac{E}{E'} \left(1 - \frac{t}{E'} \sin^2\theta \cos^2\beta\right)^{-1/2} \tag{5.4}$$

If t/E'<<1 is fulfilled, (5.4) reduces to I/I'=E/E'. In the present case it follows
that t/E'>1, i.e., the projectile is faster than the ejected Auger electrons. As a
consequence, for 30-MeV oxygen projectiles, Auger electrons are only seen at forward
angles smaller than 43°. In this range, (5.3) has two solutions with respect to E
so that the projectile Auger peak appears twice in the spectrum. (See also the un-
labeled arrow above the O^{4+} spectrum in Fig.5.3).

It should be noted that (5.3) and (5.4) apply only to Auger lines of sufficiently
small width. The Auger peaks in Fig.5.1 are composed of several lines and the kinematic shift is different for lines of different energies. Eq.(5.3) shows that the
line shift increases with increasing Auger energy. Thus, the projectile Auger peak
shifted to higher energies appears to be "stretched" in comparison with the corresponding target peak, see Fig.5.1. (Similarly, Auger peaks shifted to lower energies
are found to be "compressed"). It should be noted that the stretching effect essentially does not influence the relative widths of individual Auger lines so that intrinsic structures of the Auger peaks are not affected. This is in contrast to kinematic broadening effects which do wash out intrinsic line structures of the Auger
peak [5.23,34].

The projectile energy dependence of the various electrons peaks is shown in
Fig.5.2. It is seen that the peaks labeled by "Electron Loss (P)", "O-K Auger (P)",
and "Binary Encounter (T)" are shifted to higher energies as the projectile energy
increases. According to (5.1-3) the energies of the peaks are essentially proportional to the projectile energy. As expected, the "Soft Collision" peak and the "O-K
Auger (T)" peak do not vary in energy.

In Fig.5.3 30-MeV $O^{n+}+O_2$ spectra are plotted for various projectile charge states
n. With increasing n, fewer electrons are available for ejection from the projectile
so that the peaks with label (P) decrease in intensity. In contrast, the peaks with
label (T) increase in intensity as highly stripped projectiles ionize the target
more efficiently. Fig.5.4 shows electron spectra produced by 30-MeV O^{5+} incident on
Ar, O_2, and Ne. For the targets Ne and O_2, Auger electrons resulting from K excitation are observed, whereas for Ar, Auger electrons are found following L-shell excitation. (The Ar-K Auger peak expected at about 2.8 keV is lost in the background).
Apart from the target Auger peaks and a constant factor influencing the overall intensity, the electron spectra are seen to be quite similar. This indicates that the
general features of the electron production mechanisms are rather independent of
the collision system.

5.3 Outer-Shells

5.3.1 Target Ionization

Ejection of continuum electrons from the target atom by light-ion impact has often been studied in the past. Both angular and energy distributions were measured for several collision systems to obtain double differential cross sections (DDCS) for electron emission. Results from different laboratories and related theories were discussed in detail by RUDD [5.25]; thus, only a few recent examples will be treated.

Among studies of target ionization by light-ion impact, the collision system H^++He has received most attention. Previous measurements by RUDD and JORGENSEN [5.10], RUDD et al. [5.35], and STOLTERFOHT [5.12] were recently extended by TOBUREN [5.36], STOLTERFOHT et al. [5.37], and RUDD and MADISON [5.38] to cover the projectile energy range from 5 keV to 5 MeV. Numerical data were summarized by RUDD et al. [5.39]. Furthermore, extensive comparisons between experiment and theory were reported in [5.40-42].

In Fig.5.5 are plotted cross sections for electron production in 5-keV, 30-keV, 200-keV, 1-MeV, and 4.2-MeV, H^++He collisions as measured at different laboratories [5.35-38]. Electron observation angles are 20^0 or 22.5^0. The results show the typical features of electron spectra produced by proton impact. Near zero electron energy, the cross sections maximize due to the dominance of soft collisions. For high incident energies, the peaks produced by binary collisions are clearly visible, whereas for low projectile velocities the binary collision peak overlap strongly with the soft collision peak so that they cannot be identified separately.

Fig.5.5 shows that electron production cross sections for H^++He were measured in a wide range of projectile energies and it appears to be a difficult task to reproduce them theoretically. Indeed, no universal theory applicable to the entire energy range exists at present. Thus, only specific cases of the ionization processes have been theoretically treated in the past. For fast collisions with light projectiles, the effect of the incident ion upon the target atom may be regarded as a sudden and small perturbation. In this case the wave functions of the incident particle and those of the target electrons are not significantly affected during the collision. Thus, methods of the perturbation theory may be applied. These features of the collision allow the use of the plane wave Born approximation (PWBA) [5.2,43].

The major work in calculating DDCS's by means of the Born approximation is the determination of the square of the matrix element (form factor)

$$F(K) = |\int \psi_k \, e^{i\underline{K}\underline{r}} \, \psi_0 \, d^3\underline{r}|^2 \tag{5.5}$$

where $\underline{K}$ is the momentum transfer and ψ_0 and ψ_k are single-electron wave functions for the ground state and continuum state of the electron ejected with momentum $\underline{k}$. The

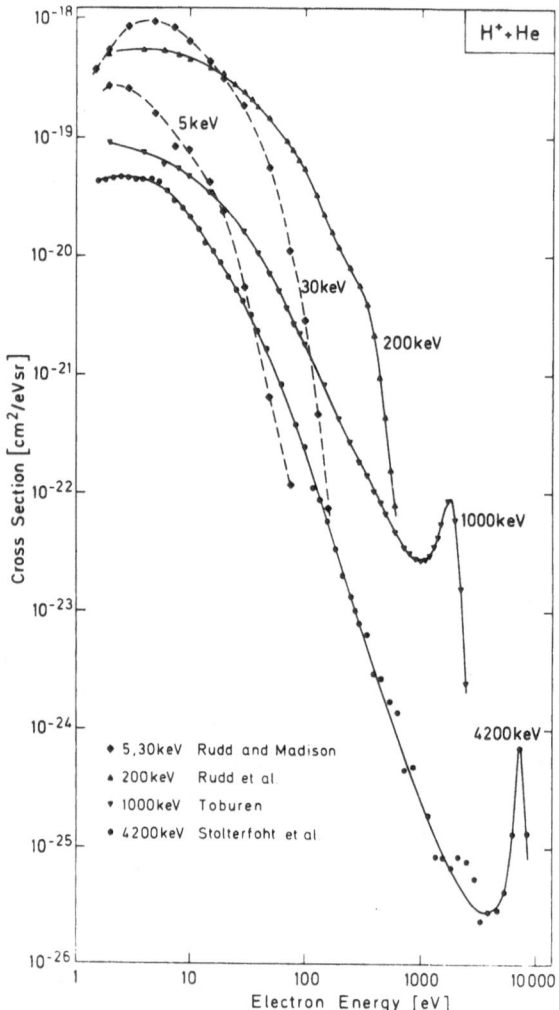

Fig. 5.5. Cross section for electron production in H⁺+He collisions. Projectile energies are 5, 30, 200, 1000, and 4200 keV. Observation angle is 22.5⁰ for 4200 keV and 20⁰ for the other energies (from RUDD and MADISON [5.38], RUDD et al. [5.35], TOBUREN [5.36], and STOLTERFOHT et al. [5.37], respectively)

integral in (5.5) can be solved analytically for bound and continuum wave functions of hydrogen [5.2,44]. These results were often scaled to determine DDCS for atomic systems other than hydrogen [5.25]. The comparison with experimental results yields good agreement at intermediate electron emission angles, whereas significant discrepancies appear at forward and backward angles. The disagreement at backward angles is primarily produced by the simplification of using hydrogenic wave functions. Recently, MADISON [5.40] and MANSON et al. [5.41] who incorporated the Hartree-Fock-Slater (HFS) continuum wave function for the first time, calculated DDCS's in good agreement with experimental H⁺+He data at backward angles. The results by MANSON et al. [5.41] are plotted in Fig.5.6 showing angular distributions of DDCS's for 81.6-eV electron production by 100-keV, 300-keV, 1-MeV, and 5-MeV H⁺ impact. The theory

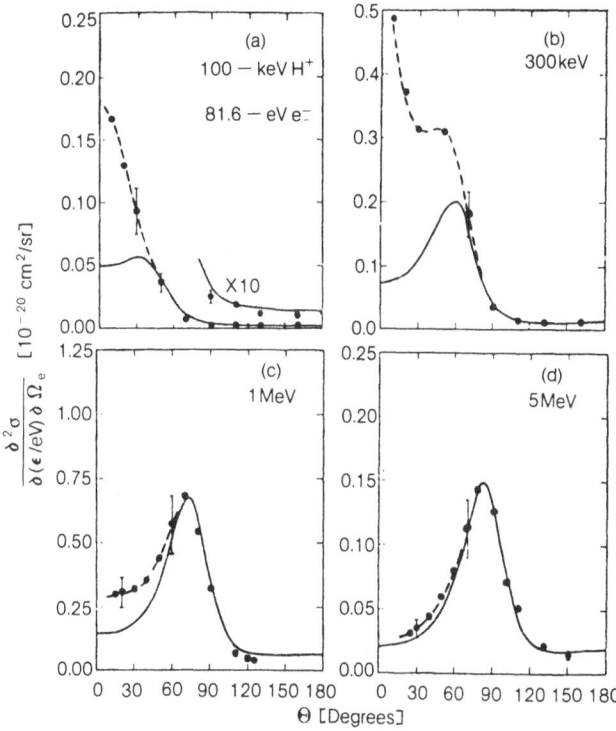

Fig. 5.6a-d. Cross section for emission of 81.6 eV electrons in H⁺+He as a function
of the observation angle θ. The experimental results for projectile energies of 100
keV and 300 keV are from RUDD et al. [5.35]. The 1-MeV and 5-MeV data were measured
at Richland and Berlin, respectively (see [5.41]). The solid curves represent calculations by MANSON et al. [5.41]

is compared with experiment and excellent agreement is found at intermediate angles
as well as at backward angles. Also, it is seen that the 5-MeV data compare well
with theory at forward angles. This shows that at sufficiently high incident energies
the first Born approximation adequately describes electron production in the full
angular ranges.

For lower incident energies the theory strongly underestimates the experimental
data at forward angles. These discrepancies are produced by charge-exchange processes
[5.45-49] not described by the first Born approximation. In the collision, electrons
are captured into low-energy continuum states of the projectile. In the laboratory
frame, these electrons are preferentially ejected at forward angles with a velocity
equal to the projectile velocity. It should be noted that electrons produced by
charge exchange into continuum processes behave similar to those resulting from
"electron loss" events described in Section 5.2. In both cases bound electrons are
transferred to low-lying continuum states of the projectile. The difference is that
the electrons originate in one case from the target and in the other case from the

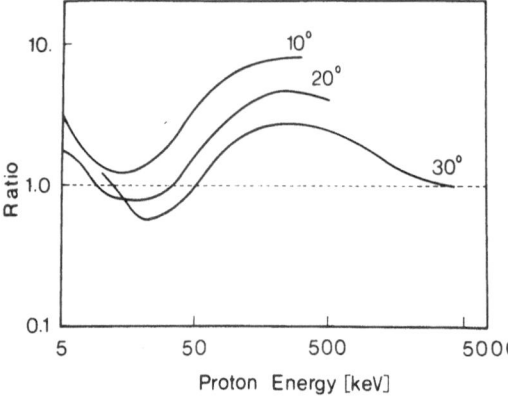

Fig. 5.7. Ratio of theoretical to experimental double differential cross sections for electron production in H^++He collision. Electron emission velocity is chosen to be equal to the projectile H^+ velocity. Observation angles are as indicated. The data are from RUDD and MADISON [5.38] including results from MANSON et al. [5.41]

projectile. It is noted that the consideration of projectile continuum states is useful only if the incident energy is sufficiently high. At low energies the colli- sion system forms a quasi-molecule for which the term "projectile continuum state" loses significance.

From the picture of the charge exchange into continuum, it is expected that dis- crepancies between experiment and theory are largest at the energy for which the ejected electron has the same velocity as the projectile. For this energy, RUDD and MADISON [5.38] plotted the ratio of experimental and theoretical DDCS for small ob- servation angles (see Fig.5.7). As expected, the ratio is largest at forward angles, e.g., 10^0. For 5-MeV H^+ and 30^0 the ratio is practically 1, showing that the process of charge exchange into continuum is not significant, as noted above. The data in- crease with decreasing proton energy and they maximize at about 300 keV. However, it is seen that theory and experiment tend to approach each other at lower incident energies. This would indicate that the Born approximation describes electron pro- duction fairly accurately at the low-energy limit.

The finding that the Born approximation applies relatively well in low-velocity $H+He^+$ collisions is rather surprising. Note that at about 40 keV the incident proton has the same velocity as the 1s electron of He. At projectile velocities smaller than the electron orbital velocity, the basic assumptions of the Born approximation break down. In this case, the interaction between the target atom and the incident particle is not a small and sudden perturbation, and the target wave functions are significantly affected during the collision. For 5-keV H^++He it is even expected that the quasi-molecular picture [5.15] is adequate to describe excitation of the electrons. A first step forward to treat the H^++He collisions in the quasi-molecular model was made by BAND [5.50], calculating charge exchange into continuum in fairly good agreement with the experiment. A full quasi-molecular calculation of ion impact ionization was carried out only for the relatively simple system H^++H [5.51]. The major reason for this limited application of the quasi-molecular model to ionization is the need for molecular continuum wave functions which are difficult to calculate. With respect to these problems, it appears to be important to verify the full range of applicability of the (first) Born approximation [5.52,53].

A sensitive test for the validity of the Born approximation is achieved from cross sections analyzed as a function of the projectile nuclear charge Z_1. The first Born approximation predicts that the DDCS is proportional to Z_1^2. The same is true for cross sections integrated over electron observation angle and energy. It should be noted that the Z_1^2 dependence of the total ionization cross section has often been discussed in the field of inner-shell ionization [5.54]. For outer-shell ionization a similar analysis is still lacking. In this case, the Z_1^2 dependence can readily be verified not only for the total cross sections but also for cross sections differential in electron observation angle and energy.

Partial information regarding the Z_1^2 dependence may be obtained from a given projectile whose incident charge state is varied. The variation of the number of projectile electrons has an effect similar to varying the projectile nuclear charge. Electrons carried with the projectile screen the nucleus and produce an effective nuclear charge Z_{eff}. However, it should be noted that Z_{eff} is not such a well-defined quantity as Z_1 since it is usually dependent on experimental parameters, e.g., the incident energy.

To verify the Z_{eff}^2 dependence of the DDCS's, the 30-MeV $O^{n+}+O_2$ data [5.22] presented in Section 5.2 are analyzed. Fig.5.8 shows cross sections $d\sigma(n)/dEd\Omega$ for ejection of electrons at 25^0 by projectiles of charge state n. It is noted that the cross sections are extrapolated in the energy region where the electron loss maximum and

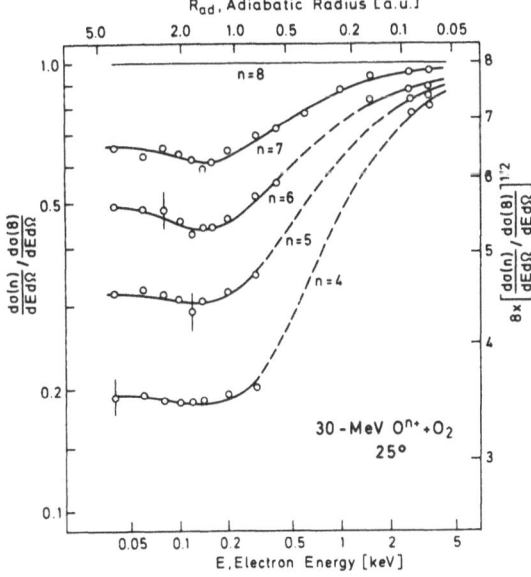

Fig. 5.8. Ratio of cross sections $\overline{d\sigma(n)}/dEd\Omega$ for electron emission at 25^0 in 30-MeV $O^{n+}+O_2$ collisions. For the adiabatic radius R_{ad} see text. The dashed lines refer to extrapolated cross sections. The data are obtained analyzing the cross sections given in Fig.5.3

the Auger peaks tend to obscure the analysis. To obtain relative cross sections, the data are normalized to $d\sigma(8)/dEd\Omega$. For n=8 the screening effects are absent and the effective nuclear charge is constant, $Z_{eff}(8)=8$.

Hereafter, the data are discussed in terms of the right-hand coordinate in Fig.5.8. Defining the reduced cross section

$$\frac{d\sigma_r(n)}{dEd\Omega} = \frac{1}{Z^2_{eff}(n)} \frac{d\sigma(n)}{dEd\Omega}$$

and the ratio

$$I_r(n) = \frac{d\sigma_r(n)}{dEd\Omega} \bigg/ \frac{d\sigma_r(8)}{dEd\Omega}$$

it follows for the right-hand scale

$$8\left[\frac{d\sigma(n)}{dEd\Omega} \bigg/ \frac{d\sigma(8)}{dEd\Omega}\right]^{1/2} = Z_{eff}(n)[I_r(n)]^{1/2} \quad . \tag{5.6}$$

Thus, the data in Fig.5.8 depend on $Z_{eff}(n)$ as well as on the ratio $I_r(n)$ of the reduced cross sections. Assuming for a moment that $I_r(n)=1$, as the Born approximation predicts, the data represent directly $Z_{eff}(n)$. It is seen that $Z_{eff}(n)$ approximates 8 at the high-energy limit, indicating that in binary collisions the projectile nucleus deeply interpenetrates the target atom so that screening effects of the projectile are not important. However, the screening effects are appreciable at lower electron energies and the data approach asymptotically a constant value at about 200 eV. Below this energy, electrons are produced in glancing (i.e., distant) collisions for which the screening effect appears to be constant.

The effective impact parameter for a collision, in which the energy W is transferred, is given by the so-called adiabatic radius $R_{ad}=v_i/W$ (see, e.g., [5.52]) where v_i is the projectile velocity. In the present case it follows approximately that W=E+I where E is the energy of the ejected electron and I is its ionization potential. In Fig.5.8 the R_{ad} is shown as derived with I=15 eV. It is seen that in the present case the effective impact parameter varies by two orders of magnitude. Accordingly, the screening of the projectile electrons is found to vary significantly.

For glancing collisions, $Z_{eff}(n)$ is expected to be equal to or slightly larger than the projectile charge state n. However, at energies <200 eV it is seen that Z_{eff} is even smaller than n. This gives evidence that the ratio $I_r(n)$ actually deviates from 1 in contradiction with the first Born approximation. Setting approximately $Z_{eff}(4)=4$, it is found that the *reduced* cross section for n=8 is 25% higher than that for n=4. This observation might be explained through "polarization" of the target electrons by means of the projectile passing by at impact parameters larger than the outer-shell radius. Similar effects were pointed out by BASBAS et al. [5.55] in calculating cross sections for inner-shell ionization as discussed in the next section. For the present case, it should be emphasized that a more definite conclusion about electron polarization would be possible, if data were available as a function of Z_1 rather than of Z_{eff}. It is suggested that measurements of DDCS's for fully stripped projectiles would reveal useful information about the applicability of the first Born approximation.

5.3.2 Projectile Ionization

In the preceding subsection, the projectile was treated as a source of a Coulomb field which causes target electrons to be ejected during the collision. Projectile electrons were incorporated only with respect to their collective effect of screening the incident nuclear charge. In this section the projectile electrons will be studied as individual particles which participate in the ionization process.

It should be pointed out that there is no principal difference between the ionization of the projectile and that of the target atom. It was noted above for symmetric collisions that the electron spectra are expected to be equal in the rest frames of the collision particles. However, considerably different energies are found for target and projectile electrons as they are simultaneously observed in one system of reference (e.g., the laboratory frame). This gives indication for strong kinematic effects on the electrons ejected from the moving projectile. As noted above, the projectile electrons are found to be concentrated within one maximum, i.e., the electron loss peak whose mean energy is equal to the projectile energy reduced by the projectile-electron mass ratio [see (5.2)].

Although the projectile electrons are usually rather intense, the electron loss peak was recognized only recently [5.56-58]. WILSON and TOBUREN [5.56] measured angular and energy distributions of electrons ejected in 0.6- to 1.5-MeV $H_2^+ + H_2$ collisions. The data indicate significant maxima at electron energies equal to the reduced projectile energy. Results showing the ratio of single differential cross sections for electron production by H_2^+ to those by H^+ of equal velocity are plotted in Fig.5.9. The H^+ data were taken as a reference to remove features characteristic for target ionization. This procedure takes advantage of the fact that H_2^+ has a similar effect on the target electrons as *two* individual protons. Indeed, the projectile electrons appear to be superimposed on a constant base line with the value of 2. (At the low energy limit the drop of the base line below 2 is probably caused by screening effects similar to those seen in Fig.5.8). After having reduced the contribution of the target electrons to a (nearly) constant background, the projectile electrons can readily be studied separately. WILSON and TOBUREN [5.56] integrated the electron loss peak and obtained corresponding cross sections for different observation angles. The results were found to be in good agreement with elastic scattering cross sections for electrons incident at a velocity equal to the projectile

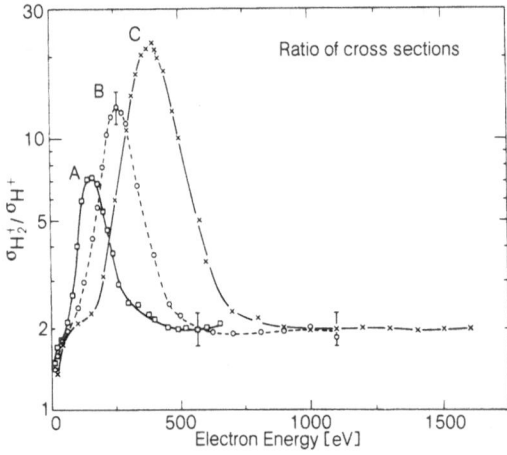

Fig. 5.9. Ratio of single-differential cross sections for incident H_2^+ to those for incident H^+ of equal velocity both on H_2. A, B, and C are, respectively, for 0.6-, 1.0-, and 1.5-MeV H_2^+ energies (from WILSON and TOBUREN [5.56])

velocity. This agreement gives some indication for the mechanisms operative in the electron loss event.

BURCH et al. [5.58] considered electrons on the incident particle to constitute merely a merged beam of electrons traveling with the velocity $\underline{v}_i$ of the incident ions. The velocity distribution $f(\underline{v}_b')$ of the electrons which results from their binding to the projectile is taken into account by adding the component $\underline{v}_b'$ to $\underline{v}_i$. Thus, electrons with velocity $\underline{v}_b = \underline{v}_i + \underline{v}_b'$ are considered to scatter elastically off the target atom which is represented by a screened Coulomb potential. The DDCS for ejection of projectile electrons is then obtained from the elastic scattering cross section folded with the velocity distribution $f(\underline{v}_b')$ and integrated over the direction of $\underline{v}_b'$. In Fig.5.10, results are shown for electron ejection at $90°$ in 23.3- to 40.8-MeV O^{4+}+Ar collisions [5.58]. The calculations were made for a velocity distribution determined from hydrogenic wave functions scaled for O^{4+}. The theoretical results are compared with (normalized) experimental data [5.58], also plotted in Fig.5.10. Despite the simple model used in the calculations, the theoretical and experimental data are found to be in rather good agreement.

Simple considerations allow for conclusions about the shape of the electron loss peak. The energy E of the ejected electron is equal to its incident energy $v_b^2/2$, as the electrons are assumed to be elastically scattered. Hence, it follows that

$$E = v_b^2/2 = (v_i^2 + v_b'^2 + 2v_i v_b' \cos\alpha)/2 \qquad (5.7)$$

where α is the angle between $\underline{v}_i$ and $\underline{v}_b'$. For $v_i \gg v_b'$, the mean energy $\bar{E}$ of the ejected projectile electrons is given by $\bar{E} \approx v_i^2/2$ in accordance with (5.2). The spread of the energy E producing the width of the electron loss peak is primarily determined by the variation of $2v_i v_b' \cos\alpha$, i.e., the spread of v_h' and α. To a good approximation, v_b' is centered around the mean value $\bar{v}_b' = \sqrt{2I_b}$, where I_b is the corresponding binding

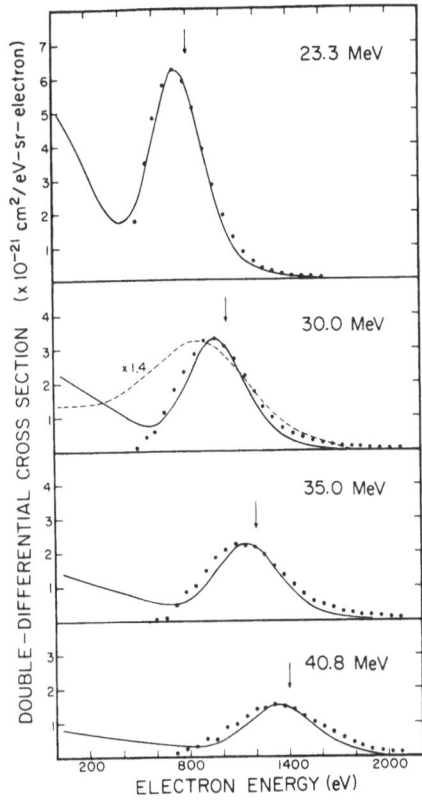

Fig. 5.10. Experimental double-differential cross sections for electron emission at 90^0 in $O^{4+}+Ar$ collisions. Projectile energies are as indicated. The data are normalized to corresponding theoretical cross sections given as solid lines. The dashed curve results from a 1s velocity distribution with a 2s binding energy (see text) (from BURCH et al. [5.58])

energy. Assume for a moment that the variation of v_b' is negligible. Then, the spread of E is determined by the variation of α within its limits of 0 to π. Hence, for the (base) width ΔE of the electron loss peak, it follows that

$$\Delta E = 2v_i \bar{v}_b' \approx 4(\bar{E}I_b)^{1/2} \quad .$$

(5.8)

This ΔE may be considered as a broadening caused by kinematic effects. The width formula (5.8) shows that the electron loss peak in Fig.5.10 is produced primarily by electrons from the 2s shell, the outer most shell of O^{4+}. Electrons from the K--shell are hardly seen as they are too much spread out over the entire spectrum.

Additional broadening is expected when v_b' varies appreciably. In Fig.5.10, DDCS's are compared for 30-MeV O^{4+} as obtained with 1s and 2s hydrogenic velocity distributions [5.58]. (To minimize differences produced by kinematic effects, the 1s velocity distribution is artificially forced to be centered around the average velocity of the 2s electron). The data show that the increased variation of v_b' causes the centroid energy of the peak to be shifted to lower energies. Also, it is seen that the results obtained with the 2s velocity distribution agree far better with experiment than those for the 1s velocity distribution. This gives some confidence that

measurements of the electron loss peak yield information about the velocity distribution of the electrons bound at the projectile [5.58].

Recent progress in the theoretical description of the electron loss peak was made
by DREPPER and BRIGGS [5.33]. They pointed out that it is very useful to analyze the
projectile ionization first in the rest frame of the projectile. There, the cross
section $d\sigma/dE'd\Omega'$ for ejection of projectile electrons is just the usual DDCS for
ionization by an incident particle, as discussed in the preceding subsection. Once
this DDCS is determined, the corresponding quantity $d\sigma/dEd\Omega$ in the laboratory frame
can simply be obtained by a velocity transformation [5.33]:

$$\frac{d\sigma}{dEd\Omega} = \frac{v}{v'}\frac{d\sigma}{dE'd\Omega'}$$

(5.9)

where $\underline{v}=\underline{v}_i+\underline{v}'$, with $v'=\sqrt{2E'}$ and $v=\sqrt{2E}$ being the velocities of the *ejected* electrons
in the frames of the projectile and the target, respectively.

Important characteristics may be understood from the frame transformation. For
instance, the process of the incident free electron with velocity $\underline{v}_b=\underline{v}_i+\underline{v}'_b$ being
elastically scattered in the field of the target atom appears to be quite familiar
in the projectile rest frame. There, the screened target atom is incident on a free
electron having a certain initial velocity distribution. The latter process is just
that which constitutes the basis of the usual binary encounter model. Hence, results
similar to that in Fig.5.10 may be obtained from (5.9) using ordinary binary encounter data for the projectile DDCS.

The projectile DDCS's can also be determined on the basis of the Born approximation. Such calculations were made by DREPPER and BRIGGS [5.33] to obtain $d\sigma/dEd\Omega$ by
means of (5.9). Results for projectile electron ejection at various angles in $He^+ + He$
collisions are shown in Fig.5.11. (As the target electrons are neglected, the data
refer to electron distributions which one would see in $He^+ + He^{++}$ collisions). At angles $\gtrsim 30^{\circ}$ the data are found to represent the usual profile of the electron loss
peak. However, at forward angles the electron distribution is very sharply peaked.
It was noted in the preceding section that the electron loss process is related to
the process of charge exchange into continuum. Indeed, the electron peaks resulting
from the two processes have the same cusp shapes at the forward direction [5.47-49].
Measurements near 0° would be useful to verify the theoretical prediction of the
cusp shaped electron loss peak.

Finally, the effect of the frame transformation is demonstrated by means of a
three-dimensional plot of the DDCS (see Fig.5.12). The plot shows the electron production cross sections $d\sigma/d^3v'$ for $He^+ + He$ collisions in the projectile frame [5.33].
The (positive) z-axis has the direction of the incident beam and the x-axis an arbitrary direction perpendicular to it. A given point (v'_x, v'_z) corresponds to the
electron velocity $v'=\sqrt{v'^2_x+v'^2_z}$ and its emission angle $\theta'=\arctan(v'_x/v'_z)$. The plot shows
a maximum which corresponds to the soft collision peak whereas the lower-lying ridge

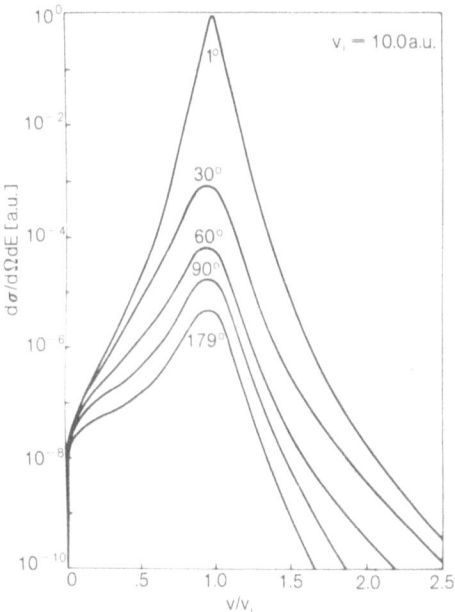

Fig.5.11. Theoretical cross sections dσ/dEdΩ for electron emission from the projectile in He⁺+He collision. The v/v₁ is the ratio of electron velocity to projectile velocity. Numbers given refer to the electron observation angle (from DREPPER and BRIGGS [5.33])

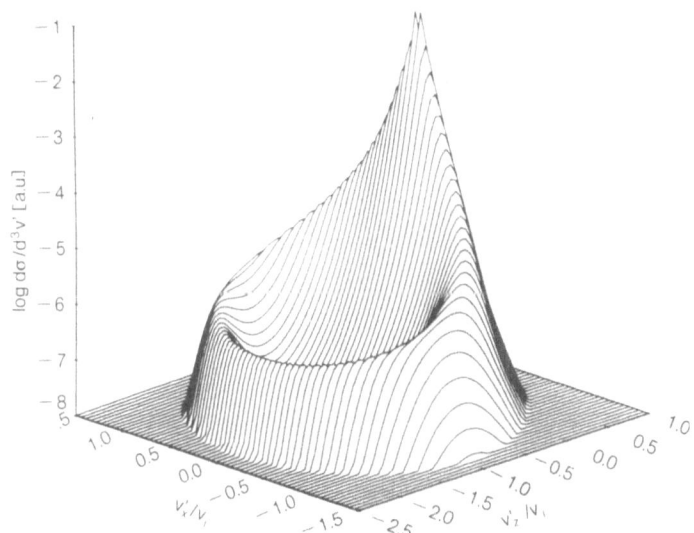

Fig. 5.12. Three-dimensional plot of projectile rest frame cross section dσ/d³v for projectile electron emission in He⁺+He collisions. Collision velocity is v₁=10 a.u. The v´z and v´x are electron emission velocities which are, respectively, parallel and perpendicular to the incident direction (from DREPPER and BRIGGS [5.33])

represents the binary encounter peak. For the transformation from the projectile
frame to the laboratory frame, the origin (0,0) is to be shifted to the valley at
(0,-1). Now, the ridge including the maximum forms a circle of radius $v/v_i=1$ around
the new origin. This circle represents the electron loss peak at $v=v_i$ for all angles. Also, it is seen that the maximum represents the dominant electron loss peak
at the ejection angle of 0°.

5.4 Inner-Shells

5.4.1 Auger Spectra

It was shown above that electron spectroscopy is a useful method to study outer-shell
ionization in ion-atom collisions. In this section, the same method is applied to
inner-shell excitation. For inner - shells there are two successive processes which
give rise to electron emission. First, during the collision, electrons are ejected
with a continuous energy distribution similar to that produced by outer-shell ionization. Secondly, electrons of discrete energies are ejected by Auger transitions
filling the inner-shell vacancy. Unfortunately, the electrons from the first process
are difficult to separate from the numerous electrons originating from outer-shells.
(An exception may be present at energies much larger than that of the binary encounter maximum). Hence, information about inner-shell excitation is usually achieved
by the detection of Auger electrons.

Generally, inner-shell vacancy states live long enough so that the vacancy "for-
gets" how it was created. Thus, Auger spectroscopy does not supply much information
about the electron initially removed from the inner-shell; the electron may be excited to upper bound states, ejected into the continuum, or captured by the projectile. However, Auger electron measurements allow for determination of the total
number of inner-shell vacancies produced in the collision. The Auger yield (Auger
electrons per vacancy) is practically 1 for the light atoms studied in this chapter.
For this case, Auger spectroscopy is an efficient method to measure total cross sections for inner-shell excitation.

Apart from this, Auger spectra can yield further information. Using kinematic
effects, the projectile Auger peak may be energetically shifted, so that it is always separated from the target peak. The peak separation allows for the individual
study of inner-shell vacancy creation in projectile and target atom. Furthermore,
Auger peaks are affected by outer-shell excitation occurring simultaneously with
the inner-shell vacancy production. The same is true for multiple ionization of the
inner-shell. Hence, Auger spectra can reveal detailed features of the collision process. Also, it is noted that Auger electrons give specific spectroscopic information
regarding the energy of the inner-shell vacancy states and their decay schemes. How-

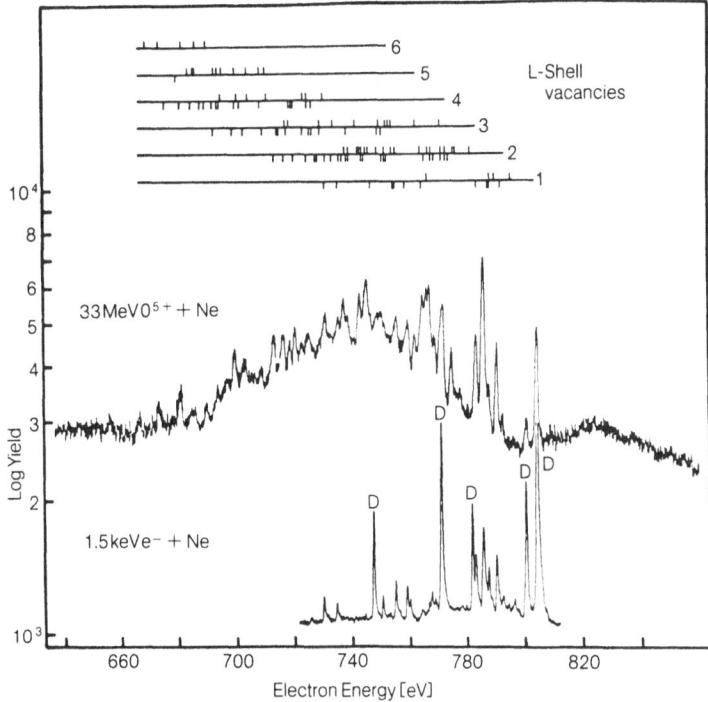

Fig. 5.13. Comparison of Ne K-Auger electron spectra produced by 33-MeV O^{5+} and 1.5-keV e^-. Normal lines are labeled D. Calculated satellite transition energies are shown on axis which label the number of L-shell vacancies (from MATTHEWS et al. [5.20])

ever, this atomic structure aspect of the Auger spectra will not be treated in detail as it lies outside the scope of this work.

Figure 5.13 shows high-resolution Auger spectra of Ne excited by 33-MeV O^{5+} and 1.5-keV e^- as measured by MATTHEWS et al. [5.20]. In the e^-+Ne spectrum, normal or diagram lines are marked with the label D. These lines correspond to K-LL Auger transitions from initial states where the outer-shell is undisturbed. Apart from that, satellite lines occur due to initial states where the L-shell is singly ionized in addition to the K-shell. The O^{5+}+Ne spectrum is dominated by satellite lines resulting from Ne multiply ionized in the L-shell. In Fig.5.13, (approximate) energies of satellites are given for each number of electrons missing in the L-shell prior to the Auger transition. It is seen that satellite lines for different vacancy states often overlap so that the analysis of the O+Ne spectrum is difficult. However, it is found that the mean energy of the satellites decreases as the number of L-shell vacancies increases. Thus, it follows that the centroid energy of the Auger spectrum gives some information about the L-shell charge state. STOLTERFOHT et al. [5.59] and MATTHEWS et al. [5.60] used the Auger spectra to estimate a mean number of 2.8 L-shell electrons missing after K-shell ionization in 30-MeV O^{5+}+Ne collisions.

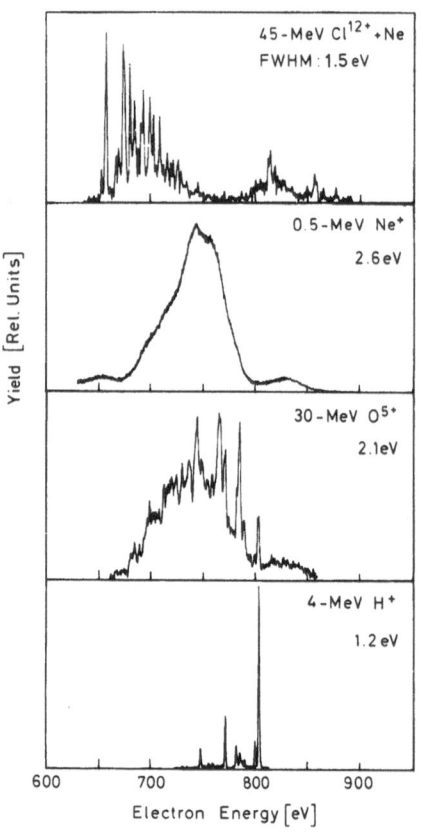

Fig. 5.14. Comparison of Ne-K Auger electron spectra excited by different projectiles. The instrumental resolution FWHM is given for each spectrum. The 4-MeV H^+, 30-MeV O^{5+}, 0.5-MeV Ne^+, and 45-MeV Cl^{12+} data are from STOLTERFOHT et al. [5.61], BURCH et al. [5.62], STOLTERFOHT et al. [5.34], and SCHNEIDER et al. [5.63], respectively

In Fig.5.14, additional Ne Auger spectra obtained at different laboratories are shown. The 4-MeV H^++Ne spectrum measured by STOLTERFOHT et al. [5.61] is practically equal to the e^-+Ne spectrum in Fig.5.13. Also, good agreement is found between the 30-MeV O^{5+}+Ne spectrum by BURCH et al. [5.62] and the corresponding data in Fig.5.13. Moreover, it is seen that 30-MeV O^{5+} and 0.5-MeV Ne^+ produce Auger spectra of similar centroid energy [5.34]. This indicated that about the same number of L-shell vacancies are produced in the two cases. However, the line structures of the two spectra are found to be rather different. (Note that the instrumental resolutions are nearly equal in the two cases). This observation is explained by kinematic line broadening which decreases with increasing projectile energy. For the 0.5-MeV Ne^++Ne spectrum, STOLTERFOHT et al. [5.34] estimated a line broadening of ~15 eV, whereas it is ~0.5 eV in the case of 30-MeV O^{5+}+Ne collisions. This shows that energetic ions are very useful for high-resolution Auger studies.

Outer-shell ionization is found to be rather dramatic in the 45-MeV Cl^{12+}+Ne spectrum measured by SCHNEIDER et al. [5.63] (see Fig.5.14). Comparison between the transition energies given in Fig.5.13 and the Cl+Ne spectrum suggests that most probably 6 L-shell electrons are removed in addition to the K-shell electron. Similar observations were previously made by BURCH et al. [5.64] using 50-MeV Cl^{15+}+Ne under conditions of limited resolution. When 1 K electron and 6 L electrons are removed, Li-like Ne is produced. At the low-energy limit, the first two lines at 652 eV and 657 eV were assigned to the $1s2s^2\, ^2S$ and $1s2s2p\, ^4P$ states [5.63]. At higher energies, lines were attributed to Be-like Ne. Above 800 eV the Cl+Ne spectrum shows structures which may be due to higher Rydberg states of Li-like Ne. Similar structures are seen in the O+Ne and Ne+Ne spectra. It should be emphasized that Li-like states of heavy particles are usually accessible only in experiments using the method of beam foil spectroscopy [5.32,65]. The observation of such states in a single Cl+Ne collision

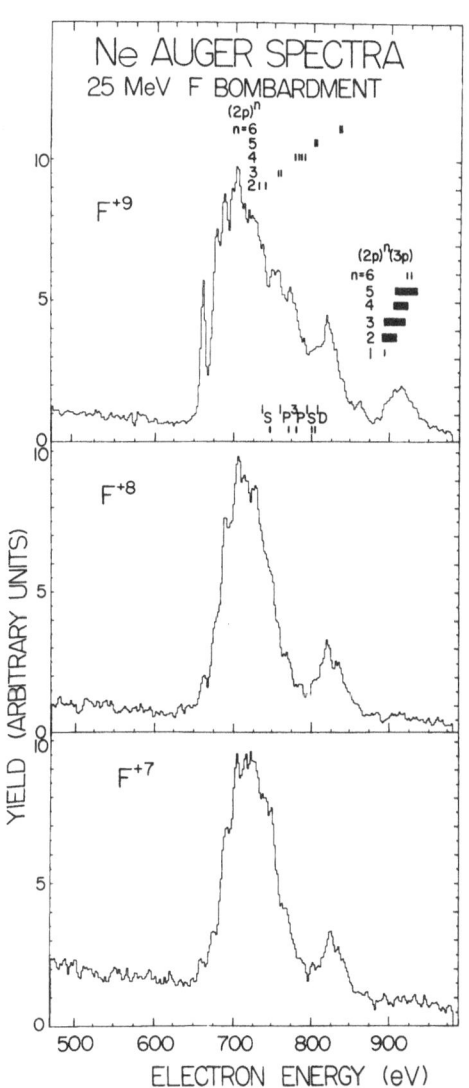

Fig. 5.15. Comparison of Ne K-Auger spectra produced in 25-MeV F^{n+}+Ne collisions for n=7, 8, and 9. Calculated transition energies for the initial configuration $(2p)^n$ and $(2p)^n 3p$ are shown. Note the difference of the spectra as F^{9+} is used instead of F^{8+} or F^{7+} (from WOODS et al. [5.66])

is rather important. This shows that a highly stripped projectile such as Cl^{12+}
passing through the K-shell radius of the target atom causes ejection of nearly all
L-shell electrons.

An interesting feature is seen in the Auger spectra produced by completely stripped
projectiles. Fig.5.15 shows Ne K-Auger spectra measured at medium resolution by
WOODS et al. [5.66] using 25-MeV F^{7+}, F^{8+}, and F^{9+} impact on Ne. For F^{7+} and F^{8+}
the shape of the spectra are similar to those observed for O^{5+} and Cl^{12+} impact
(Fig.5.14). Most of the Auger lines are found in one peak centered around ~720 eV.
At ~830 eV the spectrum shows a second peak probably due to excitation to higher
Rydberg states as noted above. However, the F^{9+}+Ne spectrum looks qualitatively different. The valley between the 720 eV and the 830 eV peaks is filled and at ~915 eV
an additional maximum is visible. These features were explained by the production
of two vacancies in the Ne K-shell [5.66]. Calculated transition energies are also
given in Fig.5.15. The data show that the spectrum filling near 750 eV can be attributed to satellite lines from initial configurations $2p^n$ and the 915 eV peak is probably due to the configurations $2p^n3p$. Now, the interesting question arises why does
F^{9+} produce two K vacancies whereas F^{8+} (or F^{7+}) does not. One explanation for this
observation is that a significant fraction of K vacancy production by highly stripped
ions occurs via electron capture into the K-shell of the projectile [5.67-70]. For
F^{9+}, two electrons may be captured into the projectile K-shell, whereas for F^{8+} the
transfer of only one electron is possible. Hence, the F^{9+} data suggests that electron capture is an important mechanism in K-shell excitation collisions.

5.4.2 Direct Excitation

It is common practice to use the term *direct excitation* in cases where the excitation
mechanism can be described within the framework of the Born approximation or the
binary encounter theory. Hence, direct excitation of an inner-shell electron is caused
by light particles incident at a velocity higher than the orbital electron velocity.
However, often, the term *direct excitation* is also used for collisions where the incident velocity is relatively low and/or the projectile is rather heavy. Such desig-
nation may be justified when an electron from a deep inner-shell is excited to a
highly lying bound state or into the continuum. Then, the corresponding transition
probability is usually small and the related theoretical treatments yield expressions
similar to those resulting from the Born approximations. Here, two limiting cases
are important. First, for sufficiently light projectiles and sufficiently high velocities, the collision partners essentially retain their atomic character during the
collision. The distortion of the atomic wave functions may be accounted for by certain corrections of the ordinary PWBA [5.55,71,72], as discussed further below. Second, for extremely low and near symmetric collisions, the atoms have to interpene-
trate each other deeply to produce electron transitions to highly lying states. As
the system is close to the united atom limit, the electron wave functions are again

atomic in character. For this case BRIGGS [5.53] derived formulas which are equivalent to the ordinary PWBA results. The only significant difference arising from the two theoretical approaches is that the separated atom wave functions in (5.5) are replaced by the united atom (UA) wave functions. Similarly, FOSTER et al. [5.73] suggested to use the UA binding energies in the BEA formulas. As the projectile approaches the target atom, the binding energies either increase or decrease. (Note that these effects cause electron promotion or demotion in the MO model, see also below). Accordingly, corresponding K-excitation cross sections decrease and increase, respectively. In this subsection the effect of increased binding will be discussed to some extent.

Figure 5.16 shows cross sections for K-shell excitation of CH_4, N_2, and Ne by 50- to 600-keV H^+ impact as measured by STOLTERFOHT and SCHNEIDER [5.74]. The data were obtained from Auger spectra integrated over electron energy and emission angle. The data agree well with similar results by TOBUREN [5.11,75] and WATSON and TOBUREN [5.76] who measured Auger electron ejection for incident energies from 0.3 to 2 MeV. In Fig.5.16 the scaling procedure by means of the K binding energy I_K follows from the binary encounter calculations by GARCIA and co-workers [5.29,77] whose results are also given for comparison. It is seen that the theoretical prediction and the experimental result agree within 20% at the curve maximum. This deviation is nearly within the experimental uncertainties of the absolute cross sections. It should be noted that the incident velocity is equal to the electron orbital velocity at the curve maximum. The data show that the binary encounter theory is well suited to de-

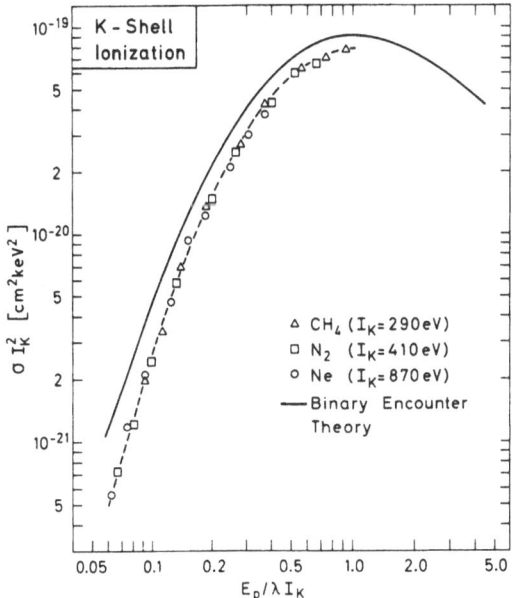

Fig. 5.16. Scaled cross sections for K-shell ionization of carbon, nitrogen, and neon. Scaling parameters are the K binding energy I_K and the mass ratio λ for protons and electrons. Experimental data are from STOLTERFOHT and SCHNEIDER [5.74]. The binary encounter calculations are from GARCIA [5.77]

scribe inner-shell excitation of light atoms by proton impact at velocities equal
to or even slightly lower than the orbital velocity of the relevant electrons. Similar results were found for the Born approximation [5.76].

With decreasing collision energy, discrepancies become significant, and at the
low-energy limit experiment and theory deviate by nearly a factor of 2. This may be
explained by distortion of the target electron shells through the field of the incident proton. BRANDT et al. [5.71] pointed out that the slow projectile being close
to the target nucleus has the effect of increasing the binding energy of the K-shell
electron. Indeed, examination of the corresponding MO correlation diagrams shows that
the target K-shell electrons are demoted. (Note that at the low-energy limit the incident velocity is smaller by a factor of 3 than the velocity of the K electron and
formation of MO's are expected). The effect of the increased binding is that the
probability for excitation of the K-shell electron decreases, in qualitative agreement with the data in Fig.5.16.

The increased binding of the K-electron in slow collisions may quantitatively be
accounted for by correction to the PWBA as shown by BRANDT et al. [5.71] and BASBAS
et al. [5.55,72]. In Fig.5.17, theoretical data are plotted along with data from
Auger measurements using helium impact for which the binding effect is expected to
be even larger than for protons. The helium data are mormalized to corresponding

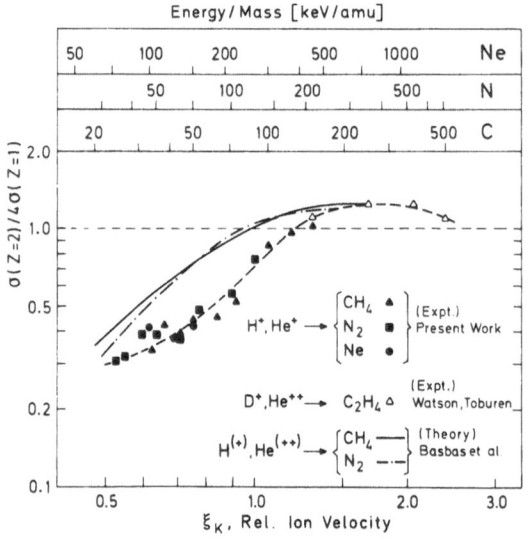

Fig. 5.17. Ratios of K-shell excitation cross sections for incident helium to those
for incident protons of equal velocity. The reduced quantity ξ_K gives essentially
the projectile velocity normalized by the orbital velocity of the K-electrons [5.55].
Projectile energies per amu are also given for the targets Ne, N_2, and CH_4. The experimental results are from STOLTERFOHT and SCHNEIDER [5.74] and WATSON and TOBUREN
[5.76]. The calculations are from BASBAS et al. [5.55]

cross sections for protons of velocity equal to the helium velocity. In this representation, the PWBA as well as the BEA predicts cross section ratios equal to 1. However, the results from the modified PWBA theory [5.55] are seen to be significantly smaller than 1 at the low-energy limit. The same is true for the experimental data; however, there are remaining discrepancies between experiment and modified PWBA theory. The decrease of the cross section ratios clearly indicates the effect of the increased binding or demotion of the K electrons.

With increase of the projectile velocity, the cross section ratio approaches 1, and at velocities larger than the electron orbital velocity this ratio even exceeds 1. The latter observation is again explained by distortion of the K electron wave function [5.55]. However, at higher velocities, K-shell excitation is preferentially produced in glancing collisions. In this case, polarization of the K electron is more likely rather than binding effects. Polarization may be described by higher order terms of the Born expansion which constitute a power series in Z_1 [5.55]. Interference between the first and the second order terms produces a contribution which is proportional to Z_1^3. When such a term is added, good agreement between experiment and theory is achieved as seen at high incident velocities (see Fig.5.17).

When projectiles heavier than protons or helium are used, strong deviations from the Z_1^2 dependence of the inner-shell excitation cross sections are expected. Using the method of x-ray detection, MACDONALD et al. [5.54] studied Ar-K excitation by bare projectiles with Z_1 ranging from 1 to 9. The velocity of the incident particles was chosen to be slightly lower than the orbital velocity of the Ar-K electron. At the high Z_1 limit the Ar-K x-ray production cross sections were found to be higher by a factor of 4 than expected from a Z_1^2 dependence. Also, MACDONALD et al. [5.67] and MOWAT et al. [5.68] measured x-ray production in 35-MeV F^{n+}+Ar and 55-MeV Ar^{n+}+Ne collisions, respectively. The results indicate a strong, nearly exponential, increase of the target x-ray production cross sections as the charge state of the projectile increases. When the projectile is highly stripped so that K electrons are missing, the x-ray production cross sections were found to be particularly large. In this case, it was suggested that capture of target K electrons into the projectile K-shell plays a significant role [5.67-70].

The x-ray production cross section is given by $\sigma_X = \bar{\omega}_K \sigma_K$ where σ_K is the K excitation cross section and $\bar{\omega}_K$ is the corresponding mean fluorescence yield. Hence, x-ray production is not only determined by the produced number of inner-shell vacancies but it is also dependent on the fluorescence yield. Variations of the mean fluorescence yield are expected, when a varying number of outer-shell vacancies is produced in the collision [5.59,60,78,79]. To verify this possibility, BURCH et al. [5.64] measured simultaneously the cross sections σ_X and σ_A for production of, respectively, x-rays and Auger electrons in 50-MeV Cl^{n+}+Ne collisions. The results, showing $\sigma_K = \sigma_X + \sigma_A$ and $\bar{\omega}_K = \sigma_X/\sigma_K$ as a function of the projectile charge state n, are plotted in Fig.5.18. It is seen that the variation of the mean fluorescence yield is rather

182

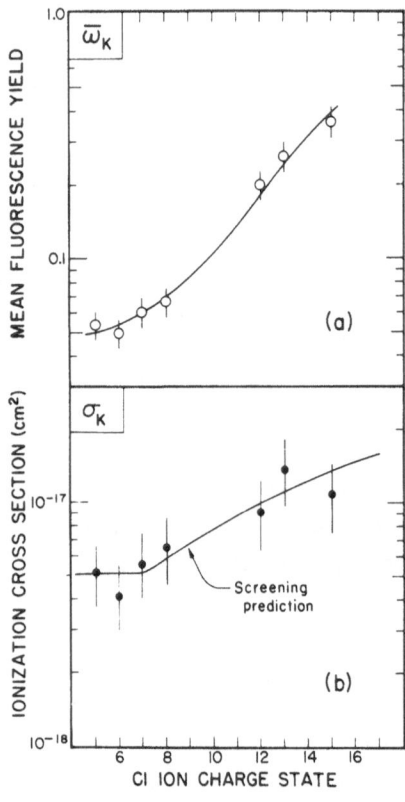

Fig. 5.18a and b. Projectile charge state dependence of the mean Ne K fluorescence yield $\bar{\omega}_K$ and Ne K excitation cross section σ_K in 50-MeV Cl^{n+}+Ne collisions. The solid curve in (b) is obtained assuming screening of the projectile nuclear charge (see text) (from BURCH et al. [5.64])

strong. This may be explained by an increasing number of Ne L vacancies produced by projectiles with increasing charge state. Hence, care must be taken with the variation of the fluorescence yield, when conclusions for projectile charge state effects on inner-shell excitation are drawn from x-ray production cross sections.

From the previous discussion, it follows that there are several effects which produce variations of the K excitation cross section with varying incident charge, i.e., 1) screening of the incident nuclear charge, 2) polarization of the target K electron, and 3) capture of a K electron. It is noted that 2) is related to 1), as the increase of the target polarization is produced by the decrease of the projectile screening. To verify 1), in Fig.5.18 the Cl^{n+}+Ne data are fitted by the expression $\sigma_K = \sigma_0(Z_1 - \alpha m_{KL})^2/Z_1^2$ where σ_0 is a constant, m_{KL} is the number of K and L electrons on the projectile and α is an effective screening number for the incident nuclear charge [5.64]. The fit yielded the value $\alpha=0.75$ which appears to be larger than the screening number one would expect for the Cl-L electrons. This may indicate that polarization of the target K electrons and/or their capture into the L-shell of the projectile play a certain role. It should be noted that the Cl ion with the maximum charge state of 15 has still a filled K-shell so that capture of target electrons into the projectile K-shell is not possible.

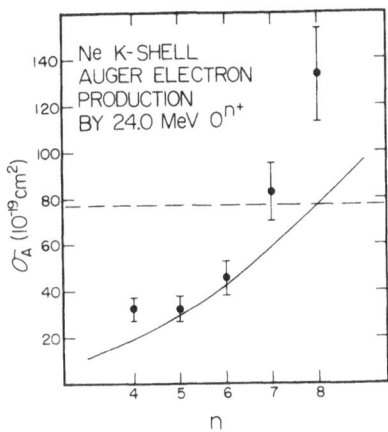

Fig. 5.19. Projectile charge-state dependence of the production cross section σ_A for Ne K Auger electrons in 24-MeV O^{n+}+Ne collisions. The broken curve is equal to $Z_1^2\sigma_p$ where Z_1 is the projectile nuclear charge and σ_p is the Ne K excitation cross section for 1.5-MeV H^+. The full curve is $n^2\sigma_p$ (from WOODS et al. [5.21])

Auger electron production following K-shell excitation by completely stripped projectiles were studied by STOLTERFOHT et al. [5.22] in 30-MeV $O^{n+}+O_2$ collisions (see Fig.5.3). Similar experiments were made by WOODS et al. [5.21] bombarding 24-MeV O^{n+} and F^{n+} on Ne. Figure 5.19 shows production cross sections σ_A for Ne K Auger electrons by O^{n+} impact as a function of the projectile charge state n [5.21]. The σ_A are assumed to be nearly equal to the corresponding Ne K excitation cross sections. In Fig.5.19 are also shown $n^2\sigma_p$ (solid line) and $Z_1^2\sigma_p$ (broken line) where σ_p is the Ne K excitation cross section for 1.5-MeV H^+ which has the same velocity as 24-MeV O^{n+}. The two curves represent lower and upper limit predictions for the cross sections if the latter scale with the square of the incident charge. (The two cases refer to the maximum and minimum screening of the incident nuclear charge). It is seen that the data measured with fully stripped projectiles exceed even the upper limit prediction of the cross sections. In this case, capture of the target K electron may considerably contribute to the K vacancy production mechanisms. However, polarization effects also tend to increase the K excitation cross section in the energy range studied. It appears that further studies are needed to verify separately the influences of the capture process and the polarization effect.

5.4.3 Electron Promotion

In sufficiently slow collisions the electrons tend to adjust their orbital energies adiabatically to the motion of the two collision partners. Thus, a quasi-molecule is transiently formed during the collision. The quasi-molecular model was used by FANO and LICHTEN [5.15] to describe inner-shell excitation in slow ion-atom collisions. The major features of the Fano-Lichten model are [5.80]: 1) Electrons occupy molecular orbitals (MO) of a diatomic system and they act independently from each other. An electron feels the other electrons only by their screening of the two nuclei involved in the collision. 2) The MO's of a many-electron system correlate the same separated atom (SA) and united atom (UA) levels as the MO's of H_2^+. 3) The principal quantum number of some MO's increases as the system moves from the SA limit to the UA limit. This phenomenon is called electron promotion, and it is the outstanding feature of the quasi-molecular model.

Condition 1) requires that the electron-electron interaction is small. This assumption appears to be well justified for inner-shell electrons which are primarily influenced by the Coulomb fields of the nuclei. In contrast, significant electron exchange forces arising during interpenetration of the atomic electron clouds were postulated by BRANDT and LAUBERT [5.81] to interpret inner-shell excitation cross sections which strongly exceed the PWBA prediction. As the electron exchange forces are closely related to the Pauli exclusion principle, the mechanism suggested by BRANDT and LAUBERT was denoted as *Pauli excitation* [5.81]. In the Fano-Lichten model, however, there is no need for postulating such strong influence of the electron--electron interaction; high probabilities for inner-shell excitation are accounted for by electron promotion and successive transitions to higher-lying orbitals. Electron promotion, in turn, is primarily produced by the varying two-center field of the moving nuclei. Moreover, the number of electrons transferred into the promoted MO is determined by statistical rules. In this picture, the Pauli principle is operative only if the higher-lying MO's are occupied so that transitions of the promoted electron are not possible. Hence, it appears that the Pauli principle has the effect of inhibiting rather than of initiating inner-shell excitation [5.80]. To illustrate this, consider a slow Ne^{10+} projectiles (carrying no electron cloud) on, say, oxygen. The Fano-Lichten model predicts that Ne^{10+} causes significant target K excitation via electron promotion. Although such an experiment has not as yet been carried out, theoretical considerations using the MO model [5.80] strongly suggest that a slow Ne^{10+} ion produces target K excitation more effectively than a neutral Ne of equal velocity. Similarly, for the extreme case of $Ne^{10+}+Ne^{9+}$ collisions, it is expected that 1s electron excitation is more probable than that in Ne+Ne collisions.

Condition 2) is also related to the dominance of the electron-nuclei interaction. Wave functions of inner-shell electrons are hydrogenic in nature and, consequently, they are similar to (adequately scaled) wave functions of H_2^+. However, there are certain differences between MO's of H_2^+ and a many-electron system. In the latter case, adiabatic MO's of equal symmetry do not cross [5.82]. The avoided crossings often complicate the theoretical treatment of the excitation process. Therefore, it is useful to look for diabatic MO's whose energy curves do cross. Due to specific symmetries of the pure two-center Coulomb field [5.83], the MO energy curves of H_2^+ do not exhibit avoided crossings. Thus, H_2^+ may serve as a useful guide to find adequately diabatic MO's. For instance, the H_2^+ system was used to find the rules for construction of the MO correlation diagrams for symmetric systems [5.15]. The correlation rules were extended to asymmetric systems by BARAT and LICHTEN [5.28] (see also EICHLER et al. [5.84]).

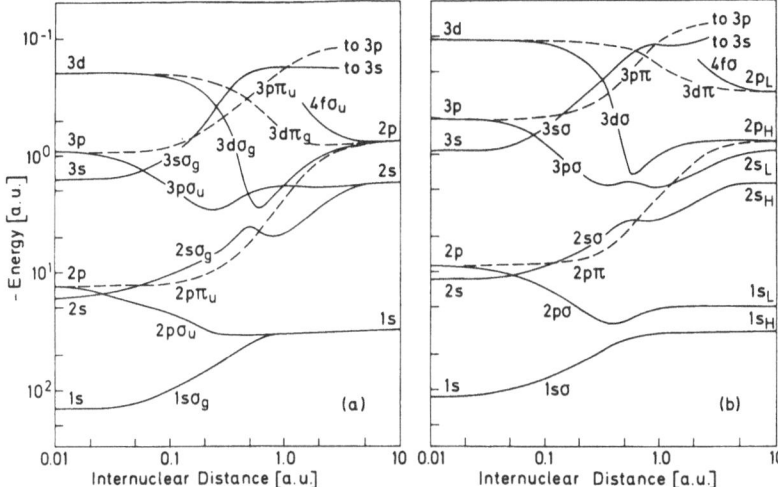

Fig. 5.20a and b. Diabatic MO energy diagrams for the systems Ne+Ne (a) and Ne+O (b).
Some higher-lying MO's are free-hand estimations. The Ne+Ne data are from LARKINS
[5.85] and the Ne+O are from TAULBJERG and BRIGGS [5.86]

Figure 5.20 shows examples for diabatic MO diagrams representing the symmetric
system Ne+Ne and the near symmetric system Ne+O. Lower-lying MO's were obtained from
HF calculations by LARKINS [5.85] and TAULBJERG and BRIGGS [5.86]. Some higher-lying
orbitals were estimated to obtain diabatic MO's. The diagrams clearly indicate the
phenomenon of promotion, e.g., see the $2p\sigma$ and the $3d\sigma$ MO's. Also, it is seen that
for asymmetric collisions the levels of the lighter partner correlate with the promoted MO's.

According to the MO diagrams, two successive mechanisms are responsible for K-
vacancy production in slow ion-atom collisions [5.87]. In the incoming part of the
collision, vacancies may be transferred into the $2p\pi_x$ MO. The $2p\pi_x$ vacancy occupation number is denoted by N_π. Furthermore, at small internuclear distances, vacancies
in the $2p\pi_x$ MO may be transferred to the $2p\sigma$ MO by a rotational coupling mechanism.
This process is characterized by σ_{rot}, the rotational coupling cross section. Under
the assumption that the two processes do not interfere, it follows that

$$\sigma_K = N_\pi \sigma_{rot} \tag{5.10}$$

where σ_K is the corresponding K excitation cross section.

Usually, σ_{rot} is calculated treating the projectile as a classical particle. The
probability for $2p\pi$-$2p\sigma$ rotational coupling transfer is determined and integrated
over the impact parameter. For symmetric collisions, the $2p\pi_u$ and $2p\sigma_u$ MO's do not
couple with the $1s\sigma_g$ and $2s\sigma_g$ MO's (because of different u-g symmetries) so that
the analysis may be based on a two-state theory. Such calculations were made by

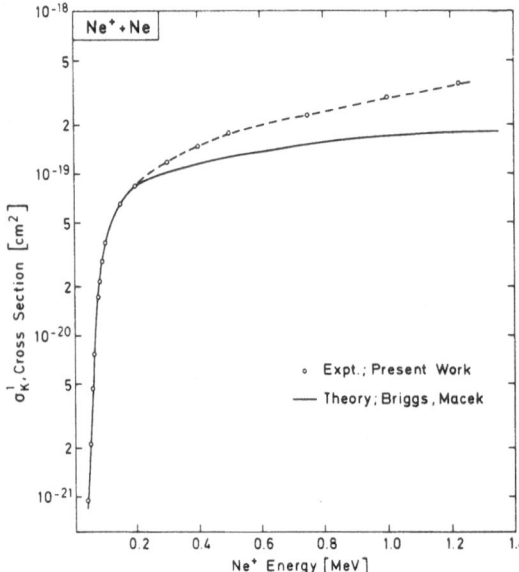

Fig. 5.21. Total cross sections σ_K^1 for K excitation in Ne⁺+Ne collisions as a function of the projectile energy. Experimental data were measured at Aarhus [5.34]. Calculations for incident energies less and greater than 200 keV are from BRIGGS [5.80] and BRIGGS and MACEK [5.88], respectively

BRIGGS and MACEK [5.88] for deuteron-deuteron collisions. A useful feature of the theory is that the results may be scaled to heavier collision systems, such as Ne⁺+Ne [5.88]. Direct calculations for Ne⁺+Ne were made by BRIGGS [5.80] who used also a screened Coulomb potential to obtain accurate particle trajectories.

In Fig.5.21 the theoretical results of BRIGGS and MACEK [5.88] and BRIGGS [5.80] are compared with Ne⁺+Ne K excitation cross sections σ_K measured at Aarhus using the method of Auger spectroscopy [5.34]. To convert the σ_{rot} to σ_K, the $2p\pi_x$ vacancy occupation number $N_\pi=1/6$ was used as discussed below. The absolute cross sections show excellent agreement (to within ±10%!) near the K ionization threshold. This observation gives good confirmation of the two-state calculation of σ_{rot} as well as for the $2p\pi_x$ vacancy occupation number $N_\pi=1/6$. However, at energies above ~200 keV the experimental data are seen to be systematically higher than the theoretical results. This deviation is primarily due to the fact that N_π is not a constant.

In the incoming part of the collision there are two processes which produce vacancies in the $2p\pi_x$ MO. First, available 2p vacancies are statistically distributed among the MO's correlating with the 2p level. This process produces a static contribution N_0 to the $2p\pi_x$ vacancy occupation number. MACEK and BRIGGS [5.89] calculated that $N_0=n/6$ and $N_0=n/3$ for symmetric and asymmetric collision systems, respectively, where n is the number of vacancies in the lower-lying 2p level prior to collision. Second, as pointed out by FASTRUP et al. [5.90], couplings to higher empty or partially filled MO's give rise to a velocity-dependent contribution N_v. (Hereafter, the projectile velocity is denoted as v). Assuming in first order that the two processes are independent, it follows that

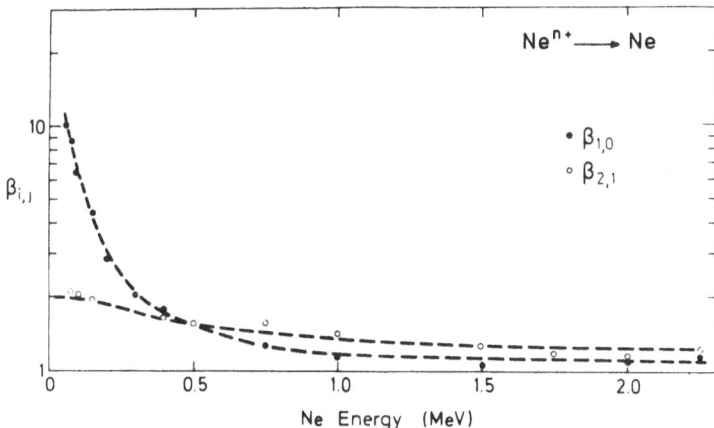

Fig. 5.22. Cross section ratio $\beta_{1,0} = \sigma_K^1/\sigma_K^0$ and $\beta_{2,1} = \sigma_K^2/\sigma_K^1$ as a function of the incident particle energy. σ_K^n denotes the K excitation cross sections in Ne^{n+}+Ne collisions. The data were measured at Aarhus [5.34]

$$N_\pi = N_0 + N_v \tag{5.11}$$

with $N_v \to 0$ for $v \to 0$. Above, it was noted that in Fig.5.21 the theoretical σ_K were obtained by setting $N_\pi = N_0 = 1/6$. Hence, the discrepancies between experiment and theory are probably due to the neglect of N_v.

More information about the $2p\pi_x$ vacancy occupation number can be obtained as the incident charge state is varied. Figure 5.22 shows ratios of cross sections σ_K^n for K excitation in Ne^{n+}+Ne collisions as measured at Aarhus [5.34] for n=0, 1, and 2. At the low-energy limit, it follows that $N_\pi = N_0$. Hence, at small energies it is expected that $\beta_{2,1} = \sigma_K^2/\sigma_K^1 = 2$. This number is in excellent agreement with the experiment, as seen in Fig.5.22. Similarly, it is expected that $\beta_{1,0} = \sigma_K^1/\sigma_K^0$ tends to infinity as the incident energy decreases. Fig.5.22 shows that $\beta_{1,0} = 10$, indicating that σ_K^1 is an order of magnitude higher than σ_K^0. This demonstrates that at K ionization threshold the cross sections strongly depend on the incident charge state. For increasing energy, however, the velocity-dependent quantity N_v gains relative importance as seen by the energy variations of $\beta_{2,1}$ and $\beta_{1,0}$. It is noted that the increase of N_v has the effect that the cross sections σ_K^2, σ_K^1, and σ_K^0 approach each other at the high-energy limit [5.34].

For asymmetric collisions the $2p\sigma$ MO correlates with the higher-lying 1s level (see Fig.5.20). Hence, K vacancies are preferentially produced in the lighter collision partner. However, there is a finite probability for vacancy transfer to the K-shell of the heavier partner in the outgoing part of the collision [5.91]. This is due to the fact that for asymmetric collisions the system loses the u-g symmetry (Fig.5.20) and, consequently, the $2p\sigma$ and $2p\pi$ orbitals may couple with the $1s\sigma$ and the $2s\sigma$ MO's. Hence, in principle, the theoretical treatment of the K excitation

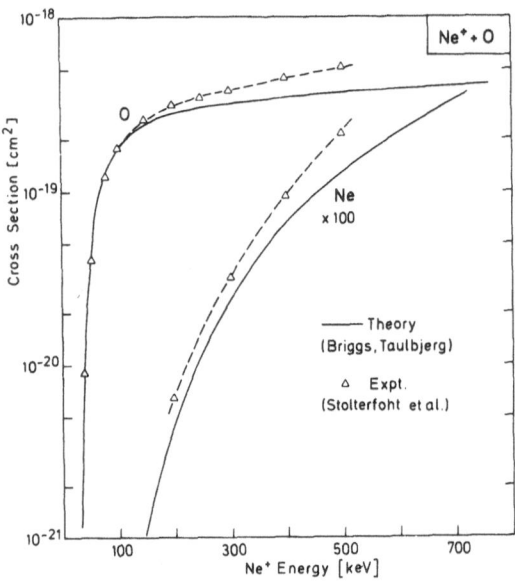

Fig. 5.23. Experimental cross sections for K excitation of oxygen and
neon in Ne^++O_2 collisions in comparison with corresponding theoretical
data. Theoretical and experimental
results are from BRIGGS and TAULBJERG
[5.92] and STOLTERFOHT and co-workers
[5.93,94], respectively

requires a four-state calculation. For Ne+O, such calculations were made by BRIGGS
and TAULBJERG [5.92] using the Hartree-Fock (HF) method. A complication for the
asymmetric system is the occurence of an avoided crossing between the adiabatic MO's
"$2p\sigma$" and "$2s\sigma$" at ~0.05 a.u. Thus, some effort was made to construct proper diabatic
orbitals [5.86]. The theoretical results are given in Fig.5.23.

Figure 5.23 also shows cross sections for K excitation of both collision partners
for Ne^+ impact on O_2 as deduced by STOLTERFOHT and co-workers [5.93,94] from Auger
measurements. It is seen that the K vacancy production cross sections for neon are
orders of magnitude smaller than those for oxygen. This indicates that the vacancy
population of the $1s\sigma$ MO is relatively weak. Similar results were found for the $2s\sigma$
MO [5.92]. Figure 5.23 indicates good agreement between theory and experiment at
the ionization threshold, whereas deviations occur at higher energies. The theoretical data were obtained using $N_\pi=N_0=1/3$. Hence, discrepancies between theory and
experiment are most probably caused by the neglect of N_v as mentioned already for
$Ne^{n+}+Ne$ collisions.

It was noted above that the cross section σ_{rot} for symmetric systems can be scaled.
Similarly, TAULBJERG et al. [5.95] presented a universal curve and corresponding
scaling rules from which σ_{rot} can be obtained for any heteronuclear system. The analysis is based on the one-electron model calculations by TAULBJERG et al. [5.96],
see also PIACENTINI and SALIN [5.97]. In the calculations, couplings to the $1s\sigma$ and
the $2s\sigma$ MO's were neglected, i.e., a two-state approximation was made. Hence, the
results must be compared with $\sigma_K=\sigma_K(L)+\sigma_K(H)$ where $\sigma_K(L)$ and $\sigma_K(H)$ are K excitation
cross sections for the lighter and heavier collision partner, respectively. [Gener-

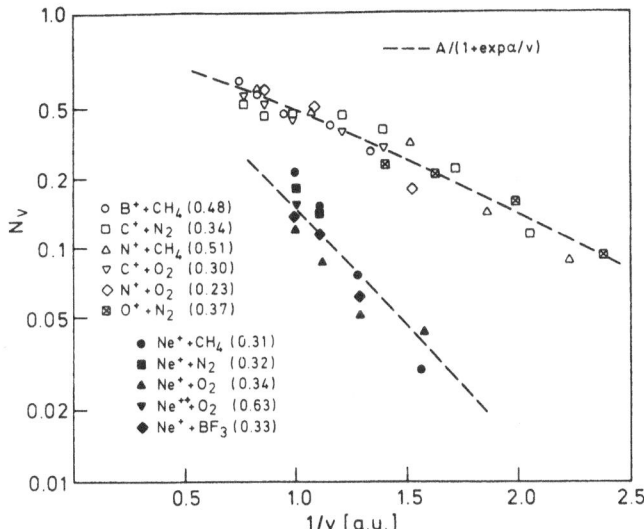

Fig. 5.24. The dynamic $2p\pi_x$ vacancy occupation number N_V as a function of the inverse projectile velocity v. The static N_0 are given as numbers for each collision system. The dashed curves are fits using $A/(1+epx\alpha/v)$ with A and α given in the text (from STOLTERFOHT and SCHNEIDER [5.94])

ally, $\sigma_K(H)$ is small so that $\sigma_K\approx\sigma_K(L)]$. The scaling results were found to be in good agreement (±15%) with previous HF calculations [5.92,98]. This agreement gives some confidence that σ_{rot} can be deduced for other (light) collision systems with relatively high accuracy. Hence, the remaining problem in deducing the K excitation cross sections is the proper determination of N_π. Unfortunately, no calculations are available for N_π, at present.

Semi-empirical N_π may be derived by combining experimental σ_K and theoretical σ_{rot}. Such an analysis was made by FASTRUP et al. [5.99] for $N_e^{n+}+N_2$ and by STOLTER-FOHT et al. [5.34] for $Ne^{n+}+Ne$. A more comprehensive study of N_π for 11 collision systems was made by STOLTERFOHT and SCHNEIDER [5.94], whose results are plotted in Fig.5.24. The data were obtained by combining $\sigma_K=\sigma_K(H)+\sigma_K(L)$ from Auger measurements [5.94] and scaled σ_{rot} [5.95]. The N_0 are deduced from zero-velocity extrapolation of N_π, and the N_V are calculated from $N_V=N_\pi-N_0$. In Fig.5.24 the N_0 are given as numbers for each collision system and N_V is shown in a semi-logarithmic plot versus 1/v. This type of plot is useful, as N_V is expected to be given by

$$N_V = A/(1+exp\alpha/v) \tag{5.12}$$

where A and α are constants. Such an expression was proposed by BARAT and LICHTEN [5.28] who considered mixing between noncrossing MO's. A similar formula was deduced by MEYERHOF [5.91] analyzing the radial coupling between the $1s\sigma$ and $2p\sigma$ MO's(see also below). In the present case, vacancy transfer to the $2p\pi_x$ MO is expected to be produced by radial coupling of the $2p\pi$ and $3d\pi$ MO's (Fig.5.20).

In Fig.5.24 the data are found to be centered around two curves given by (5.12) with the fit parameters $A=1.7$ (1.9) and $\alpha=1.2$ (2.5) for the upper (lower) curve [5.94]. It is seen that the lower-lying data belong to collision systems involving Ne^+ impact, whereas the upper-lying data are due to systems with atoms lighter than Ne. Qualitatively this may be explained by the fact that the 2p electron of Ne^+ has a particularly large binding energy (~42 eV). It is expected from Meyerhof's formalism (see below) that N_v strongly decreases as the energy gap between the $2p\pi$ and $3d\pi$ MO's increases at large internuclear distances. This gap is typically 25 eV for systems involving Ne^+, whereas it is roughly 10 eV for the other systems studied.

It was noted above that $N_0=n/3$ where n is the number of precollision vacancies in the lower-lying 2p level. For collisions involving Ne^+ impact, it follows that $N_0=1/3$ in good agreement with experiment (see Fig.5.24). However, it appears that N_0 for the lighter systems cannot be determined simply from the n/3 expression. Although the molecular nature of the target particles complicates the accurate determination of n, the data suggest that N_0 is smaller than expected from the n/3 rule. For instance, O^+ carries three vacancies into the collision, whereas N_0 is found to be only 0.39 for $O^+ + N_2$. Here, further studies are required to clarify the situation.

5.4.4 Vacancy Sharing

The K vacancy productions in both collision partners are similarly affected by the $2p\pi_x$ vacancy occupation. The K excitation cross section for the lighter particle is proportional to N_π [see (5.10)]. The same is true for the K excitation cross section of the heavier particle. Hence, the deviations between theory and experiment are expected to be similar for the two collision partners, when inaccurate N_π values are used in the calculations of the cross sections. This can clearly be seen by comparison of the data in Fig.5.23. The reason for the similar behavior of the cross sections is that the $2p\pi$-$2p\sigma$ and the $2p\sigma$-$1s\sigma$ vacancy transfers are greatly independent processes [5.92]. The $2p\pi$-$2p\sigma$ rotational coupling takes place near the distance of closest approach, whereas the $2p\sigma$-$1s\sigma$ radial coupling is operative at intermediate internuclear distances in the outgoing part of the collision. The latter coupling may be denoted as a vacancy-sharing mechanism [5.91], as the distribution of the vacancies among the $2p\sigma$ and the $1s\sigma$ MO's is usually independent of the mode of creation of these vacancies. Hence, problems arising from the uncertainties of the N_π values are avoided, when cross section ratios rather than absolute values are considered. Indeed, BRIGGS and TAULBJERG [5.92] found good agreement between the experimental sharing ratio $S=\sigma_K(H)/\sigma_K(L)$ for Ne+O and the four-state HF calculations given in Fig.5.23. Similar results were obtained by one-electron model calculations using both a three-state ($2p\pi$, $2p\sigma$, $1s\sigma$) and a two-state ($2p\sigma$, $1s\sigma$) approximation [5.96].

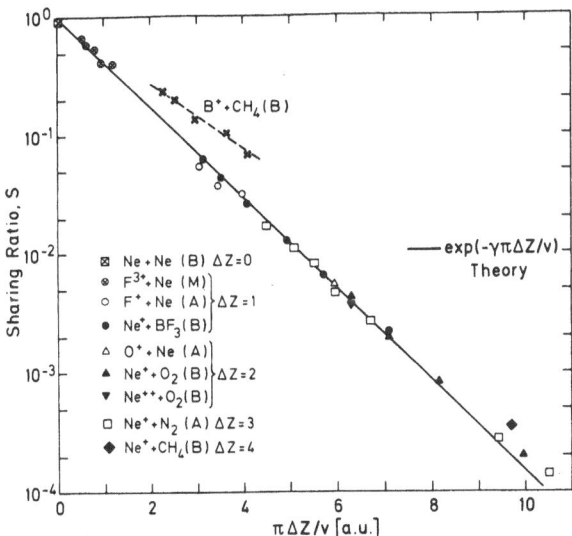

Fig. 5.25. Ratio S for K vacancy sharing as a function of $\pi\Delta Z/v$ where v is the projectile velocity and ΔZ is the difference in the atomic numbers of the collision partners. Labels A, B, and M refer to measurements at Aarhus by FASTRUP et al. [5.103], at Berlin by STOLTERFOHT et al. [5.104], and at Manhattan by WOODS et al. [5.105]. The theoretical results with $\gamma=0.87$ a.u. are obtained using the formalism by MEYER-HOF [5.91]

MEYERHOF [5.91] was first to describe theoretically the K vacancy sharing process. The theoretical results were found to be in good agreement with a large number of experimental data. The analysis is based on the charge-transfer model by DEMKOV [5.100] (see als NIKITIN [5.101]). The derivation yields the simple formula for the sharing ratio

$$S = \exp(-\alpha/v) \tag{5.13}$$

with

$$\alpha = \pi\left(\sqrt{2I_H}-\sqrt{2I_L}\right) \tag{5.13a}$$

where I_H and I_L are the K binding energies of the heavier and the lighter particles, respectively. [Note that (5.12) and (5.13) are equivalent]. As the K binding energy scales rather accurately with Z^2, it is useful to define the quantity γ by $\alpha=\gamma\pi\Delta Z$ where ΔZ is the difference in the atomic numbers of the heavier and the lighter particle. Thus, it follows [5.93] that

$$\dot{S} = \exp\left(-\gamma\frac{\pi\Delta Z}{v}\right) \tag{5.14}$$

with

$$\gamma = (\sqrt{2I_H}-\sqrt{2I_L})/\Delta Z \quad .$$

(5.14a)

Now, γ is nearly independent of the collision system. An analysis based on experimental K binding energies [5.102] shows that γ=0.87 a.u. to within ±2% for collision particles of Z=5 to 11.

In Fig.5.25 theoretical results are shown in a semi-logarithmic plot of S versus $\pi\Delta Z/v$. In this plot, the results from (5.14) yield a straight line with a slope given by γ. This line represents a universal curve for the sharing ratios. In Fig.5.25 the theoretical curve is compared with experimental results for several collision systems as measured by FASTRUP et al. [5.103], STOLTERFOHT et al. [5.93,104], and WOODS et al. [5.105]. It is seen that nearly all experimental data are in excellent agreement with the theoretical predictions. In particular, it is found that the exchange of the target and the projectile as well as the molecular composition of the target has no effect on the sharing ratio. (For example, compare Ne^++O_2 and O^++Ne). Furthermore, Fig.5.25 exhibits that there is no visible dependence on ΔZ left, once the data are plotted versus $\pi\Delta Z/v$.

It is noted that the calculated sharing ratio depends very critically on the related K binding energies [see (5.13)]. The K binding energies are slightly different for the free atom and the atom bound in a molecular compound. In most cases [5.104, 105] far better agreement is obtained between theory and experiment when binding energies are used for atoms rather than for molecules. This finding indicates that the target molecule generally breaks up in the K excitation collision. The same conclusion is suggested by the agreement of the experimental sharing ratios for, say, Ne^++O_2 and O^++Ne. Also, it is noted that the K binding energies are significantly affected by the electrons missing in the L-shell. In Section 5.4.1 it was pointed out that several L electrons are removed in the K excitation collision. However, surprisingly, it is found that in spite of considerable changes in I_H, I_L, and even I_H-I_L, the quantity $\sqrt{I_H}-\sqrt{I_L}$ is not significantly altered when the numbers of missing L electrons are assumed to be equal for the collision partners. Hence, it appears that the sharing ratio is rather independent of the degree of outer-shell ionization.

It should be emphasized that the derivation of Meyerhof's equation (5.13) implies several simplifying assumptions. For instance, the analysis is based on an *atomic* two-state expansion and the coupling matrix element is approximated by a simple expression depending exponentially on the internuclear distance. Thus, the excellent agreement between experimental and theoretical sharing ratios is rather noteworthy. It appears that Demkov's charge-transfer model [5.100] applies well in the description of the $2p\sigma$-$1s\sigma$ vacancy sharing. However, it should be noted that the validity of the approximations made in the model has not as yet been fully confirmed. Hence, it appears interesting to determine the limits of applicability of (5.13).

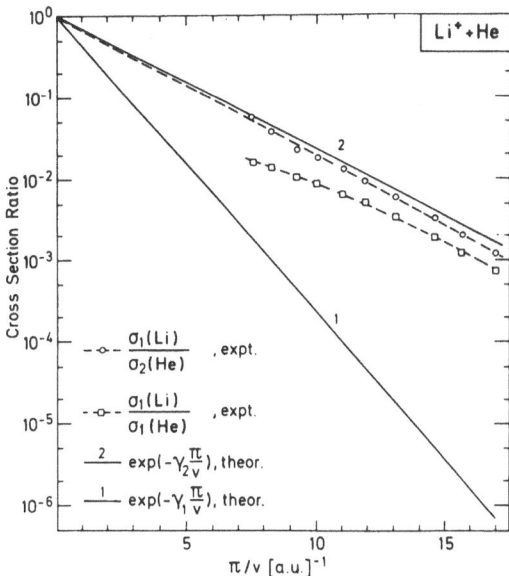

Fig. 5.26. Ratios of cross sections for 1s excitation in 6- to 30-keV Li$^+$+He collisions as a function of the inverse projectile velocity v. The σ_1(Li) and σ_1(He) denote cross section for production of, respectively, Li and He excited in the 1s
level. The σ_2(He) is the corresponding cross section for doubly excited He. The
theoretical data are calculated with γ_1=0.83 a.u. and γ_2=0.376 a.u. using the formalism by MEYERHOF [5.91]. The experimental data are from STOLTERFOHT and LEITHÄUSER
[5.106] including results from PARK et al. [5.107]

Figure 5.25 indicates systematic discrepancies between theory and experiment for
the relatively light collision system B$^+$+CH$_4$ [5.104]. These discrepancies increase
strongly for the even lighter system Li$^+$+He. Fig.5.26 shows the ratio σ_1(Li)/σ_2(He)
where σ(Li) is the cross section for single 1s excitation of Li and σ_2(He) is the
corresponding cross section for double 1s excitation of He. The data were deduced
from autoionization spectra measured by STOLTERFOHT and LEITHÄUSER [5.106] in 6- to
30-keV Li$^+$+He collisions. Also shown are σ_1(Li)/σ_1(He) where σ_1(He) is the cross
section for single 1s excitation of He as derived including results from energy loss
experiments [5.107]. Furthermore, Fig.5.26 shows the theoretical curve 1 following
from (5.13) with I_H=64.4 eV and I_L=24.5 eV, the 1s binding energies of Li and He,
respectively. Curve 1 should be compared with σ_1(Li)/σ_1(He) which, however, is seen
to be orders of magnitude higher than the theoretical values.

These discrepancies may be explained by two-electron transitions from the 2pσ MO
into the 2pπ MO [5.106]. Previous experimental studies of slow Li$^+$+He collisions
have shown that the probability of double excitation of He is relatively high [5.107-
109]. In the case of double excitation, the 2pσ MO is significantly decreased in
energy and the probability for 1sσ-2pσ electron transitions is dramatically enhanced.
For two electrons missing in the 2pσ MO, the sharing ratio may be estimated by (5.13)

using modified binding energies [5.106]. In Fig.5.26 the theoretical curve 2 is given as obtained using I_L=44.4 eV, the 1s binding energy of He(1s2p). Curve 2 should now be compared with $\sigma_1(Li)/\sigma_2(He)$ for which good agreement is achieved with the theoretical prediction. Similar results were obtained from *ab initio* HF calculations by SIDIS et al. [5.110]. This shows that Meyerhof's formalism needs revisions with respect to the possible double excitation of the 2pσ electrons. For example, the B^++CH_4 data in Fig.5.25 may well be reproduced by a revised formula when a fraction of ~0.15 double 2pσ excitation is adopted [5.106].

It should be emphasized that the proposed transition from the 1sσ MO to the completely empty 2pσ MO corresponds to a one-electron transition which might still be described within the framework of the independent electron MO model. However, to account for the enhanced transition probability, it is required to consider dynamic alterations of the MO's during the collision. Changes of the MO's are important if the collision is sufficiently adiabatic so that the electrons are able to readjust in the field resulting from the new configuration produced during the collision. Hereafter, this process will be referred to as *dynamic MO relaxation*. The dynamic relaxation results in modifications of the MO energy curves at a certain stage of the collision. For instance, to account for the 2pσ-2pπ transitions it is necessary to use MO diagrams different for the incoming and outgoing part of the collision. Such complications are avoided when state diagrams rather than MO diagrams are applied.

Figure 5.27 shows a simplified state diagram for the (LiHe)$^+$ system [5.106]. Hereafter, the states $(1s\sigma)^2(2p\sigma)^2\Sigma$, $(1s\sigma)^2(2p\sigma)(2p\pi)\Pi$, $(1s\sigma)^2(2p\pi)^2\Delta$, $(1s\sigma)(2p\pi)^2(2p\pi)\pi$, and $(1s\sigma)(2p\sigma)(2p\pi)^2\Delta$ are abbreviated by Σ, Π, Δ, KΠ, and KΔ, respectively. The Li-K excitation following single excitation of the 2pσ MO is represented by the process Σ-Π-KΠ, whereas Li-K excitation following double vacancy production in the 2pσ MO is described by Σ-Π-Δ-KΔ. It is seen that at large internuclear distances the energy gap between the states Δ and KΔ is smaller by a factor of 2 than the spacing between the states Π and KΠ. From this it is evident that the probability for 1sσ-2pσ transition corresponding to the Δ-KΔ process is significantly higher than that for Π-KΠ.

It should be noted that Li-K vacancies may also be produced by Δ-KΠ transitions. A similar process was introduced by AFROSIMOV et al. [5.111] to explain N-K excitation in slow N^++Ar collisions. This process was justified by the magnitude and energy dependence of the vacancy sharing ratio between the N-K and Ar-L shells. For instance, the experimental sharing ratio was found to be higher than the theoretical prediction resulting from MEYERHOF's formalism. It is noted that the mechanism considered by AFROSIMOV et al. [5.111] implies double electron excitation for which the process of *dynamic MO relaxation* is expected to play a certain role. Unfortunately, such considerations were not incorporated in the analysis of the N^++Ar data. However, a part from the *MO relaxation*, the mechanism proposed by AFROSIMOV et al. [5.111] gives rise to interesting questions. It should be emphasized that the Δ-KΠ process involves a two-electron transition which can only be accounted for by electron-correlation effects. If such effects are found to be important, the MO picture is no

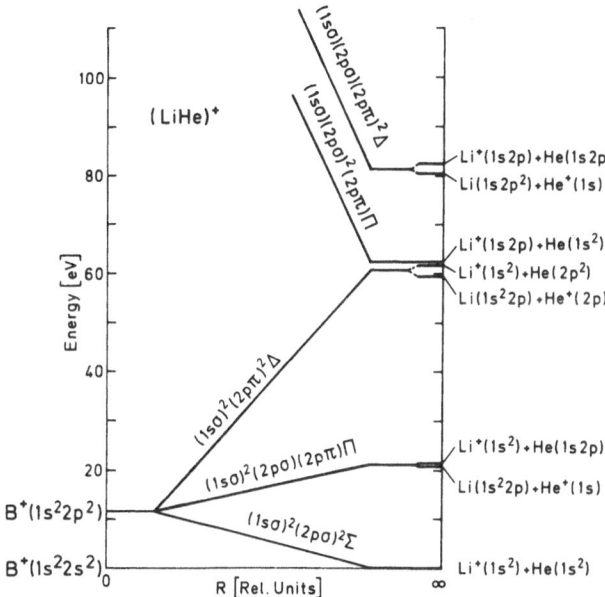

Fig. 5.27. Schematic molecular state diagram for the system (LiHe)$^+$. R is the inter-nuclear distance between the collision partners (from STOLTERFOHT and LEITHÄUSER [5.106])

longer valid and, consequently, many-electron models have to be used. For the future, it would be an interesting prospect to study the role of electron correlation in more detail, as such an analysis reveals information about the validity of the MO model.

Finally, in this section, the mechanisms of direct excitation and electron promotion are discussed by means of the collision system F+Ne for which K excitation cross sections were measured in a wide range of incident energies. Using the method of Auger spectroscopy, STOLTERFOHT and SCHNEIDER [5.94] determined the cross sections $\sigma_K(F)$ and $\sigma_K(Ne)$ for K excitation of, respectively, F and Ne in 0.1- to 0.5-MeV Ne$^+$+BF$_3$ collisions. Similar measurements were made by WOODS et al. [5.105] for 3-15 MeV F^{3+}+Ne. For the following rather heuristic discussion it is suggested to neglect possible effects arising from the molecular nature of the target particle BF$_3$ and from the interchange of target atom and projectile. In Fig.5.28 the cross section $\sigma_K=\sigma_K(F)+\sigma_K(Ne)$ is shown as a function of the incident velocity. The experimental data are compared with theoretical results following from scaled MO calculations (MOA) with $N_\pi=N_0=1/3$ [5.95]. Also shown are scaled binary encounter results (BEA) [5.77] as obtained by calculating separately $\sigma_K(F)$ and $\sigma_K(Ne)$. The effective nuclear charges $Z_{eff}=4.0$ and 3.6 were chosen for, respectively, F and Ne to obtain agreement with the experimental $\sigma_K(F)$ and $\sigma_K(Ne)$ for the incident velocity of 5.6 a.u.

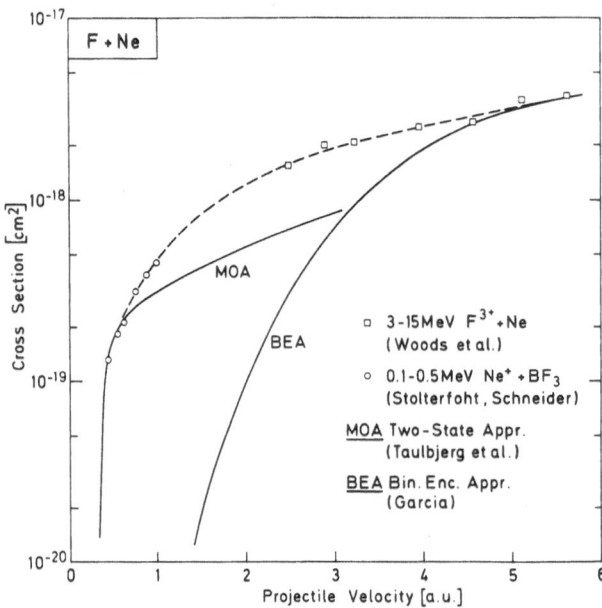

Fig. 5.28. Sum of cross sections $\sigma_K = \sigma_K(F) + \sigma_K(Ne)$ for K excitation of neon and fluorine in F+Ne collisions. Data for 3- to 15-MeV F^{3+}+Ne are from WOODS et al. [5.105]. Results for 0.1- to 0.5-MeV Ne^++BF_3 are from STOLTERFOHT and SCHNEIDER [5.94], where the measured cross sections were divided by 3 to obtain σ_K per target F atom. The theoretical results MOA and BEA refer to the scaled MO calculations by TAULBJERG et al. [5.95] and to the binary encounter theory by GARCIA [5.77]

It is noted that the incident velocity is equal to the orbital velocity of the F and Ne K electrons at 7.1 and 8.0 a.u., respectively.

Without entering into too many details, a few common features of the two theoretical descriptions will be pointed out. Figure 5.28 demonstrates that each theory is applicable in a restricted velocity range which is close to one of the limits of the overall range studied. Also, MOA as well as BEA depends on multiplicative parameters, i.e., the $2p\pi_x$ vacancy occupation number N_π and the effective nuclear charge Z_{eff}, respectively. Apart from specific cases, N_π and Z_{eff} are known only approximatively. Improvements of the MOA results are possible as the increase of N_π is adequately taken into account with increasing velocity. Similarly the BEA results may be improved by allowing for a decreased binding of the relevant inner-shell electron [5.73,112]. It is noted that these modifications tend to reduce differences between the two theoretical descriptions. Accordingly, in the intermediate energy range, the distinction between direct excitation and electron promotion diminishes. With respect to these considerations it may be fruitful to consider MOA and BEA (or PWBA) not as descriptions of different excitation mechanisms but as opposite starting points for a unified treatment of the inner-shell excitation process.

Acknowledgement. The author is indebted to Dr. J.S. BRIGGS, Dr. W. DREPPER, Prof. E. RUDD, and Prof. D. MADISON for communication of their results prior to publication. Helpful comments on the manuscript by Dr. F. HOPKINS and Dr. D. HILSCHER are gratefully acknowledged.

References

5.1 H. Bethe: *Handbuch der Physik*, ed. by H. Geiger, K. Scheel (Berlin: Springer 1933) Vol. 24, Teil 1, p. 273
5.2 M. Born: Z. Physik 38, 803 (1926)
5.3 M. Inokuti: Rev. Mod. Phys. 43, 297 (1971)
5.4 Y.K. Kim: Radiation Res. 61, 21 (1975); 65, 205 (1975)
5.5 M. Gryszinski: Phys. Rev. 138, A305, A322, A336 (1965)
5.6 E. Blauth: Z. Physik 147, 228 (1957)
5.7 D. Moe, E. Petsch: Phys. Rev. 110, 1358 (1958)
5.8 H.W. Berry: Phys. Rev. 121, 1714 (1961)
5.9 C.E. Kuyatt, T. Jorgensen, Jr.: Phys. Rev. 130, 1444 (1963)
5.10 M.E. Rudd, T. Jorgensen, Jr.: Phys. Rev. 131, 666 (1963)
5.11 L.H. Toburen: Phys. Rev. A3, 216 (1971)
5.12 N. Stolterfoht: Z. Physik 248, 81 (1971); 92 (1971)
5.13 J.S. Risley: *Proceedings of the Fourth International Conference on Atomic Physics* (Plenum Press, New York 1975) p. 487 and references therein
5.14 M.E. Rudd, T. Jorgensen, D.J. Volz: Phys. Rev. 151, 28 (1966); M.E. Rudd: Phys. Rev. Lett. 13, 503 (1964)
5.15 U. Fano, W. Lichten: Phys. Rev. Lett. 14, 627 (1965)
5.16 For a review of energy-loss experiments see Q.C. Kessel, B. Fastrup: *Case Studies in Atomic Physics*, Vol. 3, ed. by E.W. McDaniel and M.C. McDowell (North-Holland, Amsterdam 1973) p. 137
5.17 Q.C. Kessel, P.M. McCaughey, E. Everhart: Phys. Rev. Lett. 16, 1189 (1966)
5.18 R.K. Cacak. Q.C. Kessel, M.E. Rudd: Phys. Rev. A2, 1327 (1970)
5.19 D. Burch, W.B. Ingalls, J.S. Risley, R. Heffner: Phys. Rev. Lett. 29, 1719 (1972)
5.20 D.L. Matthews, B.M. Johnson, J.J. Mackey, C.F. Moore: Phys. Rev. Lett. 31, 1331 (1973); D.L. Matthews, B.M. Johnson, J.J. Mackey, L.E. Smith, W. Hodge, C.F. Moore: Phys. Rev. 10, 1177 (1974)
5.21 C.W. Woods, R.L. Kauffman, K.A. Jamison, C.L. Cocke, P. Richard: J. Phys. B7, L474 (1974)
5.22 N. Stolterfoht, D. Schneider, D. Burch, H. Wieman, J.S. Risley: Phys. Rev. Lett. 33, 59 (1974); *Nuclear Physics Laboratory Annual Report 1974* (University of Washington, Seattle 1975) unpublished
5.23 M.E. Rudd, J. Macek: *Case Studies in Atomic Physics*, Vol. 3, ed. by E.W. McDaniels and M.C. McDowell (North-Holland, Amsterdam 1973) p. 47
5.24 G.N. Ogurtsov: Rev. Mod. Phys. 44, 1 (1972)
5.25 M.E. Rudd: Radiation Res. 64, 153 (1975)
5.26 D. Burch: *Proceedings of the VIIIth International Conference on Electronic and Atomic Collisions*, Invited Lectures and Progress Reports, ed. by B.C. Cobic and M.V. Kurepa (Institute of Physics, Beograd 1973) p. 97
5.27 N. Stolterfoht: Ref. 5.26, p. 117
5.28 M. Barat, W. Lichten: Phys. Rev. A6, 211 (1972)
5.29 J.D. Garcia, R.J. Fortner, T.M. Kavanagh: Rev. Mod. Phys. 45, 111 (1973)
5.30 Q.C. Kessel, B. Fastrup: *Case Studies in Atomic Physics*, Vol. 3, ed. by E.W. McDaniel and M.C. McDowell (North-Holland, Amsterdam 1973) p. 137
5.31 G. Gerber, R. Morgenstern, A. Niehaus: J. Phys. B5, 1396 (1972); 6, 493 (1973) and references therein

5.32 H.H. Haselton, R.S. Thoe, J.R. Mowat, P.M. Griffin, D.J. Pegg, I.A. Sellin:
     Phys. Rev. 11, 468 (1975) and references therein
5.33 F. Drepper, J.S. Briggs: J. Phys. B9, 2063 (1976)
5.34 N. Stolterfoht, D. Schneider, D. Burch, B. Aagaard, E. Bøving, B. Fastrup:
     Phys. Rev. A12, 1313 (1975)
5.35 M.E. Rudd, C.A. Sautter, C.L. Bailey: Phys. Rev. 151, 20 (1966)
5.36 L.H. Toburen: *Proceedings of the VIIth International Conference on the Physics
     of Electronic and Atomic Collisions*, Abstracts of Papers, ed. by L.M. Branscomb
     et al. (North-Holland, Amsterdam 1971) p. 1120, see also Ref. 5.41
5.37 N. Stolterfoht, H. Gabler, U. Leithäuser: (unpublished) see also Ref. 5.41
5.38 M.E. Rudd, D. Madison: Phys. Rev. A14, 128 (1976), and private communication
5.39 M.E. Rudd, L.H. Toburen, N. Stolterfoht: Atomic and Nuclear Data Tables 18, 413 (1976)5)
5.40 D.H. Madison: Phys. Rev. 8, 2449 (1973)
5.41 S.T. Manson, L.H. Toburen, D.H. Madison, N. Stolterfoht: Phys. Rev. A12, 60
     (1975)
5.42 K.L. Bell, A.E. Kingston: J. Phys. B8, 2666 (1975)
5.43 H.S. Massey, C.B. Mohr: Proc. Roy. Soc. A140, 613 (1933)
5.44 M.C. Walske: Phys. Rev. 101, 940 (1965); G.S. Khandewahl, E. Merzbacher: Phys.
     Rev. 151, 12 (1966)
5.45 A. Salin: J. Phys. B2, 631 (1969)
5.46 G.B. Crooks, M.E. Rudd: Phys. Rev. Lett. 25, 1599 (1970)
5.47 K. Harrison, M.W. Lucas: Phys. Lett. 33A, 142 (1970)
5.48 J. Macek: Phys. Rev. A1, 235 (1970)
5.49 K. Dettmann, K.G. Harrison, M.W. Lucas: J. Phys. B7, 269 (1974)
5.50 Y.B. Band: J. Phys. B7, 2557 (1974)
5.51 W.R. Thorson: Phys. Rev. A12, 1365 (1975)
5.52 D.H. Madison, E. Merzbacher: *Atomic Inner-Shell Processes*, Vol. 1, ed. by B.
     Crasemann (Academic Press, New York 1975) p. 1
5.53 J.S. Briggs: J. Phys. B8, L 485 (1975)
5.54 J.R. MacDonald, L.M. Winters, M.D. Brown, L.D. Ellsworth, T. Chiao, E.W. Pettus:
     Phys. Rev. Lett. 30, 251 (1973)
5.55 G. Basbas, W. Brandt, R. Laubert, A. Ratkovski, A. Schwarzschild: Phys. Rev.
     Lett. 27, 171 (1971)
5.56 W.E. Wilson, L.H. Toburen: Phys. Rev. A7, 1535 (1973)
5.57 F.D. Schowengerdt, S.R. Smart, M.E. Rudd: Phys. Rev. A7, 560 (1973)
5.58 D. Burch, H. Wieman, W.B. Ingalls: Phys. Rev. Lett. 30, 823 (1973)
5.59 N. Stolterfoht, D. Schneider, P. Richard, R.L. Kauffman: Phys. Rev. Lett. 33,
     1418 (1974)
5.60 D.L. Matthews, B.M. Johnson, L.E. Smith, J.J. Mackey, C.F. Moore: Phys. Lett.
     48A, 93 (1974)
5.61 N. Stolterfoht, H. Gabler, U. Leithäuser: Phys. Lett. 45A, 351 (1973)
5.62 D. Burch, N. Stolterfoht, D. Schneider, H. Wieman, J.S. Risley: *Nuclear Physics
     Laboratory Annual Report 1974*, (University of Washington, Seattle 1975) unpu-
     blished
5.63 D. Schneider, C.F. Moore, B.M. Johnson: J. Phys. B9, L 153 (1976)
5.64 D. Burch, N. Stolterfoht, D. Schneider, H. Wieman, J.S. Risley: Phys. Rev. Lett.
     32, 1151 (1974)
5.65 K.O. Groeneveld, R. Mann, G. Nolte, S. Schumann, R. Spohr, B. Fricke: Z. Physik
     A274, 191 (1975)
5.66 C.W. Woods, R.L. Kauffman, K.A. Jamison, N. Stolterfoht, P. Richard: Phys. Rev.
     12, 1393 (1975)
5.67 J.R. MacDonald, L. Winters, M.D. Brown, T. Chiao, L.D. Ellsworth: Phys. Rev.
     Lett. 29, 1291 (1972)
5.68 J.R. Mowat, I.A. Sellin, D.J. Pegg, R.S. Peterson, M.D. Brown, J.R. Macdonald:
     Phys. Rev. Lett. 30, 1289 (1973)
5.69 A.M. Halpern, J. Law: Phys. Rev. Lett. 29, 1291 (1973)
5.70 F. Hopkins, N. Cue, V. Dutkiewicz: Phys. Rev. A12, 1710 (1975)
5.71 W. Brandt, R. Laubert, I.A. Sellin: Phys. Lett. 21, 518 (1966)
5.72 G. Basbas, W. Brandt, R.H. Ritchie: Phys. Rev. A7, 1971 (1973)
5.73 C. Foster, T. Hooghamer, P. Woerlee, F.W. Saris: J. Phys. B9, 1943 (1976)
5.74 N. Stolterfoht, D. Schneider: Phys. Rev. A2, 721 (1975)

5.75 L.H. Toburen: Phys. Rev. A5, 2482 (1972)
5.76 R.L. Watson, L.H. Toburen: Phys. Rev. A7, 1953 (1973)
5.77 J.D. Garcia: Phys. Rev. A1, 1402 (1970); 4, 955 (1971)
5.78 M.H. Chen, B. Crasemann, D.L. Matthews: Phys. Rev. Lett. 34, 1309 (1975)
5.79 C.P. Bhalla: J. Phys. B8, 1200 (1975)
5.80 J.S. Briggs: Rep. Prog. Phys. 39, 217 (1976)
5.81 W. Brandt, R. Laubert: Phys. Rev. Lett. 24, 1037 (1970)
5.82 J. von Neumann, E.P. Wigner: Z. Physik 30, 467 (1929)
5.83 H.A. Erikson, E.L. Hill: Phys. Rev. 75, 29 (1949)
5.84 H.J. Eichler, V. Wille, B. Fastrup, K. Taulbjerg: Phys. Rev. A14, 707 (1976)
5.85 F.P. Larkins: J. Phys. B5, 571 (1972)
5.86 K. Taulbjerg, J.S. Briggs: J. Phys. B8, 1895 (1975)
5.87 W. Lichten: Phys. Rev. 164, 131 (1967)
5.88 J.S. Briggs, J.H. Macek: J. Phys. B5, 579 (1972); 5, 982 (1972)
5.89 J.H. Macek, J.S. Briggs: J. Phys. B6, 841 (1973)
5.90 B. Fastrup, E. Bøving, G.A. Larsen, P. Dahl: J. Phys. B7, 206 (1974)
5.91 W.E. Meyerhof: Phys. Rev. Lett. 31, 1341 (1973)
5.92 J.S. Briggs, K. Taulbjerg: J. Phys. B8, 1909 (1975); 9, 1 (1976)
5.93 N. Stolterfoht, P. Ziem, D. Ridder: J. Phys. B7, L409 (1974)
5.94 N. Stolterfoht, D. Schneider: Proceedings of the Vth International Conference
      on Atomic Physics, Abstracts of Papers (Berkeley, 1976) p. 438
5.95 K. Taulbjerg, J.S. Briggs, J. Vaaben: J. Phys. B9, (1976)
5.96 K. Taulbjerg, J. Vaaben, B. Fastrup: Phys. Rev. A12, 2325 (1975)
5.97 R.D. Piacentini, A. Salin: J. Phys. B7, 311 (1974)
5.98 J.S. Briggs, M.D. Hayns: J. Phys. B6, 514 (1973)
5.99 B. Fastrup, B. Aagaard, J. Vaaben, E. Bøving: Proceedings of the IXth Inter-
      national Conference on the Physics of Electronic and Atomic Collisions, Ab-
      stract of Papers, ed. by J.S. Risley, R. Geballe (University of Washington
      Press, Seattle 1975) p. 1159
5.100 Y.N. Demkov: Zh. Eksper. I. Teor. Fiz. 45, 195 (1963). English transl.: Sov.
      Phys. - JETP 18, 138 (1963)
5.101 E.E. Nikitin: Izv. A. N. SSSR Ser. Fiz. 27, 996 (1963), The Theory of Nonadia-
      batic Transitions, in Advances of Quantum Chemistry, Vol. 5, ed. by P.O. Löwdin
      (Academic Press, New York 1970) p. 135
5.102 C.M. Lederer, J.M. Hollander, I. Perlman: Table of Isotopes, 6th ed. (Wiley,
      New York 1967) p. 566
5.103 B. Fastrup, B. Aagaard, E. Bøving, D. Schneider, P. Ziem, N. Stolterfoht:
      Ref. 5.99, p. 1058
5.104 N. Stolterfoht, D. Schneider, D. Ridder: Ref. 5.99, p. 1060
5.105 C.W. Woods, R.L. Kauffman, N. Stolterfoht, P. Richard: Phys. Rev. A13, 1358
      (1976)
5.106 N. Stolterfoht, V. Leithäuser: Phys. Rev. Lett. 36, 186 (1976)
5.107 J.T. Park, V. Pol, J. Lawler, J. George, J. Aldag, J. Parker, J.L. Peacher:
      Phys. Rev. A11, 857 (1975)
5.108 D.C. Lorents, G.M. Conklin: J. Phys. B5, 950 (1972)
5.109 R. François, D. Dhuicq, M. Barat: J. Phys. B5, 963 (1972)
5.110 V. Sidis, N. Stolterfoht, M. Barat: Proceedings of the 2nd International Con-
      ference on Inner-Shell Ionization Phenomena, Abstracts of Contributed Papers,
      ed. W. Mehlhorn (University of Freiburg, Freiburg 1976) p. 68
5.111 V.V. Afrosimov, Y.U.S. Gordeev, A.N. Zinoviev, D.H. Rasulov, A.P. Shergin:
      Ref. 5.99, p. 1066
5.112 W.E. Meyerhof: Phys. Rev. A10, 1005 (1974)

6. X-Ray Production in Heavy Ion-Atom Collisions

P. H. Mokler and F. Folkmann

With 45 Figures

X-ray emission was observed in 1895 by RÖNTGEN and was so named because of its mysterious nature at that time. It has now long been classified as electromagnetic radiation of atomic origin, and is known to be produced in ion-atom collisions from work dating back to around 1930, as reviewed by MERZBACHER and LEWIS [6.1]. Owing partly to the intervening rapid progress in nuclear physics, the interest in atomic collision processes subsided, and the subject was categorized as one of basically well-known phenomena until recently, when refined techniques led to many new and unexpected results. A major reason for the present interest in x-rays from ion-atom collisions is the complexity introduced by using heavy ions as projectiles, which in the last decade has resulted in the discovery of several intriguing effects not previously predicted. Experiment, together with application of well-accepted theories, has enabled many such effects to be understood and studied, and opened up the possibility of examining details of the ionization process which as a rule occurs prior to the x-ray emission.

Within a vast field of interesting problems, we have chosen to treat only those aspects of the ionization which are most closely related to x-ray emission induced by swift heavy ions, namely the ionization of tightly bound electrons and the appearance of specific sources of continuous x-ray emission. Inner-shell ionization has also been treated theoretically in Chapter 4 by BRIGGS and TAULBJERG, and experimentally in Chapter 5 by STOLTERFOHT, in connection with his discussion of electron emission.

To facilitate the discussion and to provide a possibility of a broader outlook, we preferentially make reference to existing review papers in the field like those of GARCIA et al. [6.2] and of KESSEL and FASTRUP [6.3] as well as in the book edited by CRASEMANN [6.4], which contains, for example, comprehensive chapters by MADISON and MERZBACHER [6.5] and by RICHARD [6.6] on inner-shell ionization and x-ray emission. Summary talks presented at recent international conferences also contain much information of current relevance, as for example those on Inner Shell Ionization Phenomena [6.7,8], on the Physics of Electronic and Atomic Collisions [6.9,10], on Atomic Collisions in Solids [6.11] and on Atomic Physics [6.12,13].

X-radiation can be clssified in terms of characteristic x-rays of sharply defined energy corresponding to a given Z, and continuous radiation extending over a broader range. The latter may originate from bremsstrahlung processes, which depend only weakly on the atoms involved, or from transitions to specific atomic or molecular states formed during a collision, and will in that case carry characteristics of these states. Consequently our discussion of ion-induced x-rays is initiated by a description of the atomic transitions leading to x-ray emission, of how x-rays are detected, and of the characteristics of well-resolved x-ray spectra. The next two sections deal with the ionization processes for direct and molecular excitation leading to characteristic x-rays, and the last two with the continuous x-ray emission.

The ionization which occurs prior to the x-ray emission may be treated in terms of inner- and outer-shells. The most tightly bound inner-shell electrons can be defined in a relative way as those electrons with mean orbital velocity u appreciably larger than the collision velocity v: u>v. The parameter $\eta=v^2/u^2$ defines the adiabaticity of the reaction. Most of the work reported here refers to the (quasi-) adiabatic region $\eta<1$ of the most strongly bound electrons, for which the inner-shell x-ray transitions are studied. There may, however, be several other and more loosely bound electrons extending in orbital velocity well into the region $\eta>1$ for the same atom. Simple Coulomb collision-induced ionization shows a maximum near $\eta=1$, and is relatively highly probable for outer-shells, which results in a high degree of depletion for outer electrons simultaneously with ejection of inner-shell electrons. The multi-vacancy state production has several consequences: the balance between x-ray emission and vacancy production, as described by the fluorescence yield, will be disturbed; the electron energy levels will be changed as the shielding changes, resulting in shifts or splittings of characteristic x-ray lines; and delayed x-ray emission may occur from metastable states [6.6,14].

For adiabatic conditions ($\eta<<1$) the electrons concerned can continuously adjust during collisions to the perturbation caused by the collision partner. Following MADISON and MERZBACHER [6.5], two general cases must be distinguished, as shown in Fig.6.1. In the case $Z_1<<Z_2$ the perturbation potential caused by the projectile (Z_1) is small compared to the (inner) electron binding energy of the target atom (Z_2). This is the domain of direct Coulomb ionization. In the case $Z_1 \sim Z_2$ the perturbation potential can be comparable to the corresponding binding energy. In this case the

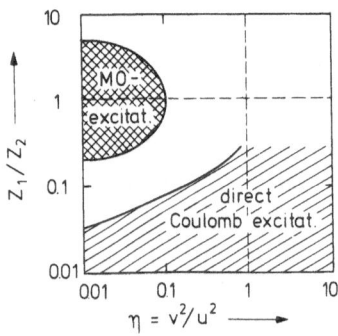

Fig. 6.1. The domain for the excitation processes, direct ionization and quasi-molecular excitation. Z_1 and Z_2 are the atomic numbers of projectile and target atoms; v and u are the collision velocity and the electron orbital velocity, respectively; η is called the adiabaticity parameter (see MADISON and MERZBACHER [6.5])

target atom and the projectile are equivalent and their role can be interchanged.
Here, a transient quasi-molecule is formed during collisions, and quasi-molecular
excitation processes prevail. The wide field of quasi-molecular radiation belongs
also to this domain, since x-rays may be emitted during the course of a collision,
and excitation and deexcitation processes are no longer always independent.

The collision dynamics can be studied in considerable detail by performing demand-
ing experiments, e.g., with coincidence requirements. A classical description of the
ion trajectory is in most cases applicable, as the de Broglie wavelength of the projectile is small compared to the shell radii of concern ($h/p<10^{-11}$ cm for energies
higher than 0.1 MeV/amu). Measurements of the scattering angle of the projectile will
then determine its trajectory, and investigations of the accompanying x-ray emission
provide insight into the details of the ionization processes.

6.1 Measurement of X-Rays

6.1.1 Atomic Transitions, Fluorescence Yields, and Lifetimes

Characteristic x-rays are emitted from atoms containing vacancies in their lowest
bound energy levels. A vacancy in the electronic shell is thus the basis for observing an x-ray quantum, which may be emitted when an outer electron fills the vacancy.
Figure 6.2 shows schematically the sequence of energy levels for an atomic electron
- or generally for one out of many electrons surrounding a nucleus. Each level is
labeled by its quantum numbers n l j, or by the shell (K,L,M,...), with a number for
the subshell, beginning with the lowest level. The state of the atom for which such
a level diagram applies is described by the number of electrons in each level.

With a less than full subshell, an electron in one of the energetically higher
levels can make a transition to it and the energy gained by this change of state
can then be liberated as a photon. The x-ray energy is the difference between the
energies of the two levels which characterize the transition. The most important
determinants of this difference are the atomic number Z of the atom and the shell of
the initial vacancy. The modest resolution of early x-ray spectrometers justified
labels only for the lower shell, with subscripts α, β,... to describe the main components as explained in Fig.6.2. Typical intensities are indicated for Z=30 and
Z=82, normalized within a shell to the strongest x-ray line, and calculated for a
statistical population of the subshells for a singly ionized atom. At the top of
the figure is a model spectrum folded with finite energy resolution. X-ray energies
and intensity ratios for subshells have been tabulated by STORM and ISRAEL [6.15].

The energy of a transition depends on the electronic state of the atom, and in
many cases the interesting feature is not so much the details of which transition
is involved as the description of the initial state as specified by the vacancies

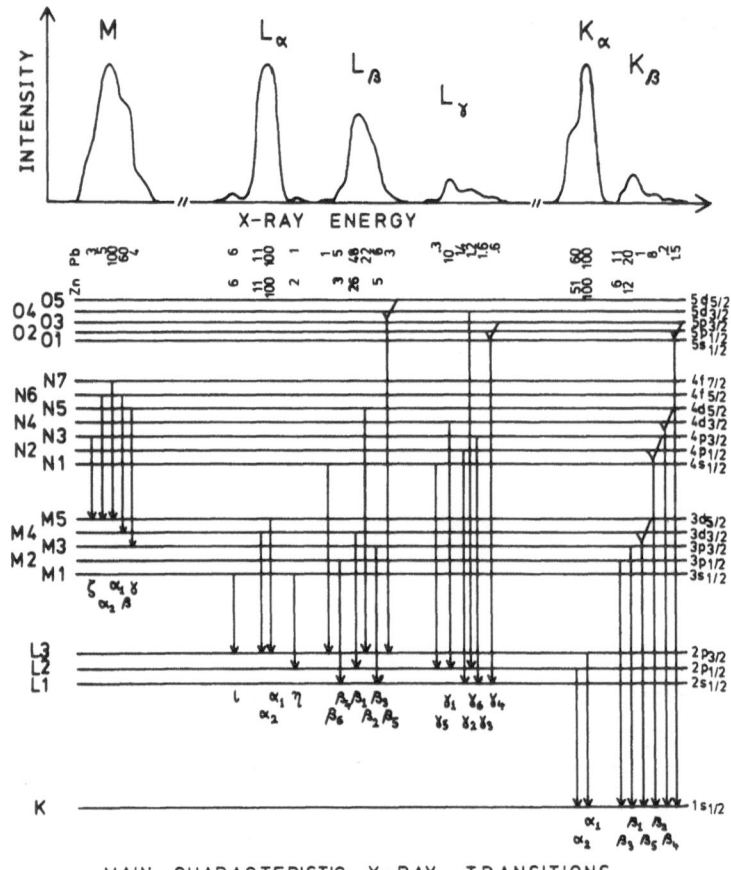

MAIN CHARACTERISTIC X-RAY TRANSITIONS

Fig. 6.2. X-ray transitions with indication of the involved atomic levels (n l j). For Zn and Pb, intensity ratios for transitions to a shell are given, assuming single vacancy states and a statistical population of the subshells, e.g., for L1, L2, L3 with ratios 2:2:4. At the top, x-rays (for Pb) are schematically shown as a function of the x-ray energy, and folded with a representative energy resolution

in each of the levels and sometimes by the associated electron angular momenta. This description relates most directly to the problem of the excitation mechanism creating these vacancies.

An atomic state can frequently decay both by x-ray emission and by other competing processes. The available energy may, for example, be used to excite another bound electron, which leaves the atom as an Auger electron of a definite kinetic energy (see Chap.5). The probability that a vacancy in a level S will decay by emission of an x-ray is called the fluorescence yield

$$\omega_S = N_S^x / N_S^{vac} \tag{6.1}$$

where N_S^X is the number of x-rays to S and N_S^{vac} the number of vacancies in S. For a uniquely specified atomic state this quantity is well defined and can be calculated theoretically [6.16,17]. For L- and higher shells, such calculations are complicated by the occurrence of Auger transitions (the so-called Coster-Kronig transitions), in which a vacancy can be transferred from one subshell to higher ones in the same shell. The number of vacancies entering in (6.1) will then contain contributions from initially ionized lower subshells.

A mean value of ω_S is observed, when the information on S is not complete. For incompletely resolved heavy ion-induced x-rays, one will often have to average over various states and subshells, and over the population of these during collisions. In Fig.6.3 the mean fluorescence yields for the K- and the complete L- and M-shells are shown, calculated for singly ionized states by BAMBYNEK et al. [6.16], assuming statistical population of the subshells. The realistic $\bar{\omega}_S$ in heavy ion collisions

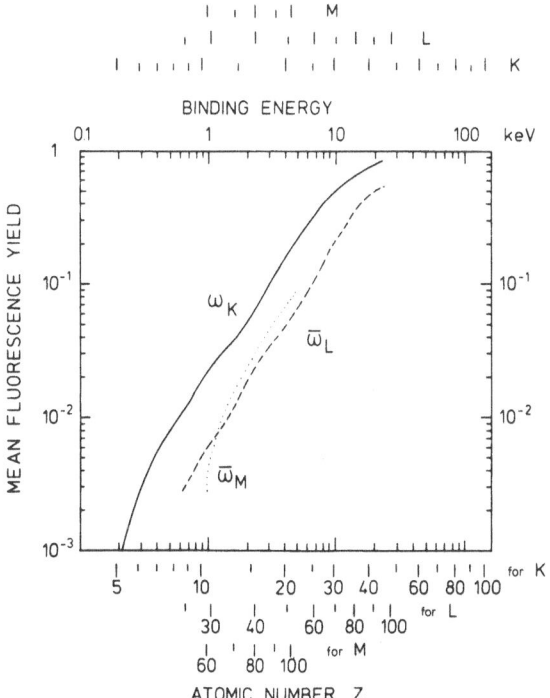

Fig. 6.3. Mean fluorescence yields for single vacancy states in the K, L, and M shells. A statistical population of the subshells, according to their j, is assumed. The fitted values from BAMBYNEK et al. [6.16] are shown as function of the experimental binding energy of the shell (U for K, L2, and M5 are used). In this representation the curves exhibit a similar shape. A scale for the Z values for each of the shells is given at the bottom of the figure

will be systematically higher than these values, because the presence of additional outer vacancies will reduce the possible Auger processes more than the radiative transitions. That $\bar{\omega}_S$ can indeed have a strong dependence on collision parameters is demonstrated in [Ref.6.18, Fig.5.18]. In that figure $\bar{\omega}_K$ is shown to increase from 0.05 to 0.36 with increasing charge of the bombarding Cl ion; the single vacancy value from Fig.6.3 is only 0.02. Simultaneously, the total ionization cross section increases, corresponding to a more violent removal of electrons and thereby produc-tion of more states with few outer electrons (high ω_K) with increasing projectile charge. For many heavy ion collisions, however, the deviation from the single vacancy value is not great. From the relationship given by (6.1), one can then express the x-ray production cross section σ^X corresponding to a shell S as

$$\sigma_S^x = \bar{\omega}_S \cdot \sigma_S^{vac} \quad . \tag{6.2}$$

The vacancy production cross section σ^{vac} can then often be approximately extracted using the standard value of $\bar{\omega}_S$ from [6.16] or Fig.6.3 rather than using a measured value. If, according to theoretical or experimentally based estimates, the deviation from Fig.6.3 is large, more reliable values require difficult direct measurements of $\bar{\omega}_S$.

The mean lifetime τ of a state can often be calculated theoretically. The tran-sition probability is expressed* as the linewidth $\Gamma = \hbar/\tau$. For radiative transitions, Γ_R is a rapidly increasing function of the x-ray energy roughly proportional to E_x^2 [6.19]. The total natural width and the corresponding lifetime are given by summa-tion over all decay modes

$$\hbar/\tau = \Gamma = \Gamma_R + \Gamma_A + \Gamma_{CK} \tag{6.3}$$

where A stands for Auger and CK for Coster-Kronig. Figure 6.4 shows the linewidths and the lifetimes for single vacancy states in the K, L_3, and M_5 shells from [6.20]. Inner-shell transitions are usually faster than transitions in outer-shells in the same atom due to the higher energy differences, except for Coster-Kronig transitions and some Auger transitions for outer-shells in light atoms. The width of an x-ray line is the sum of the natural widths of the levels involved

$$\Gamma_{S_1 \to S_2}^{x-ray} = \Gamma(S_1) + \Gamma(S_2) \quad . \tag{6.4}$$

The multi-vacancy states often encountered in heavy ion-atom collisions will have total widths and lifetimes different from those shown in Fig.6.4. Due to the smaller number of possible transition electrons, Γ_R is reduced and Γ_A will be reduced even

---

*$\hbar$=h/2π (normalized Planck's constant)

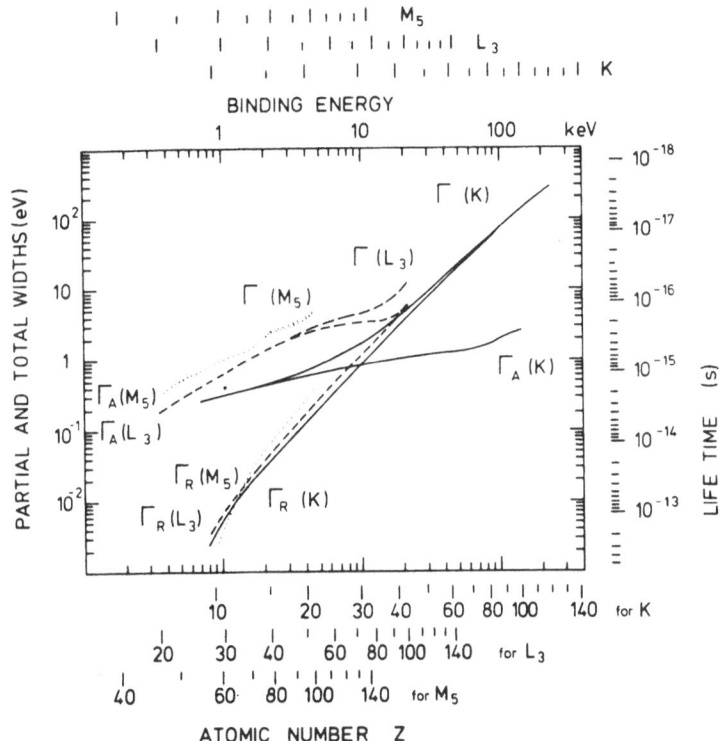

Fig. 6.4. Lifetimes and total, radiative, and Auger widths of the K, L3, and M5 levels with single vacancy states for various elements. The partial widths from KESKI-RAHKONEN and KRAUSE [6.20] and the lifetimes (right) are displayed as a function of the binding energy of the level concerned. This procedure will tend to coalesce the radiative widths on similar curves. Z values are given at the top and the bottom. Lower subshells (L1, L2, M1, M2, M3, M4) can also make fast Coster-Kronig transitions, which will give shorter lifetimes for those levels, but shift the main x-ray intensity to the levels shown

more, because the Auger process involves pairs of electrons. Hence the natural level width will be smaller and the lifetime longer than deduced from Fig.6.4, and the fluorescence yield $\omega = \Gamma_R/\Gamma$ will be larger than shown in Fig.6.3.

Ions moving in a solid can deexcite in collisions with other atoms in a way which may be as important as the spontaneous decay described so far. Many new possibilities arise, e.g., a vacancy may be filled by electron capture from a target atom (see Sec.6.2.3) as BETZ et al. [6.21] have shown. Lifetimes and fluorescence yields must thus be handled with much care for projectiles radiating inside a solid target.

6.1.2 X-Ray Detectors

The most widely used x-ray detectors are solid-state detectors, which with high efficiency give a direct energy signal, and crystal spectrometers, which have a good wavelength dispersion and usually employ a proportional counter to measure the diffracted x-rays. Since several comprehensive recent reviews exist [6.22-25], we shall only mention a few characteristics for these types of detectors.

The proportional counter is a simple and compact ionization detector usually made in cylindrical form with an axial anode wire, and filled with a noble gas (often Xe or Ar, sometimes with a quenching agent like CH_4). The electric field is so high that electrons originally produced by the photoelectric ionization are accelerated near the wire to velocities which will cause secondary ionization and give a multiplication of the measured charge. X-rays enter through a thin window, which determines the efficiency for lower energies, and which is covered with a metal layer to provide a good equipotential surface, At high photon energies the efficiency is governed by the absorption in the active gas volume, so that a high-Z gas is advantageous. The efficiency is typically higher than 0.4 for radiation between 1 and 10 keV for an $Ar(CH_4)$-filled counter, but has a discontinuity at the Ar absorption edge [6.22]. The energy resolution is determined partly by the statistics of the ionization process and partly by avalanche fluctuations, whereas amplifier noise is negligible. The charge collection statistics are determined by the average energy w required to produce an electron-ion pair, and by the Fano factor F describing the coupling to the larger number of final carriers. The full width at half maximum (FWHM) for a peak of energy E from the collection statistics alone is

$$\delta E_{(FWHM)}^{stat}/E = 2.355 \cdot (F \cdot w/E)^{1/2} \quad . \tag{6.5}$$

Table 6.1. Energy resolution of gas and solid-state detectors. The statistical contribution to the energy resolution is given by the mean electron-ion (hole) energy w and the Fano factor F from (6.5). The experimental energy resolution has the energy dependence of (6.6) for the gas detectors and of (6.7) for the solid-state detectors

Detector type	Material	w [eV]	F	δE_{FWHM}^{stat} for E=5.9 keV [eV]	δE_{FWHM}^{exp} for E=5.9 keV [eV]	Operating energy region [keV]
Proportional counter	Ar (90%) CH_4 (10%)	26.5	0.20	416	830	0.2 - 10
Proportional Scintillation counter	Xe	21.9	0.21	388	500	0.2 - 20
Si(Li) diode	Si	3.81	0.12	122	140	0.2 - 30
Intrinsic Ge diode	Ge	2.96	0.08	88	140	0.5 - 100

The noble gases have w~20-30 eV and F~0.2. For details see Table 6.1. Variations in the multiplication increase the width, so the best energy resolution expected and found is of the form

$$\delta E(FWHM)/E = C_g \cdot E^{-1/2} \tag{6.6}$$

with $C_g=10.8$ $eV^{1/2}$ for a proportional counter [6.22,26].

The gas proportional scintillation counter is a gas detector superior in some respects to the proportional counter. It is operated with lower electric field as an ionization chamber, so that no secondary ionization is possible, and the light produced in the gas by the electrons accelerated towards the anode (secondary scintillation) is detected with a photomultiplier [6.27,28]. With Xe gas, which gives large light output, the energy resolution is given by (6.6) but with $C_g=6.5$ $eV^{1/2}$, i.e., nearly half that of a proportional counter and closer to the limit in (6.5), because avalanche fluctuations are omitted. The output signals from the photomultiplier are fast [6.27], so the detector can be suited for coincidence measurements when good energy resolution is not needed. The gas volume can be large with a simple parallel plate geometry [6.28] and will thus allow high total detection efficiency.

Solid-state detectors give reasonably good energy resolution. They typically have an active volume of Si or Ge with a thickness of 3-7 mm and a surface area of 10-500 mm^2, and are cooled to 77 K, the temperature of liquid nitrogen. The advantage over gas counters is that the mean excitation energy w for an electron-hole pair is only 3-4 eV and F is low, so that the fundamental statistical contribution to the energy resolution in (6.5) is as low as indicated in Table 6.1. A schematic drawing of a semiconductor detector is shown in the lower part of Fig.6.5. To keep the effect of impurities low, the Si or Ge diode is doped with Li or an intrinsic Ge crystal is used. Using an applied bias voltage, the charge is collected and the signal amplified. The first amplification stage contains a field effect transistor (FET) kept at low temperature. A major contribution to the energy resolution comes from the noise in the amplifier which adds quadratically to the collection statistics according to (6.5)

$$\delta E/E = [(\delta E^{noise}/E)^2 + (\delta E^{stat}/E)^2]^{1/2} \quad . \tag{6.7}$$

The electronic noise δE^{noise} is ~70-110 eV for the best detectors. A careful description of the contributions to this electronic noise and of other factors affecting the detection process is given by GOULDING [6.23]. His introduction of the pulsed optical feedback in the preamplifier, in which the photosensitive FET is reset with a light emitting diode (LED), was a major step towards good resolution also at high count rates. Such solid-state detectors normally work up to ~50,000 counts/s. Inclusion of pulse pile-up rejection electronics - or equivalently by deflecting the ion beam, whenever an event is analyzed [6.29] - enables rejection of pulses sepa-

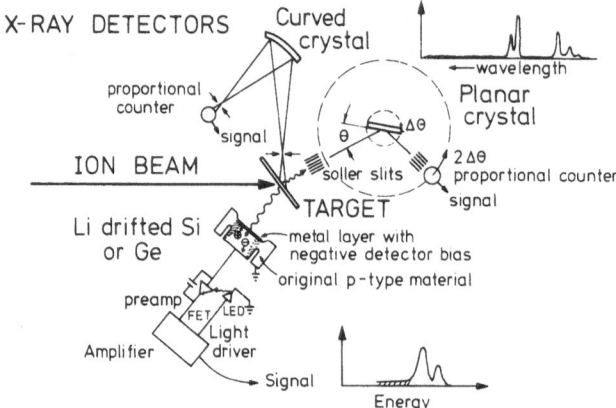

Fig. 6.5. Schematic drawing of three major types of x-ray detectors. The solid-state detector, Si(Li) or intrinsic Ge, in the lower part produces an energy signal. The crystal spectrometers at the top utilize a measurement of the wavelength of the radiation and require crystal rotation for varying the wavelength

rated more than ∼200 ns, permitting good quantitative measurements. For a pulsed beam also, externally controlled reset pulses will simplify the corrections. The good energy analysis corresponding to (6.7) and Table 6.1 is obtained at the cost of a long analysis time per pulse - ∼10-50 μs. The detection efficiency is determined by the absorption in the active layer of the diode, and is close to one over a broad ragion. At low energies the efficiency is limited by the attenuation in the window and the effect of a surface dead layer of the detector. The window has typical thicknesses of 8-150 μm Be (10% transmission for 0.7-2 keV) and can even be eliminated if the detector is in vacuum. The dead layer is 0.2-0.4 μm thick causing low energy tails on the peaks. This is pronounced at low energies, especially for Ge due to its higher absorption coefficient [6.23]. On the other hand, the greater absorption of Ge gives these detectors better efficiency at high x-ray energies. Low energy tails of peaks are predominantly produced by incomplete charge collection in the edge regions of the detector.

In addition to the normal peaks, smaller peaks sometimes occur at lower energy corresponding to escape of a Ge K (or Si K) quantum. These escape peaks are more pronounced for Ge than for Si. For further information on such semiconductor detectors we can refer to the review by BERTOLINI and RESTELLI [6.24]. To summarize, the solid-state detectors rapidly register x-rays over a broad energy region, with high efficiency, and with reasonable good energy analysis.

Crystal spectrometers are dispersive x-ray detectors, which select a certain wavelength λ according to Bragg's interference condition for diffraction from crystal lattice planes

$$n \cdot \lambda = 2d \cdot \sin\theta \qquad\qquad (6.8)$$

where d is the spacing between crystal planes, θ is the angle of incidence and reflection relative to this plane, and the integer n is the order of the reflection.
Figure 6.5 shows two types of frequently used crystal spectrometers, of which the
planar crystal type at the upper right is the simplest. The radiation reaches the
crystal through a Soller collimator consisting of parallel plates, which defines θ
and the spread $\delta\theta$ in θ, and is reflected into a proportional counter. The selected
wavelength is varied by turning the reflecting crystal through the angle $\Delta\theta$ and the
proportional counter through the angle $2\Delta\theta$. The resolution can be significantly
better than with the previously mentioned energy-analyzing detectors and is found
by differentiation of (6.8) using the relation $\lambda = 1.24(\text{nm keV})/E$:

$$\frac{\delta E}{E} = -\delta\theta \cdot \cot\theta = -\delta\theta \cdot \left\{ \left[\frac{2dE}{n \cdot 1.24(\text{nm keV})} \right]^2 - 1 \right\}^{1/2} \qquad . \qquad (6.9)$$

The angular uncertainty $\delta\theta$ depends on the structural purity of the crystal (its
mosaic structure) and on experimental parameters such as the Soller slits. Typically
$\delta\theta \sim 0.002\text{-}0.004$, and twice the lattice constant 2d is 0.874 nm for PET, 0.4027 nm
for LiF (200), and 0.236 nm for quartz ($13\bar{4}0$). Hence $\delta E/E \sim 10^{-2}\text{-}10^{-3}$ for crystals
with properly chosen 2d, which for $E \lesssim 15$ keV is clearly better than for solid-state
detectors. The disadvantage is that a crystal spectrometer has to be mechanically
adjusted for each new energy and has a very low detection probability $10^{-5}\text{-}10^{-6}$.

In a curved crystal spectrometer, as shown at the top of Fig.6.5, the efficiency
can be raised a factor 10 to 100, because the different parts of the bent crystal
can focus the source region (~ 1 mm diameter) at the entrance slit of the proportional counter. A mechanically coupled movement of crystal and proportional counter
allows simple variation of the analyzed energy. For a deeper description of crystal
spectrometers we refer the reader to the work of CACHOIS and BONNELLE [6.25] and of
KAUFFMAN and RICHARD [6.22]. For high x-ray energies (small 2d or high n), transmit-
ting crystal spectrometers can be used, where the x-rays are diffracted through the
crystal. A curved crystal may be placed close to the target concentrating the diffracted transmitted radiation to a distant point in front of a proportional counter,
or alternatively a larger bent crystal placed far from a point source will transmit
the x-rays. These types of bent crystal spectrometers are usually constructed for a
special energy region for x-rays or γ-rays of interest.

Other types of x-ray detectors include photo electron spectrometers, effective in
the region from 20 eV to 20 keV with a resolution no better than that of crystal
spectrometers and with a similar efficiency [6.22]. For ion beams special attention
may be paid to Doppler-tuned spectrometers. They utilize the fact that the Doppler
shift of a monoenergetic photon emitted from the ion depends on the angle relative
to the beam. By varying this angle with a properly chosen absorber foil in front of

a conventional x-ray detector, the transmission through the foil will drastically change, as the energy-shifted radiation sweeps over an absorption edge of the foil. The detector signal will then change with angle as a step function, since it represents the integration of the emitted radiation from the edge-energy downwards.

6.1.3 High Resolution X-Ray Data

X-rays from ion-atom collisions as a rule exhibit each of the simple diagram lines (see Fig.6.2 and [6.15]), accompanied by a number of satellite lines at higher energies, which are best studied with the good resolution of crystal spectrometers. As seen from Fig.6.6, the satellite structure grows more and more pronounced for increasing atomic number of the projectile.

The satellite lines are due to the presence of spectator vacancies in the atom at the time when an x-ray transitions fills another vacancy and are a consequence of the high degree of ionization caused by heavy ions. When an inner-shell electron is ejected in a collision there is a high probability that electrons in higher shells are also ejected by the strong Coulomb interaction of the incident ion, and these additional vacancies are often not able to decay prior to the filling of the inner- -shell vacancy. In the left part of Fig.6.6 (from WATSON et al. [6.30]) are shown x-rays of potassium labeled by the number of vacancies in the K- and L-shell, where only the K hole is filled by the K_α transition.

Estimates of the energy shifts of the satellite lines can be made from a calculation of the change in the binding energy of the levels concerned with simultaneous vacancies in higher shells. Using the simple procedure of BURCH et al. [6.32], employing hydrogenlike wave functions for both the missing electron and for the eigenfunctions of each level, we obtain the changes in binding energy for K, L, M, and N levels with simultaneous 2p (L) or 3d (M) vacancies as given in Table 6.2. The binding energy is increased because of less screening, and the increase is highest for the most strongly bound shells. The x-ray energy also increases with additional vacancies, in amounts per extra outer shell hole which are given in Table 6.3 for the main K and L transitions. The energy shifts in Tables 6.2 and 6.3 are proportional to the screened charge of the L or M electrons according to Slater's rules [6.33]. Although the values are good for giving an overall impression of the magnitude of the shifts for different transitions, they are not very accurate, especially for the higher shells (where, for example, additional 4s and 4p vacancies behave differently from 4d), and should for cases of detailed interest be replaced by more realistic Hartree-Fock calculations.

The satellites of the K lines in Fig.6.6 are in correspondence with Table 6.3, and it is noted that the K_β shift per vacancy is 2.5 times the K_α shift in the Ca x-ray spectra from McWHERTER et al. [6.31]. In that experiment, "hypersatellite" lines between the K_α and the K_β groups were observed, with an additional vacancy present in the K-shell. These states having two initial K vacancies give a bigger

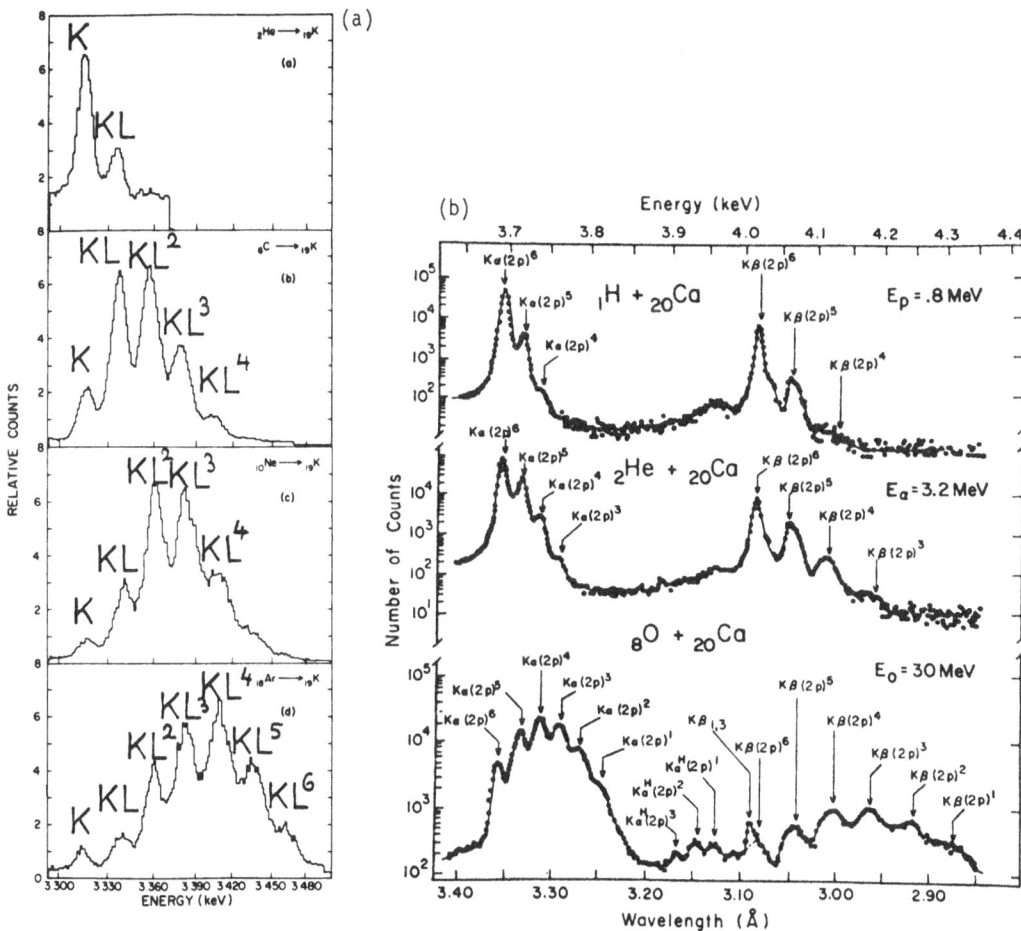

Fig. 6.6a and b. High resolution x-ray spectra measured with a crystal spectrometer, showing the satellite structure and its growth of complexity with increasing Z_1. On the left are shown the K_α satellite lines of potassium for bombardment by 1.7 MeV/amu He, C, Ne, and Ar from WATSON et al. [6.30]. The peaks are labeled by the initial vacancies in the K- and L-shells. McWHERTER et al. [6.31] measured the similar spectra of K satellites of Ca with incident H, He, and O ions as shown on the right. Both K_α and K_β lines are seen, and in between hypersatellite lines appear. In the lowest spectrum these are indicated with H and correspond to transitions with a spectator K vacancy. The peaks are labeled by the transition and the initial number of 2p L electrons

energy shift. For $Z \lesssim 30$, separate L satellites can be resolved in the K lines, whereas M shifts are too small to be resolved and only contribute to the width of the single peaks. Satellites for L x-rays are often hard to resolve, as Fig.6.18 shows, because there are many energetically closely spaced lines. In addition, coupling of the spins to form multiplet terms will produce splittings of the above-mentioned satellite lines, which have been adequately treated only for few hole or few electron systems.

Table 6.2. Binding energy corrections in the presence of an additional vacancy in the 2p level ($L_{2,3}$ shell) and in the 3d level ($M_{4,5}$ shell). Hydrogenic wave functions are used for the calculations (see BURCH et al. [6.32]). Z_L and Z_M are screened charges, estimated from the SLATER rules [6.33], which for full shells give $Z_L = Z-4.15$, $Z_{M1,2,3} = Z-11.25$, and $Z_{M4,5} = Z-21.15$

Level	nl	Increased binding per L vacancy [eV]	Increased binding per M vacancy [eV]
K	1s	$Z_L \cdot 6.61$	$Z_M \cdot 3.02$
$L_{2,3}$	2p	$Z_L \cdot 4.95$	$Z_M \cdot 2.96$
$M_{2,3}$	3p	$Z_L \cdot 2.23$	$Z_M \cdot 2.11$
$M_{4,5}$	3d	$Z_L \cdot 2.71$	$Z_M \cdot 2.40$
$N_{4,5}$	4d	$Z_L \cdot 1,24$	$Z_M \cdot 1.24$

Table 6.3. Energy difference between satellite lines. The initial and final vacancies for a transition are given according to Fig.6.2, and the shift per 2p ($L_{2,3}$) and 3d ($M_{4,5}$) vacancy is extracted from Table 6.2

Transition	Initial vacancy	Final vacancy	Shift per L vacancy [eV]	Shift per M vacancy [eV]
K_α	K	$L_{2,3}$	$Z_L \cdot 1.66$	$Z_M \cdot 0.06$
$K_{\beta 3,1}$	K	$M_{2,3}$	$Z_L \cdot 4.38$	$Z_M \cdot 0.91$
L_α	L_3	$M_{4,5}$	$Z_L \cdot 2.24$	$Z_M \cdot 0.56$
$L_{\beta 1}$	L_2	M_4	$Z_L \cdot 2.24$	$Z_M \cdot 0.56$
$L_{\beta 2}$	L_3	N_5	$Z_L \cdot 3.71$	$Z_M \cdot 1.72$

With a solid-state detector the satellite structure is not resolved, but observed as centroid energy shifts and increased widths. Only the hypersatellite group can sometimes be resolved.

Rearrangement of outer-shells prior to the decay of an inner-shell vacancy, causing dramatic changes in the satellite line pattern, is strongly influenced by the near surroundings of the decaying atom. In measurements of the satellite spectra of SiH_4 and SiF_4 gases bombarded with 53 MeV Cl ions, HOPKINS et al. [6.34] observed much higher Si K_α satellite lines for SiH_4 than for SiF_4, which means that the presence of near F atoms induced a fast filling of Si L vacancies. The results of WATSON et al. [6.35] of a systematic dependence for target satellite lines on the chemical environment in solid targets (e.g., expressed by the density of valence electrons) is another important observation in the same direction.

A variety of interesting rare phenomena can be studied with crystal spectrometers, such as the radiative Auger effect and radiative electron rearrangement, which both give lines of lower energy than the normal x-ray transitions. We refer the reader to the discussion of these effects by RICHARD [6.6,36]. We shall also call the attention to the double K transitions, where two electrons simultaneously fill two K vacancies under emission of a single photon. Although this is only a weak decay branch, the energy of the photon is so high—about twice the K_α energy—that it can be clearly seen over the background, even with solid-state detectors, as described by WÖLFLI et al. [6.37] for Ni-Fe collision systems.

6.2 Direct Ionization

6.2.1 Coulomb Excitation, Simple Description

The most prominent way of producing vacancies in ion-atom collisions is by the direct interaction of the electric field of the projectile with the target electrons. By such atomic Coulomb excitation one or several electrons can be ejected from their original bound states. This ionization model in many cases totally accounts for the observed vacancy production, especially for inner-shell ionization by light ions of atomic number Z_1 on heavier targets Z_2, or for projectiles faster than the mean orbital velocity of the ejected electrons. From Fig.6.1 and [6.5] this direct ionization applies to collisions with $Z_1 \ll Z_2$ or $v > u$:

$$[Z_2/(Z_1 \cdot 30)]^2 + (v/u)^2 \gtrsim 1 \quad . \tag{6.10}$$

This criterion is based on a comparison of measured x-ray cross sections with theoretical calculations using the above-mentioned interaction for vacancy production combined with the fluorescence yields (6.1,2) [6.1,2,38]. It is hence of interest to treat the physical assumptions entering into such calculations and to state the main results. As most of these data are obtained with light ions $Z_1 \sim 1$, we will later focus on the deviations which occur and which effects have to be included, when Z_1 increases.

For heavy ions the projectile can be treated classically, as its de Broglie wavelength is much less than the atomic dimensions of interest. Often it will only suffer a small change in direction and energy during a collision. A target electron with which it interacts is, however, to be treated quantum mechanically due to the uncertainty principle and its small mass m.

The electron probability distribution for a given state S is characterized by its quantum numbers and its experimental binding energy U. It is commonly described by a hydrogenic wave function with energy E, having a mean velocity u and radius r, all with a subscript S for the shell - or subshell - of interest (S=K, L, M, or L1, L2, L3, M1,...,M5). The projectile may be treated as a point charge, and we denote its atomic number Z_1, velocity v, energy E_1, and mass A_1M (A_1 is the atomic number and M the atomic mass unit $M=1.66 \cdot 10^{-24}g=931.5$ MeV/c^2). The plane wave Born approximation (PWBA) represents a proper treatment of such an idealized process assuming the projectile is described by an incoming and an outgoing plane wave [6.1,5,39]. Electron hydrogenlike wave functions in the target of atomic number Z_2 are frequently described by a reduced charge

$$Z_S = Z_2 - C_S \qquad (6.11)$$

where the "inner" screening correction C is taken to be the SLATER value [6.33] which for fully occupied shells is 0.3 for K, 4.15 for L, 11.25 for M1, M2, M3, and 21.15 for M4, M5. The energy of a state with principal quantum number n (equal to 1 for K, 2 for L, and 3 for M) is then

$$E_S = Ry \cdot Z_S^2/n_S^2 \qquad (6.12)$$

with the Rydberg energy Ry=13.6 eV. The mean electronic velocity is

$$v_S = (2E_S/m)^{1/2} = v_0 \cdot Z_S/n_S \qquad (6.13)$$

where $v_0=2.19 \cdot 10^8$ cm/s^{-1} and m denotes the electron mass. The average radius is

$$r_S = a_0 \cdot n_S^2/Z_S \qquad (6.14)$$

with the Bohr radius $a_0=5.29 \cdot 10^{-9}$ cm. The "atomic units" $2 \cdot Ry$, v_0, and a_0 are occasionally used later. To take into account the actual binding energy U, a screening constant θ is introduced

$$\theta_S = U_S/E_S \qquad (6.15)$$

describing the "outer screening" for giving realistic integration limits in the Born approximation. The projectile enters in the calculation with Z_1 and through its velocity v relative to v_S of the electron, with which it interacts, expressed by the parameter

$$\eta_S = [v/(n_S v_S)]^2 = \frac{m}{A_1M} \cdot \frac{E_1}{E_S \cdot n_S^2} = \frac{E_1}{A_1 \cdot Z_S^2 \cdot n_S^2 \cdot 25 \text{ keV}} \qquad (6.16)$$

PWBA results are given in [6.1,39]. For the total cross section for ejecting an electron from S one obtains

$$\sigma_S^{PWBA} = 8\pi r_S^2 \cdot \frac{Z_1^2}{Z_S^2} \cdot \frac{1}{n_S^4 \cdot n_S} \cdot f_S \ (n_S, \theta_S) \tag{6.17}$$

where the function f_S can be found in [6.39]. In [6.1,5,38] is argued that at least for small $n_K/\theta_K^2 \lesssim 1$, to a good approximation one can set

$$\frac{\theta_K}{n_K} \cdot f_K \ (n_K, \theta_K) \simeq \frac{\theta_K^2}{n_K} \cdot f_K \left(\frac{n_K}{\theta_K^2}, 1 \right) = F_K \left(\frac{n_K}{\theta_K^2} \right) \ . \tag{6.18}$$

The function F, which with $N_S = (2j+1)$ electrons in S is generally [6.40]

$$F_S = \sigma_S^{PWBA} \cdot \theta_S \cdot Z_S^4 / (Z_1^2 \cdot N_S \cdot 4\pi a_0^2) \ , \tag{6.19}$$

depends thus for the K-shell only on the parameter n_K/θ_K^2, when this quantity is not much larger than 1. Figure 6.7 shows F_K from BRANDT [6.41] and BASBAS et al. [6.38].

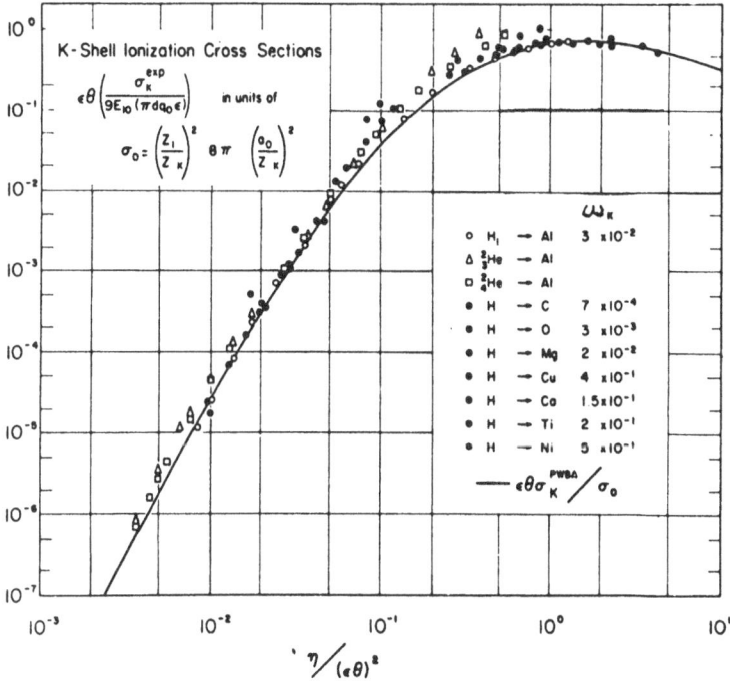

Fig. 6.7. K vacancy production from the plane wave Born approximation including the perturbations from increased binding and Coulomb deflection. The diagram from BRANDT [6.41], and BASBAS et al. [6.38] shows the function F_K in (6.18,19). The points indicate experimental results for H and He ions

F_S has a maximum value of 1 with a corresponding cross section $N_S \cdot 4\pi r_S^2 (Z_1/Z_S)^2/\theta_S$ for $v \sim v_S$ and is given in the low velocity region by the asymptotic forms [6.40], valid for $2v/(\theta_S v_S) \leq 0.1$,

$$F_K \simeq [v/(\theta_K v_K)]^8 \cdot 2^{17}/45 , \qquad (6.20)$$

$$F_{L1} \simeq [v/(\theta_{L1} v_{L1})]^8 \cdot 2^{20}/45, \qquad (6.21)$$

$$F_{L2} \simeq [v/(\theta_{L2} v_{L2})]^{10} \cdot 2^{21}/11, \qquad (6.22)$$

with F_{L3} having the same form as F_{L2}. For the K-shell, (6.20,19) give a low-energy dependence

$$\sigma_K^{PBWA} \simeq \frac{2^{20}}{45} \frac{\pi}{\theta_K^9} \frac{1}{Z_K^{12}} \frac{Z_1^2}{} \left(\frac{mE_1}{A_1 M \cdot Ry}\right)^4 a_0^2 , \qquad (6.23)$$

or E_1^4 and U_K^{-6} for $v/(\theta_K v_K) \leq 0.15$. The more complex structure of the L-shell restricts the validity of (6.21,22) to $v/(\theta_S v_S) \leq 0.05$. F_{L1}, i.e., the L_1 cross section, has a peculiar shape with two humps due to the radial node in the 2s wave function.

For various projectiles incident on the same Z_2, one sees from the general result (6.17) that the cross section scales with Z_1^2 for fixed velocity (same energy per nucleon E_1/A_1) as

$$\sigma_S^{PWBA} (Z_1, Z_2, E_1) = Z_1^2 \cdot \sigma_S^{PWBA}\left(1, Z_2, \frac{E_1}{A_1}\right) . \qquad (6.24)$$

A virtue of the PWBA is not only the results already stated but its ability to be used for further refinements. PWBA calculations can be made for cross sections differential in the energy and angle of the ejected electrons or of the scattered particle (cf.[6.42]). The description of the bound electron by hydrogenic wave functions with the screening procedure of (6.11-15) is unnecessarily restrictive, and may be replaced by Hartree-Fock wave functions (see references in [6.42]), or by relativistic wave functions as are important for the K electrons of heavy elements. The Born approximation may also treat distorted waves for the projectile, and the PWBA results constitute then a first step in these more realistic calculations.

The semi-classical approximation (SCA) takes as its starting point the classical description of the ion trajectory beginning parallel to the direction to the target nucleus in a certain distance ρ, the impact parameter. One calculates then in first order time-dependent perturbation theory the ionization probability as a function of the impact parameter [6.43-45]. Most calculations employ straight-line trajectories and hydrogenic wave functions described by the screening procedure in (6.11-15), so it is evident that one gets exactly the same results for the total cross sections as from the PWBA [6.46,47]. SCA has a simple conceptual basis and is primarily useful for impact parameter dependent calculations, but it should also be noted that SCA calculations with segmented linear trajectories (with a single break), and relativistic electronic wave function have recently been successful in accounting for the ionization of uranium by incident protons [6.48,49].

In the binary-encounter-approximation (BEA) we have still another treatment of the same collision process [6.2]. This time the initial picture is the classical encounter between the projectile and an electron, which can be treated analytically for energy transfer to an electron of a given velocity. Integration over the velocity distribution of the bound state electrons will then yield total cross sections or cross sections differential in energy and angle of the ejected electrons. The treatment is the same as the "impulse" approach, and, thanks to the agreement of the quantum mechanical and classical Rutherford cross sections, one gets the same kind of results as in the PWBA. The resulting total cross sections are very simple to estimate and obey in a commonly used approximation [6.2] the simple scaling relation

$$\sigma_S^{BEA} = \frac{Z_1^2}{U_S^2} \cdot n_S^2 \cdot g\left(\frac{mE_1}{A_1 M \cdot U_S}\right) \qquad (6.25)$$

where U is the experimental binding energy of shell S. The basis for this calculation is the description of the bound electron by a hydrogenic K (1s) velocity distribution determined by U and also the integration limit determined by U, i.e., with both inner and outer screening governed by the experimental binding energy U. The factor n_S^2 is just the weight factor for the mean number of electrons in a full shell (K, L, M,...) applying for an average value of σ over this shell. The scaling of (6.25) is a simple consequence of (6.18,16) for E_S equal to U_S (and thereby $\theta_S=1$). In the BEA procedure one can as well use more differentiated wave functions, e.g., Slater-experimental screened functions as in (6.11-15), or Hartree-Fock bound wave functions in a proper velocity representation.

6.2.2 Perturbation Improvements for Coulomb Ionization

The perturbed stationary state (PSS) approximation was used by BASBAS, BRANDT et al. [6.38,40,50] as the basis for investigating the ionization process by expanding the wave functions for a given impact parameter around the point of closest approach. They obtained total cross sections utilizing the PWBA results, and including two additional effects: 1) the binding of the electrons to the projectile during the collision, and 2) the deflection of the ion in the electric field of the target nu-

cleus. Both effects are treated as perturbations. The attraction, 1), by the projectile will (for the mostly contributing impact parameters smaller than the shell
radius) cause an increased effective binding energy of the electrons to be ejected,
and thereby a decrease in ionization cross section. The effect is described by ex-
changing θ_S with $\varepsilon_S \theta_S$ in (6.17-19) where the factor

$$\varepsilon_S = 1 + g_S \cdot 2Z_1/(\theta_S Z_S) \tag{6.26}$$

contains the velocity dependence in g_S ($\lesssim 1$), which is analytically given for K-
and L-shells in [6.38,40]. As ε_S is greater than 1, the effective value of θ_S is
increased and the ionization cross section will become smaller [see (6.18) and Fig.6.7]
by inclusion of 1).

The second effect, caused by the deviation from a straight line ion trajectory,
is determined by the reduced velocity and the increased distance from the target
nucleus at the closest approach for a hyperbolic trajectory of given impact parameter
It also gives lower cross sections, mainly due to the reduced effective projectile
velocity and is determined for s electrons by the factor $9E_{10}$ in σ occurring in
Fig.6.7 and described in [6.38,40]. Figure 6.7 displays this PSS treatment with both
the increased binding and the Coulomb deflection combined with the PWBA scaling.

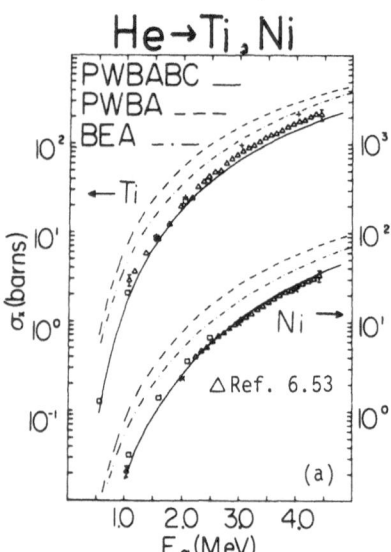
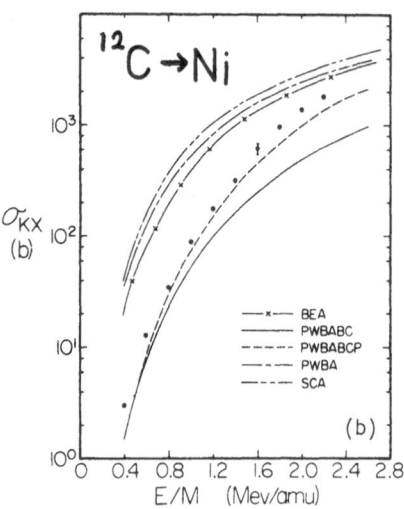

Fig. 6.8a and b. Direct ionization. Target K x-ray production cross sections for He and C
ions on Ni. Comparison is made between measurements and the binary encounter approximation (BEA), the plane wave Born approximation (PWBA), PWBA with corrections for
binding energy and Coulomb deflection (PWBABC) [6.38], and PWBABC with addition of
the polarization [6.41,52]. The results for 0.25-1.0 MeV/amu He+ on Ti and Ni are
from SOARES et al. [6.53] and for 0.4-2.2 MeV/amu ^{12}C ions on Ni from GRAY et al.
[6.54]

A third effect, the "polarization" of the electron distribution in the target atom due to the passage at large impact parameters, has also been treated [6.41,51, 52], and will by attracting the electron give lower binding energies and hence greater ionization cross sections, especially for high velocity ions.

It should be emphasized that the PSS procedure, mentioned above, incorporates new physics into the description in terms of PWBA, and dramatically improves the quantitative understanding, establishing a basis for generalizing the direct ionization results.

For light projectiles the "binding effect" represents a perturbation treatment of the attraction to the projectile. For heavy-ion impact, it leads to a description of the electrons by two center wave functions. Such molecular orbitals are discussed in Sections 6.3 and 6.4.

Experimental cross sections typical of direct ionization by heavy ions are shown in Fig.6.8 for 0.2-1.0 MeV/amu He on Ti and Ni from SOARES et al. [6.53], and for 0.4-2.4 MeV/amu C on Ni from GRAY et al. [6.54] as a function of the projectile energy. It is seen from Fig.6.8 that the unperturbed descriptions of the Coulomb excitation by PWBA, SCA, or BEA give high values. The measured data are well accounted for, especially in the low velocity region, when the effect of binding energy and Coulomb deflection is added to the PWBA treatment (PWBABC) in the PSS approach by BASBAS et al. [6.38]. When the polarization effect is included (PWBABCP), the high velocity data may also be understood.

6.2.3 Impact Parameter Dependent Direct Ionization

For $Z_1 \ll Z_2$ and for $\eta < 1$, the measured x-ray production cross sections are in reasonably good agreement with the different model calculations on inner-shell ionization. This holds true for the K-shell and at least partially for the total L-shell ionization. For the L-subshell ionization the different model calculations do not give the same answers. A crucial test for the various models can be given by differential measurements, i.e., coincidence measurements between scattered projectiles and emitted x-rays. Such coincidence measurements yield finally the ionization probability as a function of impact parameter, $P(\rho)$.

It should be mentioned that scattering angle and impact parameter can be calculated from each other as long as Bohr's criterion is fulfilled, cf. [6.43]

$$\kappa = 2Z_1 Z_2 e^2/(\hbar v) \gg 1 \quad . \tag{6.27}$$

(The relative uncertainty in the scattering angle $\Delta\theta/\theta$ is determined by $\sim (\kappa)^{-1/2}$; normally $\kappa \gg 1$). For each scattering angle or impact parameter, different regions of the electron distribution of concern contribute to the resulting ionization. Hence, a structure in the electron distribution should be visible in measured $P(\rho)$

curves. For example see the "knee" in the excitation function for the total L_1 sub-shell ionization, cf. Section 6.2.1.

For the various shells and subshells, theoretical values for the $P(\rho)$ curves can be calculated, primarily within the SCA model. This model is inherently appropriate for such calculations because the total cross sections are found by integrating cal-culated ionization probabilities for the various trajectories over all impact para-meters

$$\sigma_{tot} = 2\pi \int_0^\infty P(\rho)\rho d\rho \quad . \tag{6.28}$$

Thus $P(\rho)$ is available as a byproduct of total cross section calculation within the SCA model. Here a scaling law applies for straight-line paths and hydrogen-like electron wave functions [6.45,55],

$$P(\rho) = (Z_1^2/U_S) \cdot f(x,\xi) , \tag{6.29}$$

using the notation $x=\rho/r_{ad}$, $\xi=r_{ad}/r_S$, where r_S is the S-shell radius and r_{ad} the adiabatic distance, given by the inverse of the minimum momentum transfer leading to ionization

$$r_{ad} = \hbar v/U_S , \tag{6.30}$$

where U_S is the electron binding energy. Tabulated $P(\rho)$ values were recently published for K-, L-, and M- (sub-) shells, giving a reasonable basis for comparison with ex-periments [6.44]. Nevertheless, for a detailed comparison with experiments, these standard values should be replaced by SCA calculations applying more realistic wave functions (e.g., incorporating relativistic effects), using hyperbolic trajectories (deflection effects), and, eventually, applying a correction for the higher effective nuclear two-center charge felt by an electron during collisions (binding effects). Compared to a heuristically introduced tangential approximation [6.50], perturbed stationary states calculations may give better results [6.55] due to a proper inclu-sion of the screening effect.

Meanwhile, $P(\rho)$ values have also been calculated by modified BEA models [6.56]. The basis for such calculations is questionable within an impulse approximation due to the artificial way of introducing impact parameters, and the results should be applied cautiously. Concerning the PWBA it should be noted that a partial wave treat-ment can also yield $P(\rho)$ values. Such a treatment seems to be inadequate in parti-cular for heavy ions due to the high angular momenta involved.

Experimentally many coincidence measurements have been made for K-shell ioniza-tion (see Fig.6.9 and also the reviews in [6.63]). This is especially true for the domain of direct Coulomb ionization $Z_1 \ll Z_2$, $\eta < 1$ (see Fig.6.1). In this domain the

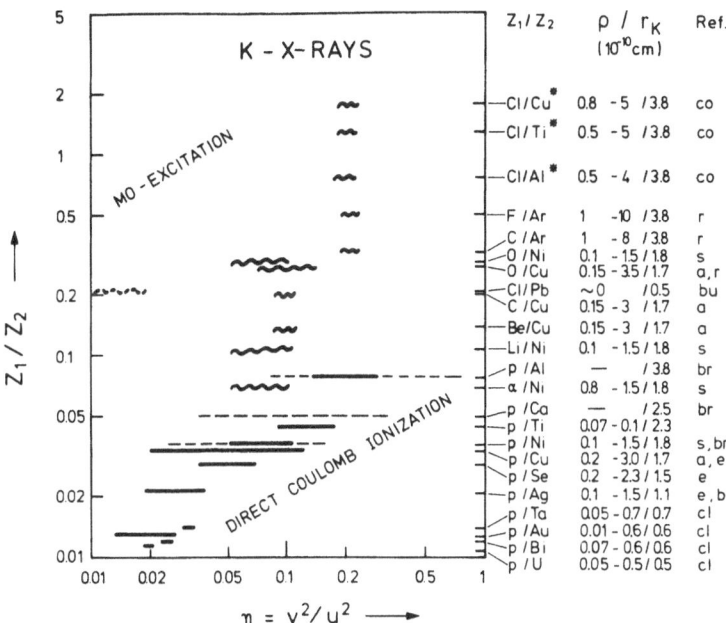

Fig. 6.9. Summary of impact parameter measurements for the K-shell ionization for the various projectile-particle combinations (Z_1/Z_2) and scaled energies (η). The impact parameters ρ scaled by the K-shell radius are given. The asterisks (*) indicate inverted collision systems $(Z_1 \leftrightarrow Z_2)$. The abbreviations refer to the following references: co, [6.57]; r, [6.58]; s, [6.59]; a, [6.55]; bu, [6.60]; br, [6.61]; e, [6.62]; cl, [6.49]

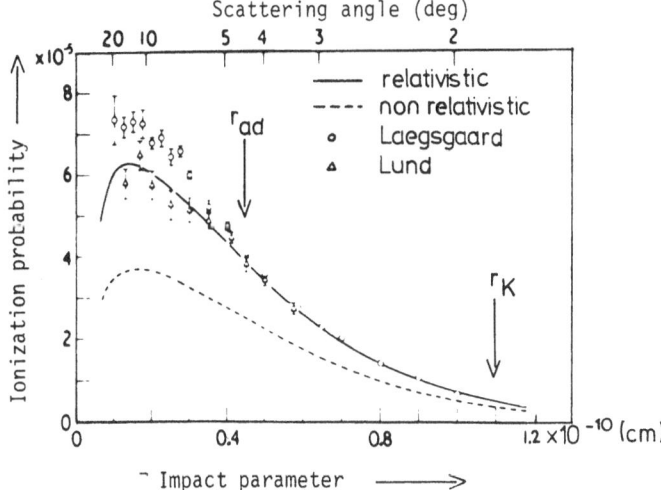

Fig. 6.10. K-shell ionization probability for 2 MeV proton impact on Ag. The curves are SCA calculations with corrections for deflections from straight-line paths using both nonrelativistic and relativistic wave functions (dashed and full lines, respectively) (see also [6.48])

experimental results, usually with protons as projectiles, can be compared with calculated $P(\rho)$ curves. In Fig.6.10 the results for 2 MeV proton bombardment of Ag are
given as an example. Applying the necessary corrections to the straight-line approximation, the SCA calculations yield generally satisfactory agreement with the experimental data concerning shape, amplitude, and energy dependence. For a comparison
the Ag-K radius r_K and the adiabatic radius r_{ad} are plotted in the figure.

According to SCA calculations, the differential (K-) ionization cross section—
given by $2\pi\rho P(\rho)$— peaks at about r_{ad}. Hence, impact parameters near r_{ad} contribute
most to the total cross section. This condition prevails for the whole direct Coulomb
ionization domain.

The validity of the calculations for the different models may be tested more effectively by studying the impact parameter dependence for subshell ionization for
higher shells. Here, the wave functions can have a radial structure, as for example
is associated with the node of the 2s wave function. SCA calculations do show the
corresponding structures for the $P(\rho)$ curves [6.44]. Unfortunately, only one x-ray
coincidence measurement exists for L-subshell ionization [6.64]. SCHMIDT-BÖCKING
and co-workers investigated the differential Lα and Lβ x-ray emission for proton
bombardment on a lead target; cf. Fig.6.11. Due to Coster-Kronig transitions the
different x-ray lines are not necessarily directly related to vacancy production in
the different subshells. Nevertheless, the Lα emission seems to describe more or
less the L_3 ionization of lead target atoms. In spite of the fact that the peak in
the $P(\rho)$ curve at about $r_L/2$ is roughly reproduced by the SCA model, there are discrepancies especially at low velocities and large impact parameters. Experimentally
unknown quantities (Coster-Kronig and fluorescence yields) as well as simplifications
in the calculation may contribute to these discrepancies. Knowing the differential
L- (sub-) shell ionization rates, one can calculate the probability for occurrence
of the different K-satellite lines. Such calculations may predict the experimentally
observed features of single-K and multiple-L excitation as observed for heavier projectiles.

Outside the direct Coulomb ionization domain a few K-shell experiments bridge the
gap to the MO domain (see Fig.6.9). Considering the SCA scaling law for $P(\rho)$, ANDER-
SEN et al. [6.55] give a scaling prescription partially applicable also for heavier
projectiles; see Fig.6.12. This prescription takes into account the impact parameter
dependent binding effects caused by the heavier projectile. Due to the increased
binding the scaled $P(\rho)/Z_1^2$ values fall appreciably below the proton curve for high-Z_1
values.

Measurements with Cl projectiles on various targets penetrate into the MO domain
at relatively high collision velocities (see [6.57]). In these experiments the projectile K radiation was measured. In Fig.6.13 the Z_2 dependence of $P(\rho)$ for Cl K
x-ray emission is shown for different ρ values. Here the role of target and projectile are interchanged (inverse collision system). A scaling with Z_2^2 does not apply.

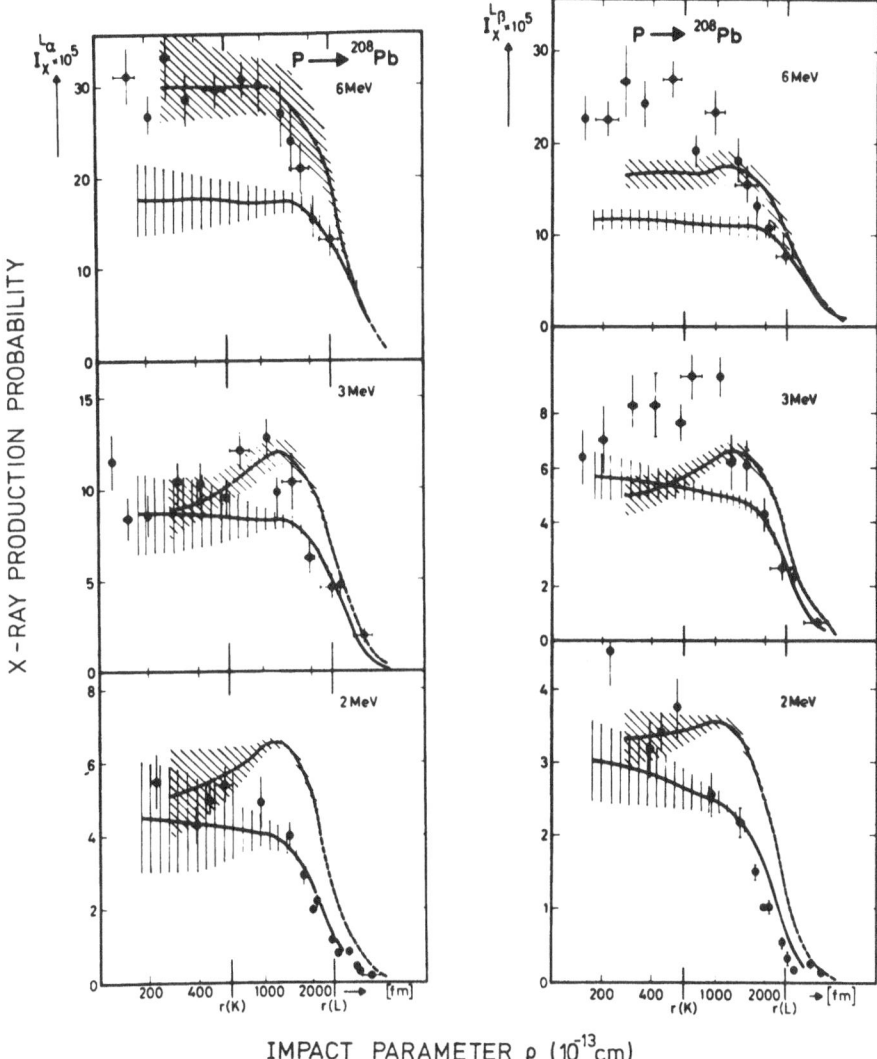

Fig. 6.11. Impact parameter dependence of Pb L_α and Pb L_β x-ray emission at 2,3, and 6 MeV proton bombardment. Solid and dashed curves give BEA and SCA calculations, respectively. The hatched regions are due to the uncertainties in fluorescence yield and in Coster-Kronig transitions (see [6.64])

There seems to be a resonance for near symmetric collision systems $(Z_1 \sim Z_2)$, which is a typical feature of the MO domain.

More elaborate measurements in the MO domain have been made for lighter systems, such as Ne-Ne [6.63]. Due to their small adiabaticity parameter $(\eta \lesssim 0.01)$, they are not shown in Fig.6.9. Because of their special character these measurements will be discussed later. This subsection should not be closed without mentioning the first

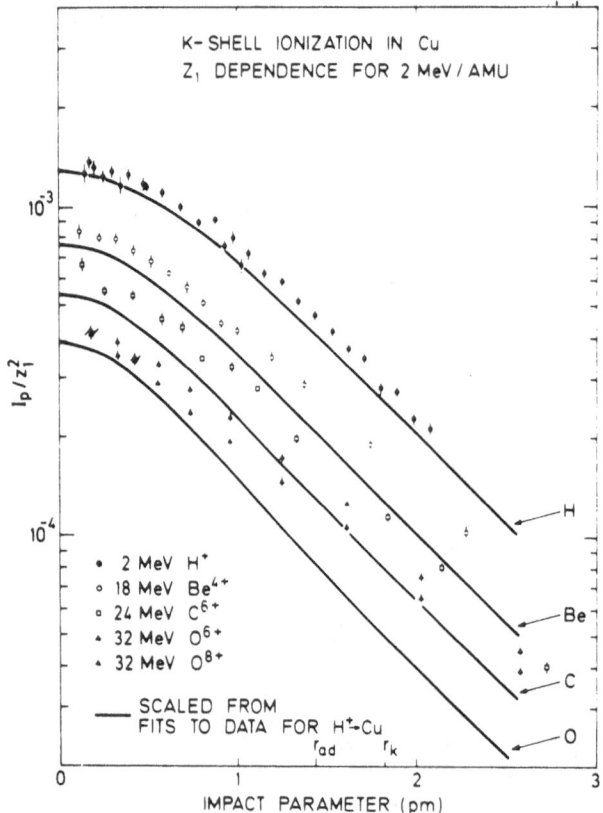

Fig. 6.12. Ionization probability for Cu-K x-rays, scaled by Z_1^2, as a function of impact parameter for 2 MeV/amu projectiles. The curves give the scaling prescription of the Aarhus group [6.55]

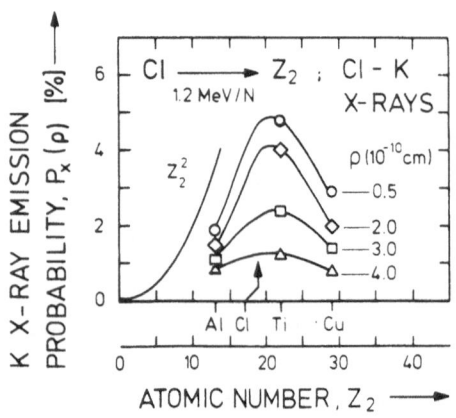

Fig. 6.13. K x-ray emission probability for Cl projectiles as a function of the atomic number Z_2 of the collision partner (inverted system) for different impact parameters. The smooth curves through the data points are only guides to the eye. A Z_2^2 dependence is indicated by the curve at the left side (cf. [6.57])

x-ray particle coincidence measurement of STEIN et al. [6.65] done for the near symmetric system I-Te. Up to now, these are almost the only differential measurements
for higher shells (L-shell) in the MO domain.

6.2.4 Electron Capture

The Coulomb interaction of the projectile with the target electrons may not only
induce transitions of the electrons to the continuum or to other bound states of the
target atom. Target electrons can also be captured into empty orbitals of the projectile, and this process may, especially for heavy ions of high charge state, contribute significantly to the vacancy production for the target atom.

Results from MACDONALD et al. [6.66] are shown in Fig.6.14, where K x-ray emission
both from the projectile and from the target from collisions of 3.0 and 3.77 MeV/amu
Cl ions incident on a SiH_4 gas target, is included. As a function of the number of
electrons carried by the projectile, the cross sections for emission of K x-rays
from the Cl projectile (circles) and from the Si target atom (squares) are displayed.
Characteristic projectile K x-rays observed with 0 or 1 initial electrons simply reflect capture to outer-shells, followed by transitions to the K-shell. It is noted
that the higher energy gives the lower projectile cross section, a trend which is
characteristic of electron capture by high velocity ions [see (6.31)]. For projectiles with more than two electrons the x-ray cross sections for target and projectile
are equal and independent of projectile energy. The reason is that the direct ionization for these incident velocities close to the mean orbital velocity of the ejected
electrons is near the maximum, seen for example in Fig.6.7. For target excitation,
a substantial increase of the cross section for the few electron projectiles is noted.

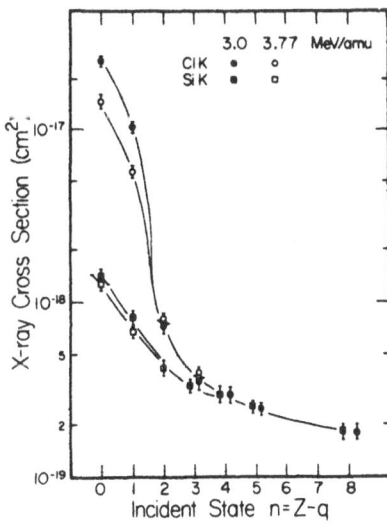

Fig. 6.14. Target and projectile K x-ray production as function of the number of electrons
carried by the Cl projectile. Cl^{q+} ions of
energy 3.0 and 3.77 MeV/amu (filled and open
symbols) bombard a SiH_4 gas target. The K x-
-rays of Si (squares) fall below those of Cl
(circles), and the higher energy gives the
lower cross section. The figure from MACDONALD
et al. [6.66] shows that electron capture is
important for highly stripped projectiles

This higher contribution, especially pronounced for projectiles with zero or one electron, can be assigned to capture of target K electrons by the projectile, and also is lower for the higher projectile energy as was the case for capture of all target electrons to outer projectile orbitals.

A simple calculation of the electron capture can be made in the first Born approximation (PWBA), which for hydrogenic wave functions gives the Brinkmann-Kramers cross section [6.67,68]

$$\sigma^{BK} = C_{BK} \cdot \frac{n_S^2 \cdot E_S^{5/2} \cdot U^{5/2} E^4}{[4 \cdot E \cdot U + (E-U+E_S)^2]^5} \tag{6.31}$$

with $C_{BK} = m \ v_0^2 \cdot a_0^2 \cdot 2^{17} \cdot \pi / 5 = 6.27 \cdot 10^{-14}$ keV cm^2 for the capture of an electron with binding energy U in the target to bound states in the shell of quantum number n_S and binding energy E_S in the projectile. The projectile velocity v enters through the kinetic energy E of an electron with the same speed $E = m \cdot v^2 / 2 = E_1 \cdot m / (A_1 M)$. The result in (6.31) is averaged over all initial states in a full shell for the target and summed over all final bound states in the projectile with principal quantum number n_S and binding energy E_S, which may be found from the Slater rules, cf. (6.11,12), for the highly stripped projectiles of interest. From (6.31) it is seen that the highest degree of electron capture is obtained for a minimum value of the denominator. For given U and v (i.e., E) it means that the strongest capture goes to bound states in the projectile of binding energy E_S closest to U-E.

The theoretical result in (6.31) is however systematically higher than experimental values, typically by a factor 3 to 10 [6.66,67], and often an overall scaling factor is included in (6.31) to reproduce measured cross sections and their dependence on v, n_S and U. A reason for this scaling procedure is that the second Born approximation gives σ^{BK} again, but reduced by a factor around 3 with a weak v-dependence [6.67]. Still, the present theoretical understanding is not satisfactory, and recent investigations show that the Coulomb deflection [6.69] and inclusion of binding energy corrections, as by the PSS description [6.67], improve the estimates.

6.2.5 Complex Excitation

Direct ionization depends not only on energy and atomic number of the projectile, as the simple theory predicts, but also on the ionic charge state, and it will furthermore be a function of several other factors which are sometimes pronounced if the excitation of a solid target is observed - and compared to that of a gas target. One category of more complex ionization is simply related to direct ionization by additional processes occurring in the bulk of the target, such as recoil of atoms, enhancement by photons and electrons, and beam excitation to well-defined electron

configurations. Another involves more complicated ionization over a sequence of excitations which can either follow in subsequent collisions in a solid or happen in
one collision.

Recoiling target atoms may contribute to the target x-ray yield for thick solid
targets. This process, described by TAULBJERG et al. [6.70,71] is important if the
incident ion causes only weak ionization, but can give a large momentum transfer to
a target atom, which then by succeeding symmetric collisions between the recoiling
atom and other target atoms has a high probability for inner-shell ionization. This
is the case for the Ne on Al data from [6.71] and for the low-energy part of the O
on Al data from [6.51], where the intense MO excitation for Al on Al, described in
Section 6.3, makes the recoil contribution to the Al K x-ray yield dominant, 10 to
100 times the direct ionization.

Simple enhancement can be caused by photons, e.g., x-rays, generated in or near
the target and by secondary electrons or electrons following the projectile, which
ionize the target atoms as well as the primary ion beam. The contribution is normally
below or at the percent level of the main ionization, but can easily reach 10%-40%
of the direct ionization for a strong flux of x-rays above an absorption edge, or
for high bombarding energies which generate energetic secondary electrons [6.30,72].
Geometrical factors affect the photon enhancement, such as the thickness and the
close surroundings of the target. Photon sources are normally seen in the x-ray
spectra, whereas secondary electrons are not. It should therefore be noted that
electron ionization is important if the beam velocity is of order of or exceeds the
mean orbital velocity of target electrons studied. Based on calculated direct excitation of secondary electrons for a solid Al target, the enhancement grows from
1% to 26% for E_1 varying from 2 to 8 MeV/amu. Both enhancement effects will give
unshifted x-ray lines (see Sec.6.1.3) and may thereby be distinguished from ordinary
heavy-ion excitation by a study of a high resolution x-ray spectrum obtained with a
crystal spectrometer [6.72].

The change of charge state and excitation of a heavy ion, when it penetrates a
solid, will affect its ionizing power, and often a nontrivial target thickness dependent x-ray yield results. A simple description, which is valid for target ionization by ions with $Z_1 \simeq 9$-20, is based on a separate treatment of ions with one or
two K vacancies, and with different ionization cross sections. With the boundary
condition of a given incident charge state, beam excitation and deexcitation cross
sections affect the degree of K-shell beam excitation and introduce a target thickness dependent ionization. This three-component model for target x-ray excitation
[6.73] successfully describes the K ionization of Cu by 1.7 MeV/amu F, Al, Si, and
S ions of initial charge states Z_1, Z_1-1, and less than Z_1-1. The complex exponential dependence on the target thickness saturates in these cases to a constant mean
cross section at 20-150 $\mu g/cm^{-2}$ and can both increase and decrease, according to the
incident charge state. For cases with only few double K vacancies in the projectile,
a simpler two-component model describes the state of the ion moving in the solid and

its ionization as shown by BETZ et al. [6.21] and GRAY et al. [6.74]. The x-ray decay of the projectile can be accounted for by a similar procedure but is more complicated, with x-rays emitted both in and after the foil with different lifetimes.

Higher shells of the projectile may also be ionized and deexcited in a solid and can affect the x-ray production, giving additional contributions to the simple direct excitation. Examples of survival of outer vacancies from one collision to a later collision, where they enhance the inner-shell excitation, are mentioned in Sections 6.3 and 6.4.

A statistical description of the ionization process has applications both to direct ionization with many electrons participating and for multi-step excitation processes. In these cases the multi-electron atomic system is characterized by a diffusion equation and can be treated as a Thomas-Fermi gas. BRANDT and JONES [6.75] have applied this alternative description for a sequence of vacancy excitations to the impact parameter dependence of K-shell ionization, and have described the observed systematics by means of a reasonable diffusion constant. The interaction of the complete electron cloud accompanying a heavy ion with that of a heavy atom in the target is another interesting problem which calls for a statistical description. In addition to direct excitation by such an encounter, collective excitations may be anticipated [6.76].

6.3 Quasi-Molecular Excitation, $Z_1 \sim Z_2$

6.3.1 The Electron Promotion Model

For the direct Coulomb ionization domain $Z_1 \ll Z_2$, the various theories predict a Z_1^2 dependence for the inner-shell ionization cross section of target atoms ($\eta \ll 1$). For heavier projectiles $1 < Z_1 \ll Z_2$, corrections due to the increased electron binding must be applied [6.38]. This binding effect is due to the additional nuclear charge (Z_1) acting (quasi-adiabatically) on the target electrons during collision. For $Z_1 \sim Z_2$ the distortion energy may even become larger than the initial binding energy. Hence, simple binding corrections may not apply and, experimentally, quite new effects are found:

1) extremely large cross sections for small η;
2) cross sections enhancements for certain Z_1/Z_2 ratios ($\sim n_1/n_2 = 1/1, 1/2, 2/1, \ldots$) [6.77];
3) abnormal intensity ratios for different x-ray lines (subshell differentiation) [6.78];
4) line shifts to higher x-ray energies and/or satellite lines.

Examples of these different effects are given in Figs.6.15 to 6.18. Effect 4) is easily explained by multiple ionization of the less tightly bound electrons caused

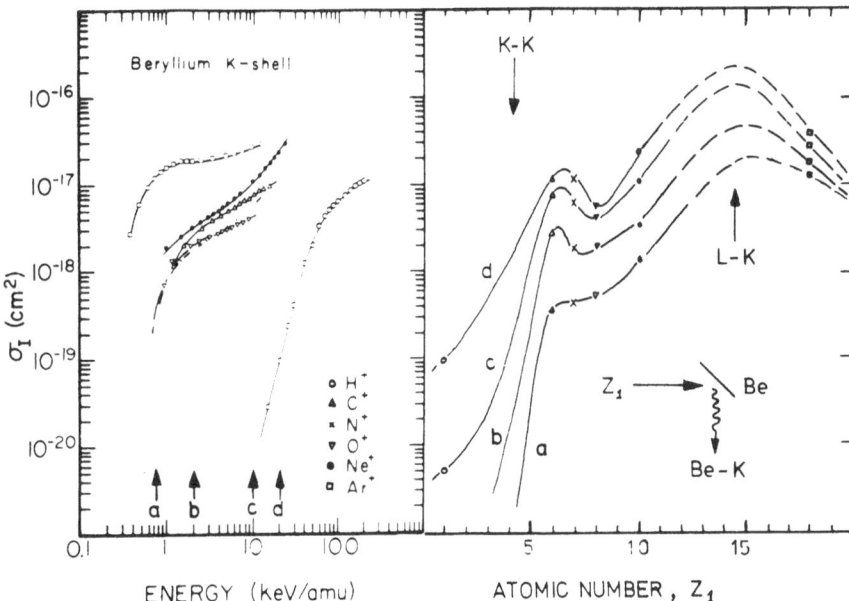

Fig. 6.15. K-shell excitation cross sections for solid Be target bombardment by different incident ions (see [6.2]). On the left-hand side the energy dependence of cross sections is given. On the right-hand side the cross section dependence on the atomic number Z_1 of the projectile is given

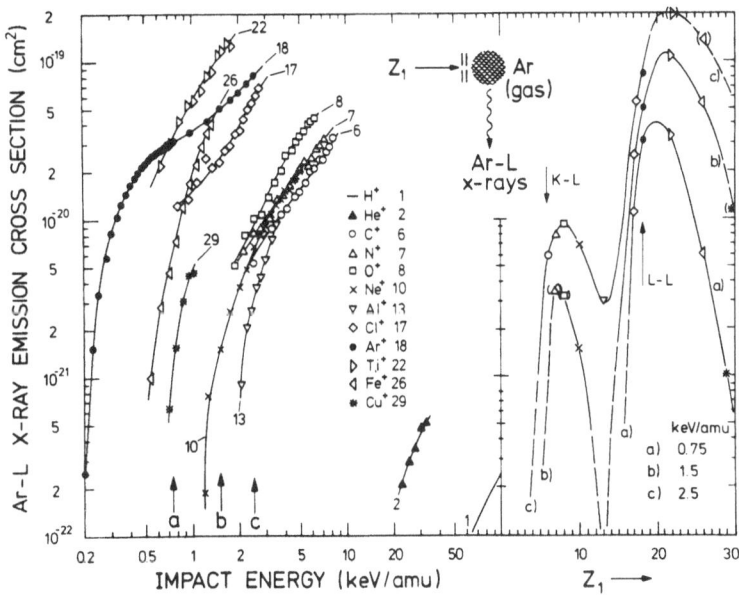

Fig. 6.16. L-shell excitation cross sections for Ar (gas target) bombarded by different incident ions (see [6.79]). On the left-hand side the energy dependence of the cross sections is given and, on the right-hand side, its dependence on the atomic number Z_1 of the projectiles is given

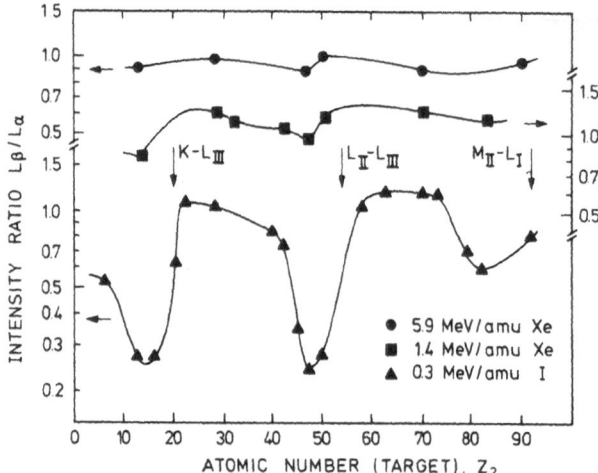

Fig. 6.17. Intensity ratio of the principal L lines (L_α and L_β) of Xe projectiles as a function of the atomic number Z_2 of the solid target atoms, shown in a semi-logarithmic plot. The low-energy data are for I projectiles taken from [6.78]. The Xe data are from [6.80]

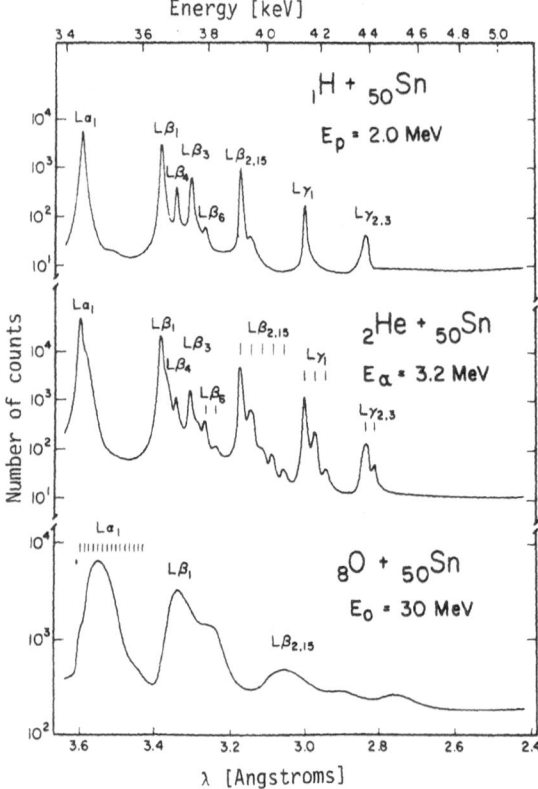

Fig. 6.18. Fine structure of characteristic x-ray spectra for p, α, and ^{16}O bombardment of a tin target. The spectra were measured by OLSEN et al. [6.81] with a crystal spectrometer and show the L_α, L_β, and L_γ groups of Sn

by the high effective charge of the collision partners. Hence, experimental line
shifts or the intensity distribution of the satellite lines may give information on
the average excitation - mainly of the next higher shell - after the collision. In
contrast, the other effects cannot be described by such simple explanations. The
electron promotion model of FANO and LICHTEN may be applied [6.82]. This model gives
a good qualitative description of most of the experimental findings, but, except in
certain special cases, no general quantitative predictions are provided. Though this
model is certainly also important in explaining effect 4), the so-called subshell
differentiation has not been explained consistently over the whole Z_1/Z_2 range.

For $\eta \ll 1$ and $Z_1 \sim Z_2$, both target atom and projectile have inner-shell electrons.
Due to the high orbital velocities compared to the collision velocity, these electrons can continuously adjust for the perturbation caused by the collision partner,
and a collision molecule with slowly varying internuclear distance R is transiently
formed for inner-shells. For small internuclear distances $R \sim 0$, the electron configuration approaches asymptotically that of the united atom (UA) system with an effective
atomic number $Z_{eff} = Z_1 + Z_2$. In Fig.6.19a the variation of electron wave functions with
internuclear distance is indicated for K-shell electrons of two identical collision
partners. Based on similar considerations for the one-electron case FANO, LICHTEN,
and BARAT [6.82] derived general rules for correlating separated atom (SA) levels
with UA levels (cf. Fig.6.19b). They neglected completely electron-electron exchange
forces and refer to their correlations as (one-electron) diabatic correlations. Hence,
the diabatic correlation rules given in Fig.6.19b are correlations for one electron
in the field of two point charges (H_2^+-like correlations). Such diabatic correlation

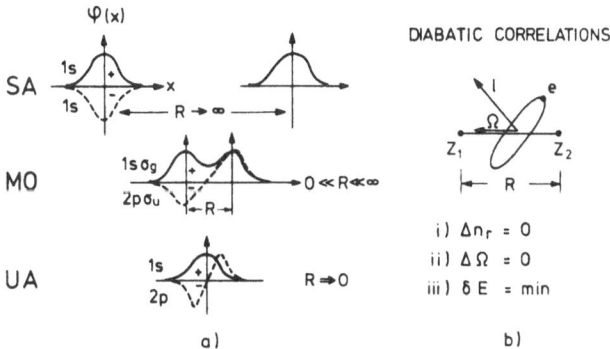

Fig. 6.19. (a) Symmetric and antisymmetric addition of two 1s electron wave functions
for two identical separated atoms (SA). At decreasing intermediate internuclear distances R, where molecular orbital (MO) wave functions occur, the united atom region
(UA) is reached, ending with 1s and 2p atomic wave functions.
(b) Diabatic correlation rules (H_2^+-like correlations). Here $n_r = n-\ell-1$ is the radial
quantum number for the SA and UA system; Ω is the projection of the angular momentum
(ℓ) on the internuclear axis, δE=min means that the deepest bound possible states
in the SA's and UA correlate with each other

234

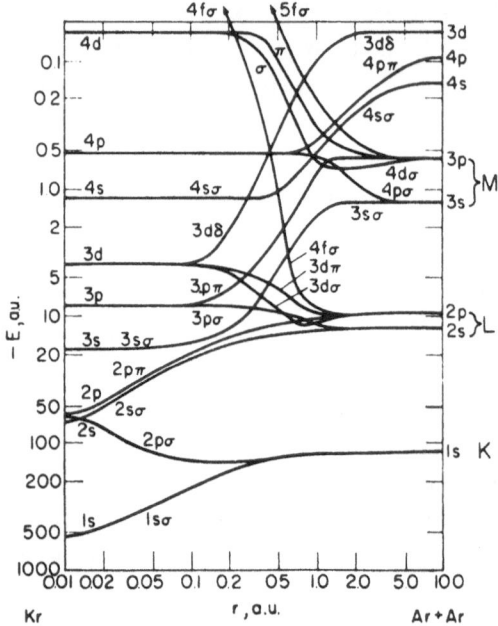

Fig. 6.20. Diabatic level diagram for the Ar-Ar collision system [6.82]

diagrams do not account for detailed dependences of molecular orbital (MO) levels on the internuclear distances R for intermediate R. Here, multi-electron level calculations are necessary. Recently, such calculations have become feasible for even very heavy systems like I-Au or U-U [6.83-86]. Nevertheless, pure one-electron diabatic correlation diagrams do already show the prominent features for inner-shell excitation: electrons are promoted or demoted in binding energy during collisions, as in the classical Ar-Ar collision system. The corresponding correlation diagram is shown in Fig.6.20. During collision, electrons are promoted to binding energy regions where they can more effectively be ionized. The vacancies created there in the quasi-molecule can be transferred during separation down to the SA levels via diabatic MOs.

For instance, the high L-ionization cross section for Ar-Ar collisions can be explained by electron promotion via the $4f\sigma$ level (cf. Fig.6.16 and 6.20). In this special case the $4f\sigma$ promotion of two Ar-L electrons enters into a binding energy region where the electrons can cross into available empty levels. In this way the electrons couple to open reaction channels. For heavier collision systems, no open reaction channels need exist. In this case a closed channel situation arises, and direct ionization of the promoted electrons will prevail (cf. also [6.87]). For intermediate collision systems or at higher impact energies, entrance channels can dynamically be open. Vacancies occurring during the incoming part of the collision may be transferred down to a binding energy region where they overlap in binding energy with the promoted electrons. Solid-state excitation may also contribute to dynamically open channels. For a dynamically open channel problem, direct ionization and electron crossing into the partially empty channel will compete.

During separation the created vacancy will be transferred down to SA levels. On its way, the vacancy created can be lost into ascending levels at crossings diminishing the inner-shell ionization cross section. Generally, one may state—independent of the entrance channel situation—that high electron promotion yields high ionization cross sections. This holds true for both collision partners, independent of which one is the projectile or target atom. Within this electron promotion model not only the large cross sections 1) but also the Z resonances 2) can be explained. For matching binding energies of both collision partners the correlations may swap, causing a steep electron promotion (cf. Figs.6.15 and 6.16). Due to this high electron promotion, cross section enhancement is found around shell matching collisions ($Z_1/Z_2 \sim 1/1$, 2/1, 1/2, ...). This statement is independent of the entrance channel situation and is true for both target and projectile ionization. Both points agree qualitatively with experiments.

To get quantitative agreement with experiments these qualitative descriptions must be replaced by more elaborate considerations. The basis of each such consideration should be a reliable level calculation. For instance, recently, systematic level calculations revealed a possible deviation from the correlation rules, shown in Fig.6.20 for the subshells [6.86]. This deviation may partially contribute to the subshell differentiation 3).

A second fundamental point to be considered is electron or vacancy transitions between MO levels. Theoretically, this procedure results in a fully coupled multi-channel system of equations [6.5]. Due to the complexity of most collision systems a complete unified treatment is not possible. Until now only isolated crossings have been studied. The few prominent crossing types so far investigated are summarized in Table 6.4. The first two crossings involve radial couplings important mainly at larger internuclear distances [6.88,89], while the last two types are of rotational character effective mainly at small internuclear distances, where the role of rotation of the internuclear axis is large [6.90,91].

Based on the above considerations the game of disentangling the ionization processes is discussed in the following subsection. The rules of the game are provided by the framework of (diabatic) level diagrams using a two-dimensional classification of open, dynamically open, and closed entrance channels; and radial and rotational couplings, respectively.

6.3.2 K-Shell Ionization

In Fig.6.21 a number of different cases for K-shell ionization are summarized. The discussion is restricted to collision systems with $2Z_1 > Z_2 > Z_1/2$. These systems have similar correlation diagrams: for $Z_h < 2Z_\ell$ the K electrons of the heavier (Z_h) and

Table 6.4. The prominent crossing types, where $\Delta\Omega$ is the change of the projection of the angular momentum on the internuclear axis R, V_{12} the matrix element for the coupling, E_i the level separation energy, I_i the binding energy in the corresponding atoms, ρ the impact parameter, v the impact velocity, and Z the atomic number. For the crossing types 1,2,3, and 4 see [6.88], [6.89], [6.90], and [6.91], respectively

No	Crossing	Type	Transition probability P	$|\Delta\Omega|$	Effective region
1	Landau--Zener		$\exp\left\{-\dfrac{\hbar}{2\pi}\cdot\dfrac{1}{v}\cdot\dfrac{|V_{12}|^2}{d/dR(E_1-E_2)}\right\}$	0	large R
2	Demkov		$\mathrm{sech}^2\left\{\dfrac{\sqrt{2\pi}}{v}\cdot\dfrac{I_1-I_2}{\sqrt{2m_0 I_\epsilon}}\right\}$	0	large R
3	Russek		$\exp\left\{-\dfrac{\hbar}{2\pi}\cdot\dfrac{b\cdot v}{R^2}\cdot\dfrac{|V_{12}|^2}{d/dR(E_1-E_2)}\right\}$	1	small R
4	Briggs--Macek		$P_Z(\rho,v)\sim P_1(Z\cdot\rho,\dfrac{v}{Z})$	1	small R

radial couplings (rows 1, 2); rotational couplings (rows 3, 4)

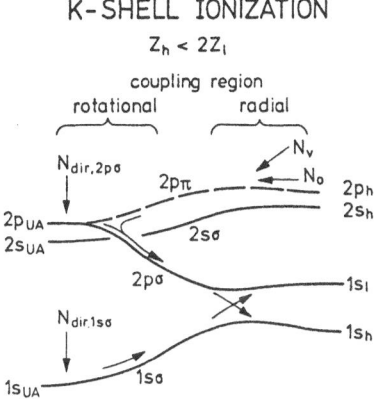

K-SHELL IONIZATION

$Z_h < 2Z_l$

coupling region

rotational   radial

$N_{dir,2p\sigma}$

$2p_{UA}$
$2s_{UA}$

$N_{dir,1s\sigma}$

$1s_{UA}$

R = 0 — INTERNUCLEAR DISTANCE → R = ∞

a) K-ionization cross section ratios		
region	1sσ excitation	σ_h / σ_l
$Z_h \not\approx Z_l$	direct ionization	$\gtrsim 4000$ *
$Z_h \sim Z_l$	sharing	$\sim \exp(-\pi \cdot \Delta Z / v)$

* depending on exit channel

b) 2pσ excitation		
region	exit channel	dominant term
$Z_h \gtrsim 10$	open	$N_0 \cdot \sigma_{rot}$
$10 \lesssim Z_{h,l} \lesssim 20$	dynamically open	$N_v \cdot \sigma_{rot}$
$Z_{h,l} \gtrsim 20$	closed	$N_{dir,2p\sigma}$

Fig. 6.21. The graph shows the different contributions to K-shell ionization of the heavy (Z_h) and the light (Z_l) collision partner for $Z_h < 2Z_l$. In the adjacent tables the relevant contributions for the different Z regions are summarized. Here N_0 is the number of static vacancies, N_v is the number of vacancies created dynamically at larger internuclear distances R, N_{dir} is the number of vacancies created by direct ionization processes (dominantly at smaller R), and V in atomic units

the lighter (Z_l) collision partner correlate via the 1sσ and 2pσ MO level to the 1s and 2p UA levels, respectively. For symmetric collision systems $Z_1=Z_2$, the 1sσ and 2pσ MO levels are shared by the degenerate K levels of both particles. In that case parity is a good quantum number and no interaction between the 2pσ odd and 2sσ even states is expected. Similar arguments may hold true for near symmetric collision systems $Z_h \sim Z_l$. For $Z_h \neq Z_l$ it is more difficult to make parity arguments [6.92]. We assume no (radial) coupling between the 2sσ and 2pσ MO's for the Z region investigated here ($Z_h < 2Z_l$). This will not hold for very heavy collision systems. In contrast we consider the coupling between the 2pπ and 2pσ levels (rotational coupling at small internuclear distances) and between the 2p and 1s levels (radial coupling at large internuclear distances).

For sufficiently asymmetric collision systems $Z_1 \neq Z_2$, the radial coupling between the relatively distant 2pσ and 1sσ levels can be neglected, i.e., the K-shell ionization for both collision partners can be treated independently. Except for very low-Z particles, we have a closed entrance channel problem (no 2pπ vacancies) to be treated approximately by the direct ionization formalism. And indeed, a surprisingly good agreement of measured and correctly scaled K ionization cross sections with the general BEA curves is found for the low velocity region [6.93,94]. For example, FOSTER et al. [6.93] used UA values for scaling: $Z_{eff}=Z_1+Z_2$, and 1s UA and 2p UA binding energies for heavier and lighter particles, respectively. Such a scaling predicts a cross section ratio for the relevant levels of $\sigma_{2pUA}/\sigma_{1sUA} \sim 4000$ corresponding to the K-shell ionization cross section ratio of light to heavy collision partner (cf. Table (a) in Fig.6.21).

For nearly symmetric collisions $Z_1 \sim Z_2$, the $1s\sigma$ and $2p\sigma$ levels approach at larger internuclear distances and the radial coupling between the levels cannot be neglected anymore, due to the vanishing ionization probability of electrons in the 1s UA level compared to those in the 2p UA level. The $1s\sigma$ MO level can be assumed as completely occupied compared to the probability of having a vacancy in the $2p\sigma$ MO level during the outgoing part of the collision. At large internuclear distances, such a $2p\sigma$ vacancy will be shared between both MO's. The dependence of the $1s\sigma$ and $2p\sigma$ level on internuclear distances does not show a Landau-Zener type crossing. Their interaction may possibly be approximated more accurately by a Demkov-type coupling as assumed by MEYERHOF et al. [6.95]. Both mentioned types of crossings yield the same approximate asymptotic transition probability:

$$P_{2p\sigma \to 1s\sigma} \sim \exp(-\beta \frac{\Delta Z}{v}) \tag{6.32}$$

with $\Delta Z = |Z_1 - Z_2|$, v the collision velocity in atomic units, and β a constant around π. Such a dependence is also obtained from consideration of the width or overlap of the MO levels caused by the short collision time due to the uncertainty principle.

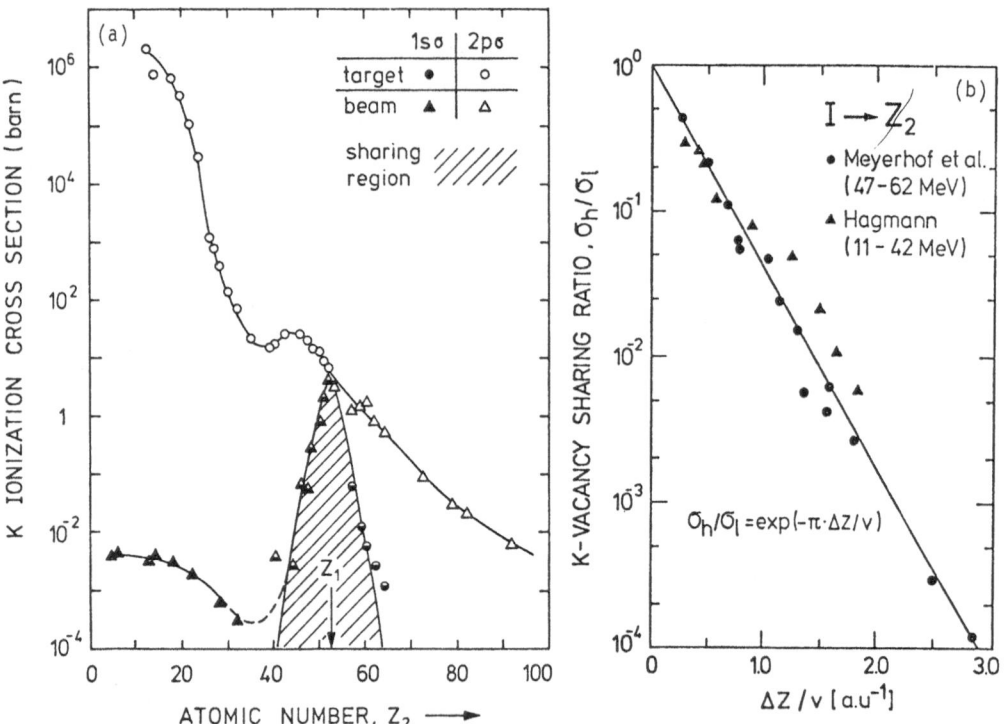

Fig. 6.22a and b. K-shell ionization for I ions impinging on various targets Z_2. (a) The K-shell ionization of both collision partners as a function of Z_2 for 47 MeV impact energy. The sharing region is indicated by the hatched area. The data are taken from [6.95]. (b) The cross section ratio of both particles $\sigma_{K-h}/\sigma_{K-\ell}$ as a function of $\Delta Z/v$. Here $\Delta Z = |Z_1 - Z_2|$ and v is the collision velocity in atomic units [6.95,96]

With $P_{2p\sigma \to 1s\sigma}$ from (6.32), the cross section ratio for K-shell excitation of both collision partners is given by

$$\frac{\sigma_{K-h}}{\sigma_{K-\ell}} = \frac{P_{2p\sigma \to 1s\sigma}}{1-P_{2p\sigma \to 1s\sigma}} \sim P_{2p\sigma \to 1s\sigma} \quad . \tag{6.33}$$

For small values of $P_{2p\sigma \to 1s\sigma}$ the cross section ratio is directly given by $P_{2p\sigma \to 1s\sigma}$ itself. A semi-logarithmic plot of K-shell ionization cross section ratios against $\Delta Z/v$ gives thus a straight line through 1 with a slope around $-\pi/\ln 10$. In Fig.6.22 this is demonstrated for a closed channel problem. STOLTERFOHT et al. [6.97] report on similar agreement using Auger electron data for open $2p\sigma$ MO channels. Agreement is also expected for dynamically open channel problems. Such systems have been described by FASTRUP [6.87], resulting in a velocity dependent probability for a vacancy in the $2p\pi$ level, $N_{2p\pi}(v)$. This velocity dependence of $N_{2p\pi}(v)$ is possibly determined by a distant radial coupling to outer levels during the incoming part of the collision. Hence, for $N_{2p\pi}(v)$ a sharing model applies. In Table (b) in Fig.6.21 the entrance channel problems are summarized for the various Z regions.

6.3.3 Impact Parameter Dependence

The K-shell ionization cross section ratio is independent of the entrance channel situation for the sharing region $Z_1 \sim Z_2$, but not outside this region. In both cases the absolute cross sections depend on the entrance channel situation, particularly for the K-shell ionization of the lighter particle. For closed channel problems the $2p\sigma$ ionization can be estimated using modifications of the direct ionization theories (see Sec.6.3.2). For open or dynamically open problems, vacancies can be additionally transferred to the $2p\sigma$ MO level via rotational coupling at small internuclear distances.

BRIGGS et al. studied the impact parameter dependent transfer behavior of a $2p\pi$ vacancy to the $2p\sigma$ level in a coupled two-channel approximation [6.91,92,98]. The transfer probability is given in Fig.6.23a. At low collision velocities the transfer is drastically reduced by the Coulomb deflection. At higher collision velocities an additional peak occurs at small internuclear distances, yielding a transfer probability of 1 at higher impact energies. This peak is situated at 90^0 in the center of mass system ($\sim 45^0$ in the lab. system) and it is caused by the nonadiabaticity of the rotation of the internuclear axis. The incoming electronic wave function will not change by a 90^0 scattering for a $2p\pi \to 2p\sigma$ transition; only the internuclear axis will be rotated and thus will indicate a transition. This kinematic effect was later referred to as a slippage effect [6.99]. But this kinematic peak contributes little to the total cross section shwon in Fig.6.23b. The cross section shows a threshold behavior due to the Coulomb deflection and increases beyond this region approximately proportional to $v^{2/3}$. Scaling laws apply for both the differential and total cross section [6.91]:

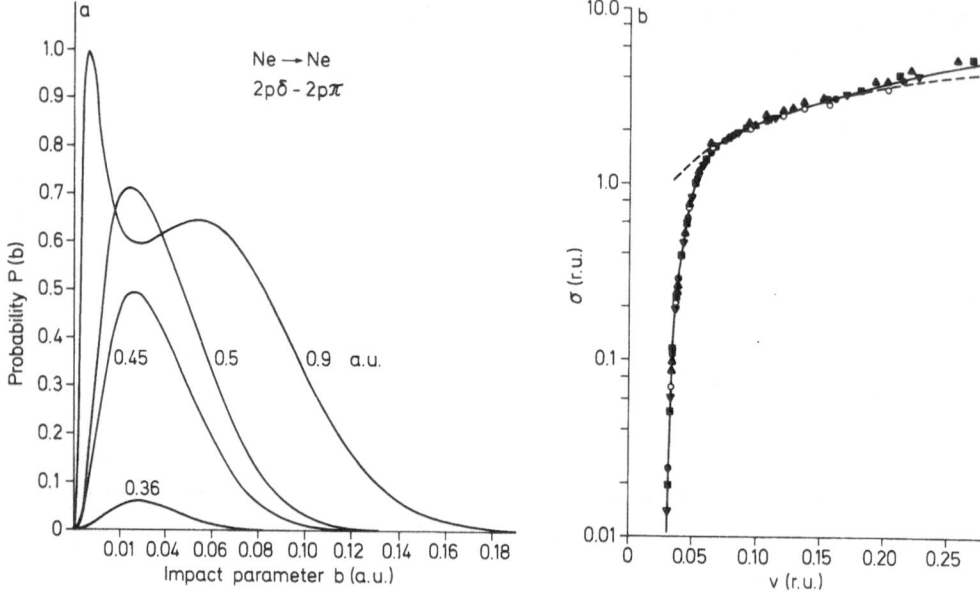

Fig. 6.23a and b. Theoretical calculation on the 2pπ→2pσ rotational coupling for Ne-Ne collisions [6.92,98]: (a) Transfer probability $P(\rho,v)$ as a function of the impact parameter ρ for different impact velocities (in atomic units). (b) Total cross section [integrated over all ρ, $\sigma(v)=2\pi\int\rho P(\rho,v)d\rho$]. The dashed curve is for straight line paths in H-H collisions

$$P_{Z,2p\sigma}(\rho,v) \sim P_{1,2p\sigma}(Z\cdot\rho,v/Z) \tag{6.34}$$

$$\sigma_{Z,2p\sigma}(v) \sim Z^{-2}\cdot\sigma_{1,2p\sigma}(v/Z) \quad . \tag{6.35}$$

Experimentally, surprisingly good agreement was found for Ne^+-Ne collisions (open channel) [6.63]. In Fig.6.24 the impact parameter dependence for the Ne-K excitation is shown. Good agreement is also found for total cross sections, especially in the low-velocity region. The calculations were also extended to heteronuclear collision systems $(Z_1 \neq Z_2)$ giving universal curves for differential and total cross sections, $P_{2p\pi\to2p\sigma}(\rho)$ and σ_{rot} (cf. [6.63,98]). For the few measurements done at fairly adiabatic velocities for heteronuclear systems (Ne-Na, Ne-O), there seems also to be good agreement with theory. It should however be pointed out that an absolute comparison of measured and calculated values will sensitively depend on the product of fluorescence yield ω times the probability $N_{2p\pi}$ of having a 2pπ vacancy. Both terms can depend on impact energy and may additionally depend on impact parameter. Early measurements seem to indicate such dependences for ω [6.63].

At higher collision velocities and for fairly asymmetric collision systems, e.g., 30 MeV F on Ar, impact parameter dependent K ionization probabilities were investi-

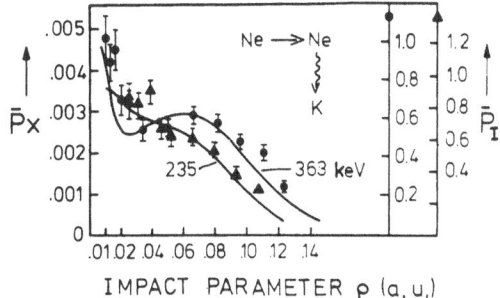

Fig. 6.24. Impact parameter dependence $P_x(\circ)$ of the K x-ray production for Ne-Ne collisions at bombarding energies of 363 and 235 keV. The theoretical curves give the $2p\sigma$ ionization probability $P_I(\circ)$, right-hand scales. The scaling factors (fluorescence yield times 2p--vacancy probability, $\omega \cdot N_{2p\pi}$) are adjusted slightly different for both bombarding energies (cf. [6.63])

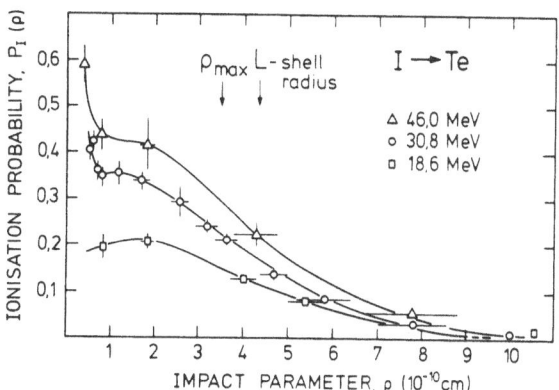

Fig. 6.25. Impact parameter dependence of the L-shell ionization probability of I and Te, $P_{I-L}(\rho)$. The L-shell fluorescence yield is assumed to be 0.14. The adiabaticity parameter is in the region between 0.01 and 0.03. Smooth curves are drawn through the data points (cf. [6.65])

gated by the Kansas group [6.58]. But these measurements are on the rim of the MO domain and cannot be explained satisfactorily by the direct ionization theory nor by the rotational $2p\pi \rightarrow 2p\sigma$ coupling model. Further measurements are required especially for heteronuclear cases, particularly at low collision velocity, for moderately asymmetric systems. For such cases it should be noted that, within the sharing region, the K ionization of both particles should show approximately the same impact parameter dependence for the $2p\pi \rightarrow 2p\sigma$ vacancy transfer. This similarity is caused by the spatial separation of the crossing regions $2p\pi \rightarrow 2p\sigma$ and $2p\sigma \rightarrow 1s\sigma$ from one another and by a very weak ρ dependence of the $2p\sigma{\rightarrow}1s\sigma$ coupling for the region of interest. Deviations will occur for higher impact energies, as both coupling regions tend to interfere there.

For higher shells no calculations and only a few measurements of the impact para-meter dependence are available. The first x-ray ion coincidence measurements were done for the L-shell ionization of the almost symmetric collision system I-Te (Z_1=53, Z_2=52) at $\eta \sim 0.01 \ldots 0.03$ [6.65]. Referring to Fig.6.25, $P_L(\rho)$ shows approximately the same features as the impact parameter dependence for the K-shell ionization. In particular we note that at small impact parameters and high impact energies a peak seems to occur. This behavior was originally interpreted as due to the onset of the direct ionization mechanism in competition there with the molecular excitation. Later an attempt was made to interpret the peak as possibly caused by recoil effects [6.100]. But in view of the similarities to K-shell ionization - the peak at small internu-clear distances might be a kinematic peak - the $P_L(\rho)$ curves may be interpreted in terms of an asymptotic rotational coupling. From correlation diagrams one sees that $4f\pi{\rightarrow}4f\sigma$ and $3d\delta{\rightarrow}3d\pi$ transitions may be responsible for the $P_L(\rho)$ curves. No quanti-tative comparison is however possible at this time. For heteronuclear cases almost no impact parameter dependent measurements are available for L-shell ionization.

6.3.4 L- and M-Shell Excitation

Due to the higher level densities for L and higher shells, it is difficult to sepa-rate isolated crossings from experiments concerning these shells. Possibly, a more statistical treatment may be appropriate. Nevertheless, the consideration of open, dynamically open, or closed channels is helpful. As mentioned in Section 6.3.1, the Ar-L-shell ionization in Ar-Ar collisions may be treated as an open channel problem. Due to the steep rise of the $4f\sigma$ level a well-defined critical promotion radius exists, r_{pr} (see Fig.6.20). Assuming an ionization probability of 1 for collisions with internuclear distances smaller than r_{pr} and 0 outside this region, the total ionization cross section can be calculated, cf. Fig.6.26 [6.101]. For this estimate it is additionally assumed that the diabatic vacancy transfer probability is 0 into levels crossing the $4f\sigma$ level on separation. For this model the energy dependence of the cross section at low collision velocities is determined by the Coulomb deflec-tion, and shows a steep threshold behavior; for higher collision velocities a plateau is expected. Such a behavior is confirmed exactly in Auger electron measurement, whereas x-ray data do not show such a well-defined plateau (see Fig.6.26). This dif-ference is caused by a variation of the fluorescence yield ω [6.102]. Due to the small absolute value of ω ($\sim 10^{-3}$), a variation of ω will strongly influence only the x-ray data. From the measured threshold behavior the promotion radius r_{pr} can be extracted. This value coincides with the corresponding ones determined by Q-value measurements [6.3]. r_{pr} can be estimated by:

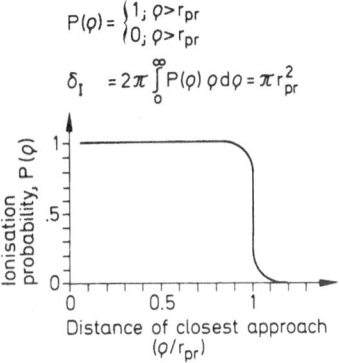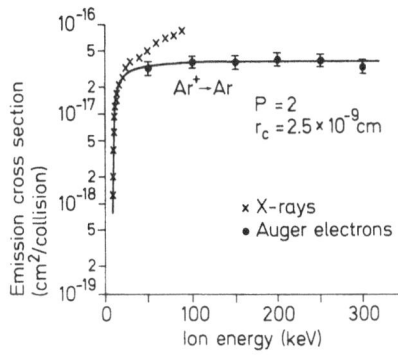

Fig. 6.26. Ionization cross sections as a function of impact energy for a steeply rising MO level coupled to open channels. As an example the Ar-L ionization is shown. (The x-rays are normalized to the electron cross sections.) The insert on the left side shows schematically the assumed dependence of the ionization probability as a function of the distance of closest approach (see [6.79,101])

$$2r_{UA}\big|_{n=4} \leq r_{pr} \leq \frac{1}{2} \cdot r_{SA}\big|_{n=2} \quad . \tag{6.36}$$

Above the threshold the ionization cross section is given by

$$\sigma_I = \pi r_{pr}^2 \tag{6.37}$$

and agrees well with the measured one. This formula can be generalized by introducing the probabilities α and P_{trans}, where α is the probability of having an open reaction channel (times the corresponding coupling probability) and P_{trans} is the probability of not losing the created vacancy on separation:

$$\sigma_I = \alpha \cdot P_{trans} \cdot \pi r_{pr}^2 \quad . \tag{6.38}$$

From such considerations, for instance, the M x-ray emission cross section for fast superheavy atoms produced in nuclear reactions was estimated, giving rise - theoretically - to an extremely high detection probability [6.103].

For closed channel conditions the direct ionization must be considered. As in open channel situations the ionization cross section may be approximated by

$$\sigma_I = (1-\alpha) \cdot P_{trans} \cdot \bar{P}_{dir} \cdot \pi r_{dir}^2 \quad . \tag{6.39}$$

Here $\bar{P}_{dir} \cdot \pi r_{dir}^2$ is the direct ionization cross section for the united atom level concerned, where r_{dir} is the minimum value of r_{pr} and the adiabatic radius $r_{ad} = \hbar v/U$, and $\bar{P}_{dir}$ is the average ionization probability near r_{dir}. These values are difficult

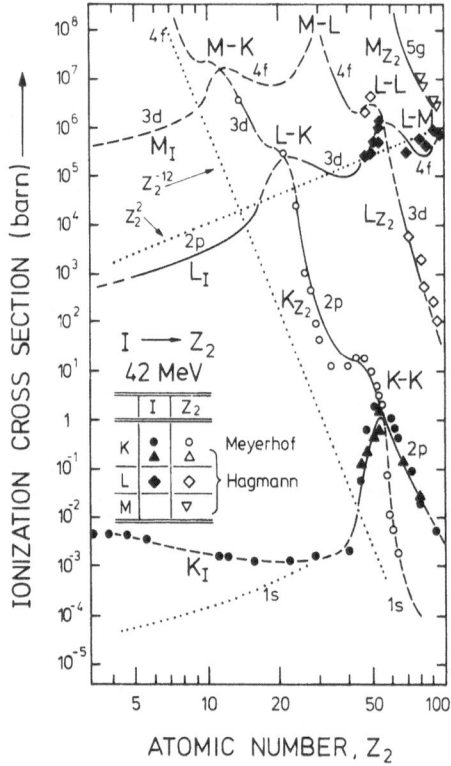

Fig. 6.27. Ionization cross sections for
1/3 MeV/amu I collision with various target
atoms (Z_2) (cf. [6.104]). The data points
are from [6.95,96]

to estimate. Applying the scaling laws of the direct ionization theories for the
UA levels concerned and using a few measured values, ionization cross sections can
be estimated over wide Z_1, Z_2, and energy regions. In Fig.6.27 such an estimate is
given for 1/3 MeV/amu I collisions on various targets. The graph shown agrees well
with the estimate given in 1972 by ARMBRUSTER et al. [6.104], which was based on
the above considerations. For such impact energies we are in the quasi-adiabatic
velocity region, e.g., for the I-L electrons η has a value of 0.04. The cross sections follow roughly the Z_1^2/Z_2^{12}-law of the direct ionization theories (6.23) (dotted
straight lines in the figure). On this general dependence a cross section enhancement because of level matching is superimposed. This enhancement is caused by the
promotion mechanism in combination with the sharing mechanism. It should be noted
that in principle the sharing mechanism - in particular its width - should not depend on the shells considered.

In summary, for the higher shells it may be stated that the ionization cross
sections may reliably be estimated if the exit channel situation and the correlation
diagram are known. For very heavy systems (closed channels) it is helpful to have
a few measured points to establish the absolute cross sections in terms of Z_1, Z_2,
and E_1 coordinates.

6.4 Quasi-Molecular Radiation

6.4.1 General Considerations

From the Section 6.3 it is seen that the inner-shell ionization can be fairly well understood in the framework of the electron promotion model. Electrons - or equivalently vacancies - can be traced in the collision molecule as they follow diabatic levels through several crossings. Thus one may have a vacancy in a well-defined inner level of the quasi-molecule during collision. If the collisional lifetime for the quasi-molecule τ_{coll} is long enough compared to the total lifetime of the vacancy in the quasi-molecule τ_{MO}/ω_{MO}, quasi-molecular radiation can be observed. Hence, by studying quasi-molecular radiation, one may get such information on quasi-molecular level separation as is important for extremely heavy collision systems [6.99].

For a given vacancy the emission probability for this MO radiation is given by

$$P_{MO} = [1 - \exp(-\omega_{MO} \cdot \tau_{coll}/\tau_{MO})] \sim \omega_{MO} \cdot \tau_{coll}/\tau_{MO} \quad . \tag{6.40}$$

For $\tau_{coll} \ll \tau_{MO}$, P_{MO} is directly determined by the time ratio: $P_{MO} \sim \omega_{MO}\tau_{coll}/\tau_{MO}$. It is difficult to get reliable estimates for the times. The (binding) energy of the vacancy state U_B depends strongly on internuclear distances, causing varying transition energies and, hence, varying lifetimes $\tau_{MO}(R)$ [and fluorescence yields $\omega_{MO}(R)$]. Hence the above formula must be considered differentially in R along the possible trajectories. According to [6.105] this differential emission probability is given by:

$$P_{MO}\big|_{E_X} = [(dU_B/dR) \cdot v_R \cdot \tau_{MO}/\omega_{MO}]^{-1} \quad . \tag{6.41}$$

This MO emission probability at a certain transition energy, $E_X(R)$, is inversely proportional to the radial velocity v_R and to the slope dU_B/dR of the molecular level. Hence, high emission probabilities are expected:

1) for regions where the levels are flat ($dU_B/dR \sim 0$);

2) for tangential collisions ($v_R = 0$; i.e., the distance of closest approach coincides approximately with the value of R concerned);

3) for high binding energies (for dipole transitions, e.g., τ_{MO}/ω_{MO} varies with about U_B^{-2}).

In order to yield MO x-ray spectra, $P_{MO}\big|_{E_X}$ must be integrated over all impact parameters:

$$d\sigma_{MO}/dE_X = 2\pi \int_0^\infty f \cdot P_{MO}\big|_{E_X} \rho d\rho \tag{6.42}$$

where f gives the fraction of available vacancies at the internuclear distance concerned. The integration along each trajectory makes f correspond to a sum of the

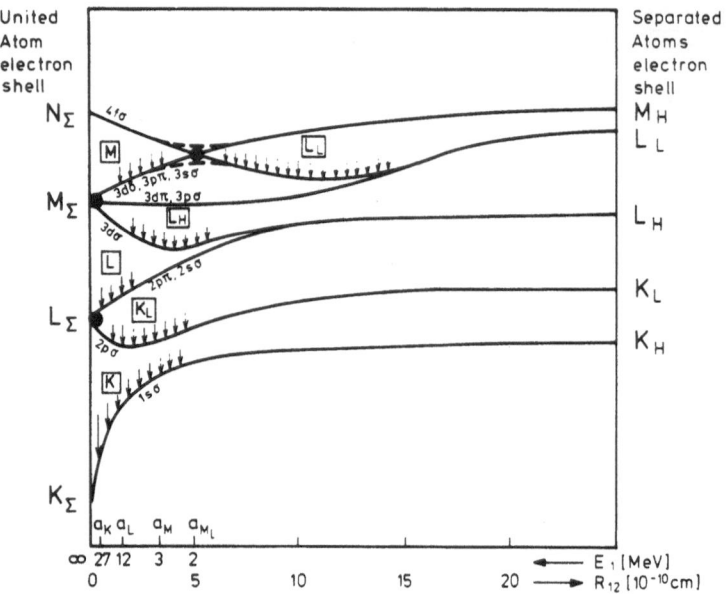

Fig. 6.28. X-ray transitions into quasi-molecular levels thought to be observed in various experiments [6.108]

values for the incoming and outgoing parts of the collision [6.106]. The integration gives emission cross sections roughly proportional to the third power of the R of concern, $\sigma_{MO}|_{E_x} \sim R^3$; this results from point 2) and from (6.41):

$$P_{MO}|_{E_x} \sim 1/v_E \sim d\tau_{coll}/dR \sim R.$$

Consideration of all these points and of realistic level diagrams leads one to expect continuous x-ray spectra for the MO radiation. Slight structures in the spectra may be observed for small R, where the levels approximate the (higher) UA binding energies over a range of values of R - this is called the quasi-atomic approximation - and also for larger R at level minima. Due to the opposing effects of promotion and subshell splitting such minima are often found near the quasi-atomic region. The structures expected in static approximations will be broadened by the dynamics of the collision (Heisenberg broadening) [6.107]. These structures can often be hidden by competing effects giving rise to continuous x-ray spectra, e.g., nuclear and electron bremsstrahlung and electronic pulse pile-up.

Quasi-atomic transitions expected in experiments are indicated by arrows in the schematic level diagram in Fig.6.28. Typical K-, L-, and M-MO radiation will be discussed in the following subsection in connection with several different phenomena. Collision broadening effects will be treated in connection with the molecular K radiation; the effects contributing to the vacancy formation of the f factor in (6.42) will be discussed using the molecular L radiation; the quasi-atomic approximation will be discussed together with molecular M radiation.

6.4.2 The Double Collision Mechanism - The L-MO Radiation

Quasi-molecular L radiation was found by SARIS et al. [6.109] in 1972 for low energy Ar-Ar collisions. Using a solid Ar target produced by high dose Ar bombardment on Si, an x-ray band in the 1 keV region was found (see [6.110] and Fig.6.29). The low energy cutoff of the band is caused by absorption in the detector windows. Inspecting the Ar-Ar level diagram of Fig.6.20, this x-ray band is attributed to transitions into the $2p\pi$ MO level. Inspection of the actual values given in Table 6.5 for these collisions shows that the observed transitions involve mainly internuclear distances where the $2p\pi$ level energy is strongly dependent on R. Hence, with increasing collision velocities, resulting in smaller possible R values, higher transition energies should be observed. Indeed, the x-ray structure shifts to higher energies, indicating its molecular origin, as seen in Fig.6.29. For higher impact energies the $2p\pi$ vacancy will be rotationally coupled to the $2p\sigma$ level, giving rise to the observed K excitation.

As indicated in Table 6.5, roughly a few percent of the available $2p\pi$ vacancies will decay during collisions at small R. But the fluorescence yield enhancement and

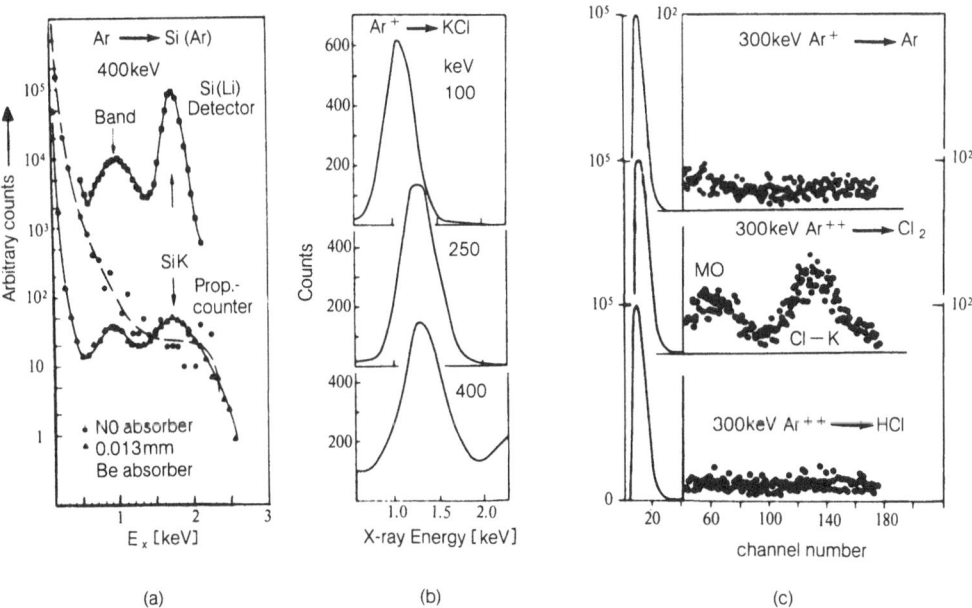

(a) (b) (c)

Fig. 6.29a-c. L-MO x-rays for Ar-Ar collisions (cf. [6.110]). (a) Proportional counter and Si(Li)-detector x-ray spectra for 400 keV Ar bombardment on Ar target atoms implanted in Si. In the absorber-corrected spectrum (dashed curve), the "band" character of the MO radiation vanishes. (b) Proportional counter "band" spectra for 100 to 400 keV Ar impact on KCl. The MO band shifts to higher x-ray energies with increasing impact energy. (c) X-ray spectra for 300 keV Ar impact on gaseous targets.. The spectra are normalized to equal Ar-L x-ray emission intensity (low energy peak)

Table 6.5. Parameters pertinent to 200 keV Ar›Ar collisions yielding a united krypton atom system, where v and u are the collision and orbital velocities, d the distance of closest approach in head-on collisions, $\tau_{coll}=r_{Kr-L}/v$ the collision time,
and $P_{MO}\sim\tau_{coll}/\tau_{Kr-L}$ the MO emission probability

Ar → Ar 200 keV	Projectile L-shell	United atom L-shell	Collision dynamics
Velocities (10^8 cm/s)	$u_{Ar-L} \sim 8$	$u_{Kr-L} \sim 16$	$v \sim 1$
Distances (10^{-10} cm)	$r_{Ar-L} \sim 10$	$r_{Kr-L} \sim 5$	$d \sim 5$
Lifetimes (10^{-16} s)	$\tau_{Ar-L} \sim 0.4$	$\tau_{Kr-L} \sim 4$	$\tau_{coll} \sim 0.05$
Fluorescence yield	$\omega_{Ar-L} \lesssim 10^{-3}$	$\omega_{Kr-L} \sim 0.03$	$P_{MO} \sim 0.01$

lifetime reduction for small R will enhance the high energy side of the MO spectra.
These facts determine the band properties. Nevertheless, there must be an initial
vacancy in the 2pπ level since it is unlikely that this vacancy is formed during the
same collision; it must be available already at the beginning of the collision. According to Section 6.3.4 and Fig.6.26, high L excitation is expected for the ions
traversing the solid target. The fraction f_2 of Ar ions having on the average an L
vacancy in the solid is determined by the product of the collision frequency producing L vacancies times the atomic lifetime of the excited projectiles τ_{Ar-L},

$$f_2 = \sigma_{I_{Ar-L}} \cdot \frac{d}{A_2 \cdot M} \cdot v \cdot \tau_{Ar-L} \qquad (6.43)$$

where $\sigma_{I_{Ar-L}}$ is the cross section for Ar-L ionization, d the density (mass per unit
volume), $A_2 \cdot M$ the atomic mass for the target (so $d/(A_2 \cdot M)$ is the number of atoms per
unit volume), and v the ion velocity. Using appropriate values, approximately, 10-20%
of the Ar ions (at 200 keV) have an initial L hole in the solid after passage through
a thin built-up layer. Such a vacancy can be transferred in a so-called second collision into the 2pπ level, giving rise to the MO radiation. The band intensity is in
qualitative agreement with this double collision mechanism.

The fraction f of L excited projectiles depends linearly on the target density d,
(6.43). Such a dependence is observed experimentally [6.110]. The density dependence
is demonstrated most dramatically by the comparison of spectra from solid and gaseous
targets (see Fig.6.29c). For collisions with Ar or HCl targets, no MO and no K x-rays
are observed, whereas for a Cl_2 target there is already a small contribution of such
x-rays indicating a double collision mechanism within the Cl_2-molecule. In considering the geometrical probability for a double collision within the Cl_2-target molecule,
the MO radiation should be suppressed by a factor of $\sim 10^3$ compared to solid targets,
a factor roughly consistent with the experiments.

In summary, the double collision mechanism for the quasi-molecular radiation is
well established for Ar-Ar collision. This holds true also for neighboring collision

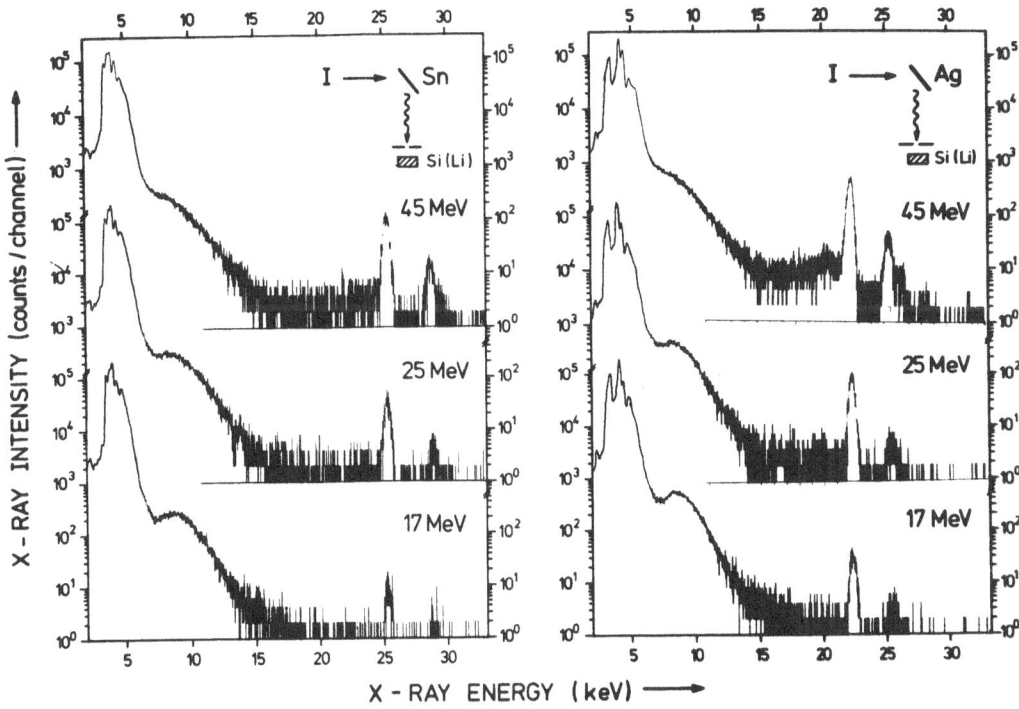

Fig. 6.30. X-ray spectra resulting from 17 to 45 MeV I bombardment of thin Ag and Sn targets, respectively. For all spectra a 40 μm Al absorber was used, producing a "band" structure for the quasi-molecular radiation at low impact energies [6.111]

systems, especially if the projectile is the lighter particle, cf. [6.110]. In successive collision systems (increasing Z_{eff}), the MO band shifts to higher x-ray energies due to the increased binding in the quasi-molecule. For heavier collision systems MO-L radiation was observed in the region of $Z_{eff} \sim 100$, i.e., for I-Ag, I-Sn, I-Te collision systems (see Fig.6.30 and [6.111,112]). These systems are not so thoroughly investigated as the Ar-Ar system and it is not clear whether a single or a double collision mechanism contributes most to the quasi-molecular emission. Due to the rapid reduction of lifetimes for increasing Z_1 (and Z_{eff}), the single collision mechanism will dominate for very heavy collision systems ($f_2/f_1 \sim Z_{eff}^{-4...6}$), [6.106]. A single collision mechanism (f_1) implies that a vacancy created in a collision will decay in the same collision. Normally f will be a sum over the two modes, $f=f_1+f_2$. For these heavy collision systems it is not clear which MO levels contribute most to the MO x-ray spectra. In Fig.6.31 a corresponding level diagram calculated by FRICKE et al. [6.113] is given for comparison.

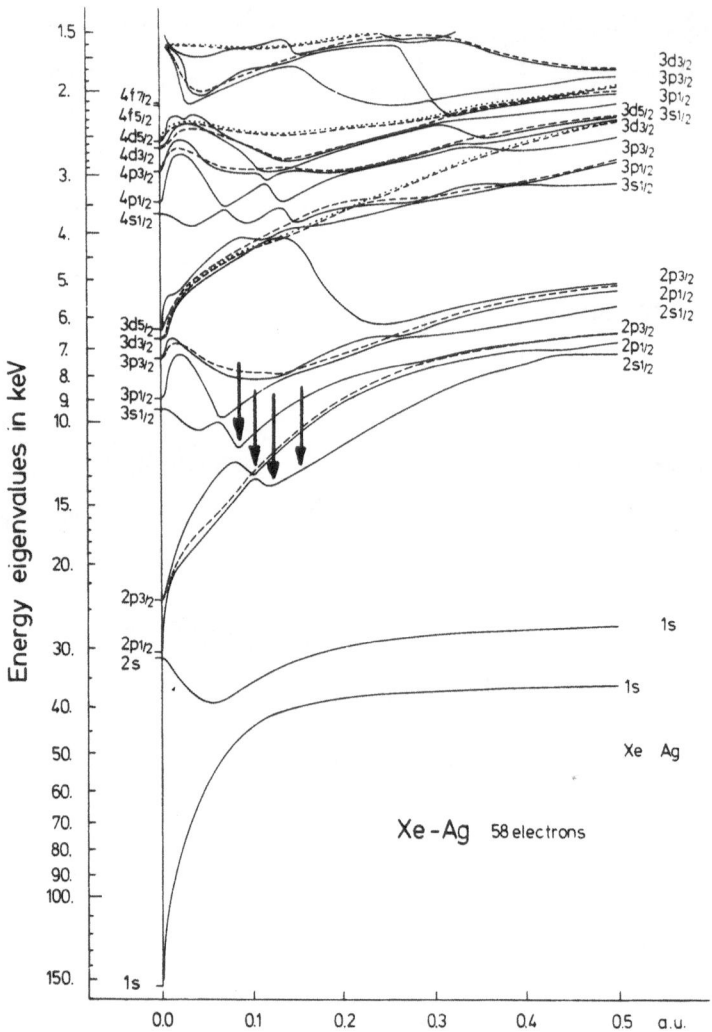

Fig. 6.31. Level diagram for the Xe-Ag collision system according to relativistic, *ab initio* multi-electron calculations of FRICKE et al. [6.113]

6.4.3 The Quasi-Atomic Approximation - The M-MO Radiation

It is difficult to extract quantitative level separation data from the MO spectra.
This is especially true for the steeply descending high-energy tail in the spectra.
To get more quantitative information, examination of structures in the spectra is
helpful. There may be some indication of such structures, for instance in low energy
I→Ag, Sn collisions, as in Fig.6.30. As indicated in Section 6.4.1, slight structures
might be produced by R-independent regions of MO levels. Such flat level dependences
may predominantly be found for small internuclear distances, that is in the quasi-
atomic approximation.

A quasi-atomic approximation can be assumed for an inner-shell if the internuclear distance at the time of emission is roughly $\lesssim r_{n-UA}/2$, i.e., an electron in the shell n sees an approximate effective charge $Z_{eff}=Z_1+Z_2$. Considering the nuclear repulsion of the collision partners, a certain threshold energy is necessary for such small distances of closest approach, resulting in a condition on the collision velocity: $v > 2 \cdot 10^{-2} u_{UA}$ (u_{UA} orbital velocity in the UA system). On the high energy side there is also a limit, that is, the existence of a quasi-molecule means the orbital period of an electron in the UA must be smaller than the collision time $\tau_{orbit\ UA} \lesssim \tau_{coll}$ (adiabaticity criterion). This results in an additional condition on the velocities which is independent on the UA shell considered: $v \lesssim 0.16\ u_{UA}$. Thus quasi-atoms are formed for collisions within the condition

$$0.02 \lesssim v/u_{UA} \lesssim 0.16 \quad . \tag{6.44}$$

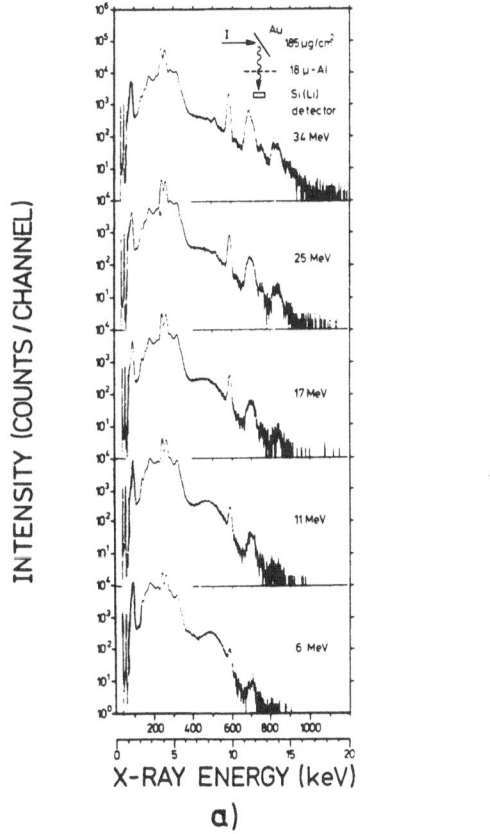

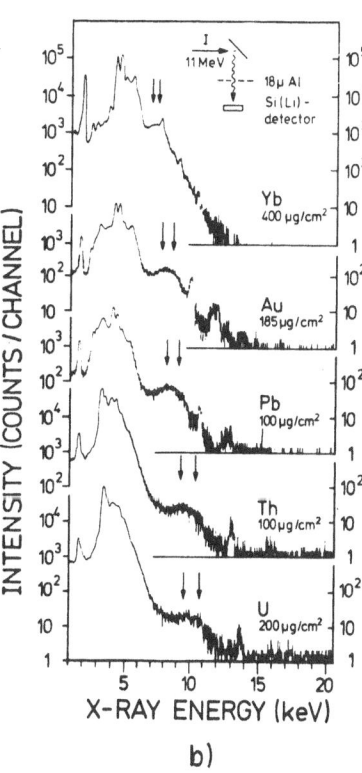

a)

b)

Fig. 6.32a and b. Quasi-molecular M radiation (see [6.111]). (a) X-ray spectra for 6 to 34 MeV I→Au collisions. (b) X-ray spectra for 11 MeV I bombardment on thin Yb, Au, Pb, Th, and U-targets. The arrows give the Mα and Mβ transition energies for the corresponding superheavy atoms

Table 6.6. Parameters pertinent to 11 MeV I→Au collisions yielding a Z=132 united atom system. (For symbols see Table 6.5)

I → Au 11 MeV	Projectile L-shell	United Atom M-shell	Collision dynamics
Velocities (10^8 cm/s)	u_{I-L} = 5.8	u_{132-M} ~ 9.0	v = 0.3
Distances (10^{-10} cm)	r_{I-L} = 3.8	r_{132-M} ~ 3.7	d = 0.9
Lifetimes (10^{-16} s)	τ_{I-L} = 2	τ_{132-M} ~ 0.1	τ_{coll} ~ 0.01
Fluorescence yield	ω_{I-L} = 0.1	ω_{132-M} ~ 0.3	P_{MO} ~ 0.01

The collision time τ_{coll} ($\sim r_{UA}/v$) is largest near threshold, thereby giving the highest emission probability P_{MO} for quasi-atomic radiation (see (6.40) and [6.114]). Due to the possible small f factors, as in (6.42), the corresponding cross sections will be small. Nevertheless the structure in the MO spectra should be most pronounced near threshold.

M-MO radiation was first observed by MOKLER et al. for I→Au collisions [6.115]. Examining this collision system, one expects quasi-atomic M radiation for 2 to 100 MeV impact energies for the I ions. In Fig.6.32a such spectra are shown. The broad peak in the spectra near 8 keV is attributed to the M radiation from superheavy quasi-atoms with Z_{eff}=132. The structure is most pronounced at small impact energies, as expected. For 11 MeV impact energy, actual numbers for the I-Au collision systems are given in Table 6.6. From these numbers one expects to find quasi-atomic M x-ray emission for this impact energy. Here it should be noted that the peak structure seen at low impact energies is definitely not produced by absorber effects. From fitting the MO radiation to an exponential background and a broad peak structure, the peak width is found to increase approximately linearly with the collision velocity, indicating a collision time broadening [6.111,114]. In this connection it is interesting to note that the peak position coincides approximately with the calculated $M\alpha$, $M\beta$ transition energies of a superheavy atom with Z=132. Such a coincidence is also found for neighboring [6.111] or inverted [6.116] collision systems, cf. Fig. 6.32b, giving strong evidence for the existence of the quasi-atomic x radiation.

One disturbing aspect of this interpretation is the rather high x-ray emission cross section, in the region of 30 b for 11 MeV impact energy, originally providing support for speculation on an induced emission mechanism [6.114] (see Sec.6.4.5). From inspection of realistic level diagrams for the I-Au collision system, one finds M-MO levels to be fairly independent of R up to relatively large internuclear distances. In Fig.6.33, a level diagram calculated within a variable screening model is reproduced [6.86]. It shows the same features as the earlier and probably more reliable diagram of FRICKE et al. [6.85] resulting from relativistic *ab initio* multi--electron calculations. The 3d 3/2 3/2 level [in the figure denoted as 2(3/2) level] shows a flat minimum around $4 \cdot 10^{-10}$ cm, giving a large effective region for "quasi-

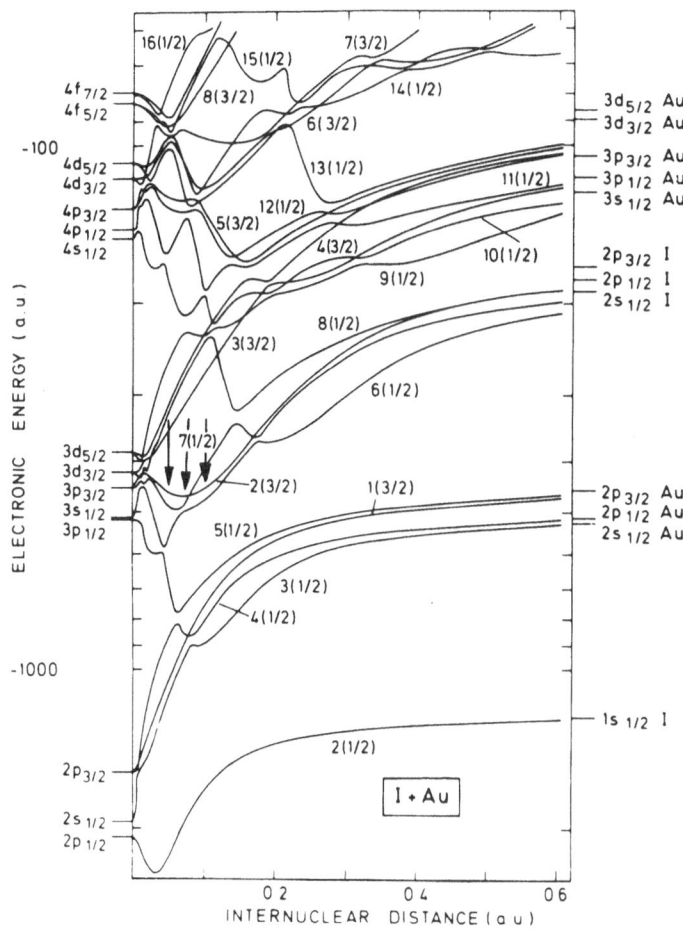

Fig. 6.33. Level diagram for the I-Au collision system according to calculations using the variable screening model of EICHLER and WILLE [6.86]

atomic x-ray emission". It may be this fact in combination with the unknown transition rates within the quasi-molecule which contributes to the observed cross sections.

It should be pointed out that for a long time a double collision mechanism was
assumed to be responsible for the M-MO x-rays. However, in bombarding a mercury vapor
target with Xe ions, an equivalent MO x-ray emission rate was observed, normalized
on the L x-ray emission rate of the ions (see Fig.6.34). This fact supports a single
collision mechanism [6.108,117].

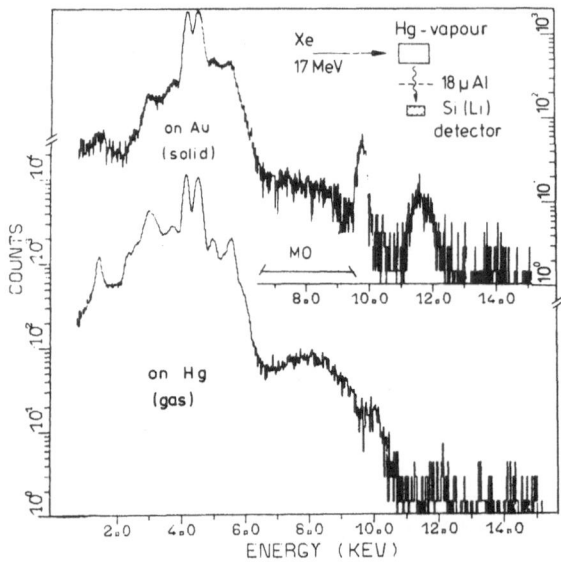

Fig. 6.34. M-MO x-ray spectra for
Xe→Au (solid) and Xe→Hg (gas) collisions at 17 MeV. The MO band is
even more pronounced for the gaseous target, whereas the target L
radiations seem to be suppressed
[6.117]

6.4.4 Collision Broadening Effects - The K-MO Radiation

Quasi-molecular K radiation was first observed by MACDONALD et al. for low energy
C-C collisions [6.118]. Due to window absorption and fluorescence yield enhancement,
an x-ray band around 1 keV was found, which shifted to higher x-ray energies with
increasing impact energy. Corrections for absorber effects yield a strongly decreasing K-MO x-ray spectrum with increasing endpoint energy. This fact seemed to be con-
nected with the strongly descending $1s\sigma$ level. Hence, the endpoint energies were
approximately transformed into binding (or transition) energies near the distances
of closest approach in head-on collisions. Similar procedures were used for neighboring and also for recoil collisions, e.g., Si-Si or Cu-Cu recoil collisions during
bombardment by Ne ions, as in Fig.6.35 [6.119].

It should be mentioned that for collision systems of first row elements - for
example in C-C collisions - the double collision mechanism dominates in the K-MO
x-ray emission. For second and possibly third row elements, high f_2 values for the
double collision mechanism occur for recoiling collisions. In a first "high-energy"
collision an effectively excited recoil is created; in a second, "slow" collision
the vacancy (of the recoil) is transferred into the quasi-molecule which decays with
a high probability. A double collision mechanism seems to be experimentally proved
up to Br-Br collisions. For example MEYERHOF et al. [6.120] report on a density dependent MO x-ray yield for 30 MeV Br-Br collisions, giving evidence for a double
collision mechanism. For heavier collision systems or for higher impact energies the
single collision mechanism should start to dominate.

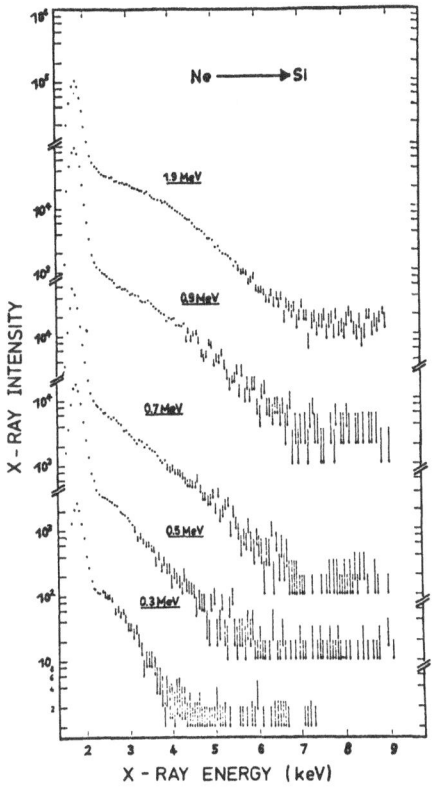
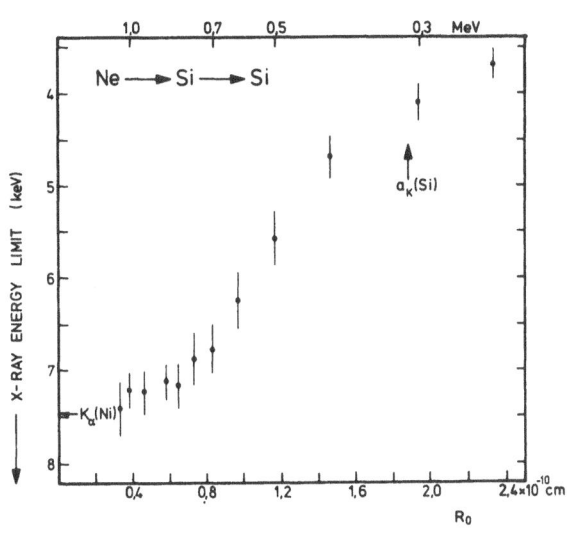

Fig. 6.35. K-MO radiation from Si-Si recoil collisions following bombardment with
0.3 .... 1.9 MeV Ne ions. The endpoint energies are transformed into transition
energies at the distances of closest approach (for head-on collisions), possibly
giving information about the 1σ level for the Si-Si collision molecule [6.119]

For the light collision systems, some efforts were made to extract endpoint energies from the x-ray spectra, and to correlate them to the UA 1s binding energy. From
the viewpoint of the quasi-static approximation given in Section 6.4.1, such a correlation seems to be justified. There are, however, two severe critisisms contra-
dicting such a correlation:

1) Competing processes, such as nuclear-nuclear bremsstrahlung, room backgrounds,
contributions from nuclear reactions, etc., may simulate wrong endpoint energies.

2) The dynamics of the collision will give rise to a Heisenberg broadening, resulting in x-ray spectra with no endpoint. For instance, even for the low energy C-C
collision the endpoint energies found already exceed the UA K-binding energies. This
collision broadening effect is clearly demonstrated in the K-MO x-ray spectra for
Ni-Ni collisions shown in Fig.6.36, cf. [6.121]. Beyond the UA limit (denoted by K_α
and K_β) exponentially decreasing spectra are seen. The half intensity "width" of

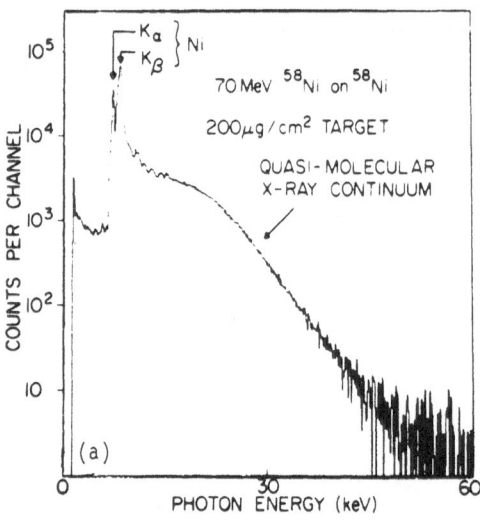
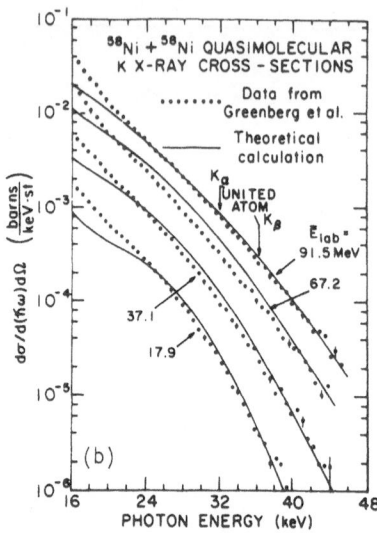

Fig. 6.36a and b. K-MO x-ray spectra for Ni-Ni collisions. (a) Total spectrum for 70 MeV impact energy. (b) The K-MO x-ray tail for 17 to 91 MeV impact energies. Theoretical spectral shapes including collision broadening are given for comparison, cf. [6.121]

the exponential tail seems to be proportional to the square root of the collision velocity [6.122].

The $(v)^{1/2}$ dependence of this half width seems not to be in close agreement with the v dependence of the total peak width of the M-MO radiation (see previous subsection). Considering the quite different energy level R dependences (steeply rising or flat minimum) this difference might be understandable.

For heavy collision systems investigated in Berkeley and Dubna, background effects predominantly from nuclear-nuclear bremsstrahlung may dominate the MO tail at high x-ray energies (see reviews of MEYERHOF [6.123] and of KAUN et al. [6.124]). In Fig.6.37 x-ray spectra for Nb-Nb collisions are shown as an example. Evidently, the x-ray continuum consists of two different parts - c_1 and c_2. Examination of correlation diagrams shows that well-pronounced minima are found for the $2p\sigma$ level at relatively large internuclear distances for these collision systems. As pointed out by HEINIG et al. [6.125], the c_1 continua are caused by transitions into the $2p\sigma$ level around minima. And indeed, the c_1 spectra shift to higher x-ray energies with increasing Z_{eff}. The shift of this c_1 structure is shown in Fig.6.38, together with the dependences of the L-MO and M-MO structures. In all three cases a $\sim Z_{eff}^2$ dependence is found for the x-ray energies of concern. For all three cases a shallow minimum of the corresponding level seems to be responsible for the structure in the MO spectra. Accordingly, for the steeply descending $1s\sigma$ levels no structures are found in the corresponding MO spectra. Hence, it is difficult to extract information

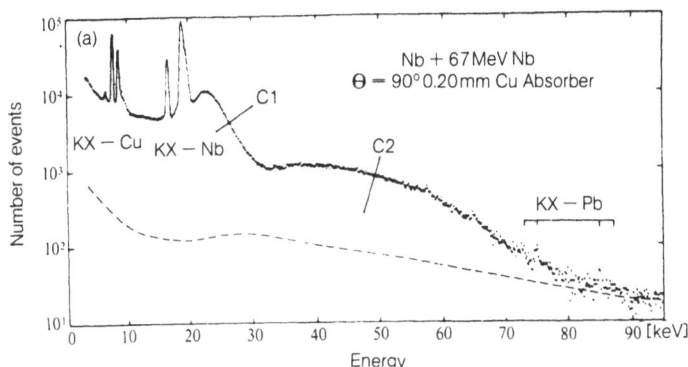

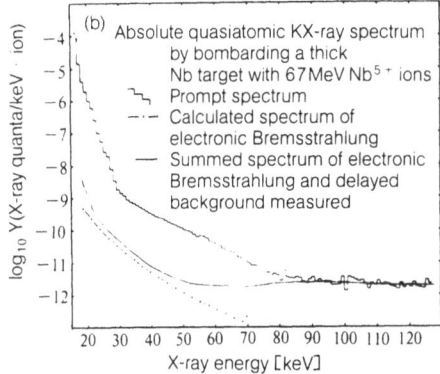

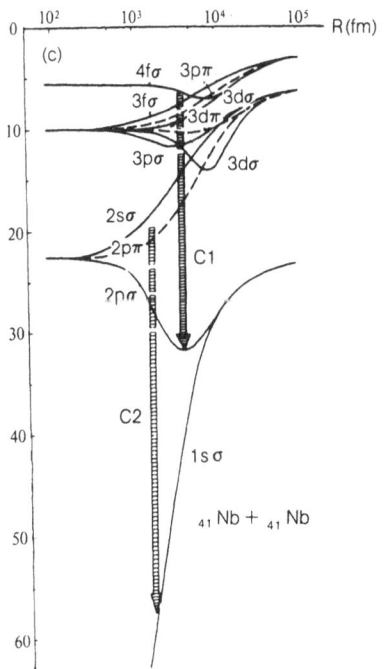

Fig. 6.37a-c. K-MO x-ray radiation for Nb-Nb collisions [6.124]: (a) X-ray spectrum for 67 MeV impact
energy. (b) Absorber corrected K-MO spectrum. (c)
Corresponding level diagram

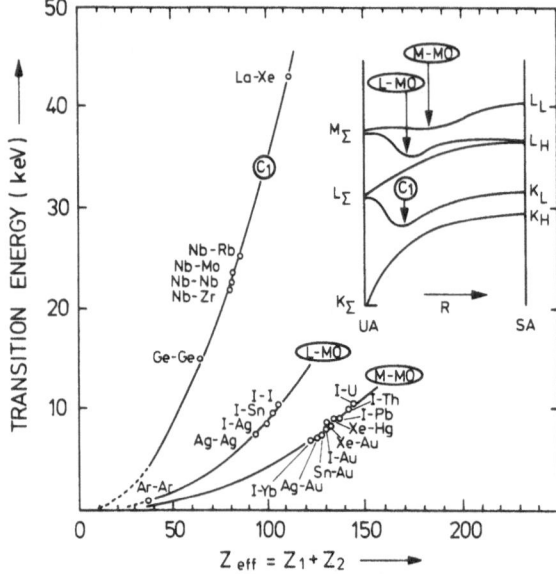

Fig. 6.38. Mean x-ray energies for the structures observed for the various MO radiations as a function of $Z_{eff}=Z_1+Z_2$

on $1s\sigma$ levels out of the K-MO spectra. It may be there is a chance to achieve such information from emission characteristic or anisotropy measurements. For all inves- tigated systems a nonisotropic emission was found for the MO radiation, with emission peaked perpendicular to the beam axis. For very heavy collision systems, overcritic- al fields may occur within the quasi-molecule and the $1s\sigma$ level may dive into the oc- cupied negative continuum (Dirac sea) [6.99]. Here quite new effects such as positron emission may occur.

6.4.5 MO Emission Characteristics and Anisotropies

To understand the large M-MO x-ray production for I-Au collisions, a speculation that an induced emission mechanism acted at small internuclear distances was invoked [6.115]. Such a mechanism may cause an anisotropic x-ray emission as experimentally observed first by KRAFT et al. [6.126]. In Fig.6.39 emission characteristics and anisotropies are shown for 11 MeV I-Au collisions. The M-MO emission characteristics show a dipole contribution, peaked perpendicular to the beam direction. For other types of MO radiation anisotropic emission is also found, with the preferential emis- sion direction perpendicular to the incoming ion beam (see summary given in [6.127]). In Fig.6.40 additional examples are given for the K-MO radiation. In all cases, the anisotropy maximum is situated near the expected UA transition energies, giving some hope for quasi-atomic spectroscopy.

Theoretically a rotationally induced emission mechanism giving rise to such an- isotropies was postulated by GREINER's group [6.129]. In this picture, Coriolis

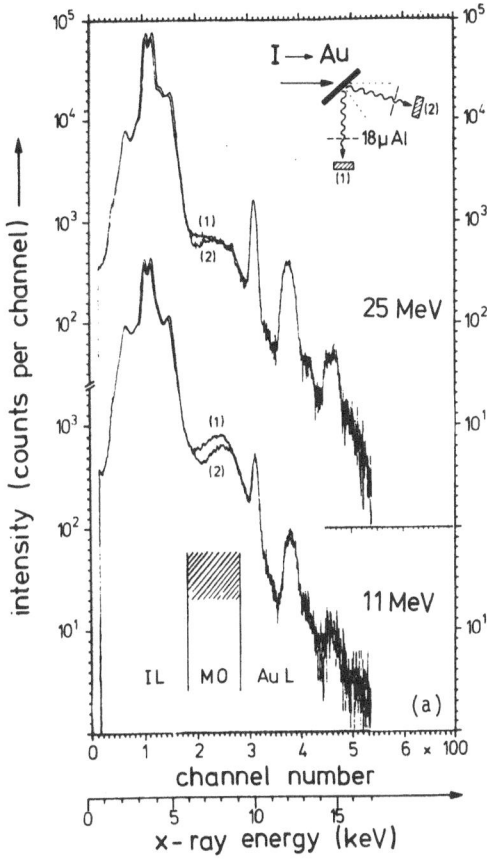

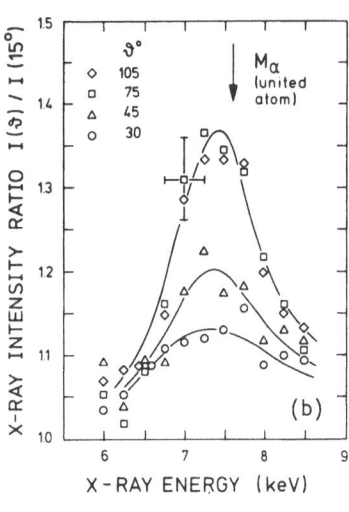

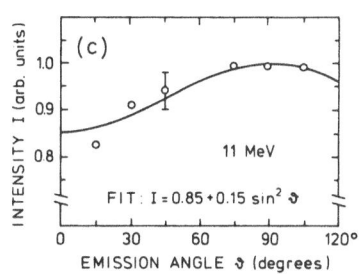

Fig. 6.39a-c. Emission characteristic for M-MO radiation from 11 MeV I-Au collisions [6.126]. (a) X-ray spectra taken under 15° and 90° degree observation angle with respect to the beam direction; the spectra are normalized to equal beam intensity. (b) Differential anisotropy distribution I(θ)/I(15°) for detection angles θ=30°, 45°, 75°, and 105°. (c) Angular distribution for the total M-MO radiation between 7.1 and 9.5 keV. The data points of (b) and (c) are corrected for Doppler effects using the center of mass velocity

forces acting on the electrons of the quasi-molecule, partially rotating within the scattering plane, induce radiative transitions. This yielded a high MO emission cross section, preferentially for small internuclear distances in the quasi-molecule, and emission directions perpendicular to the beam axis. The anisotropy should increase with increasing impact energy; see dashed curves in Fig.6.41. Such a trend was observed only for Al-Al and Si-Al, which might be due to contributions from radiative electron capture. For all other experiments the opposite trend was found. In view of the general discrepancy between experiment and theory, the rotationally induced emission mechanism is probably not the principal mechanism responsible for the anisotropies found. Moreover, it is claimed by several authors that this theory needs some modifications, e.g., [6.130].

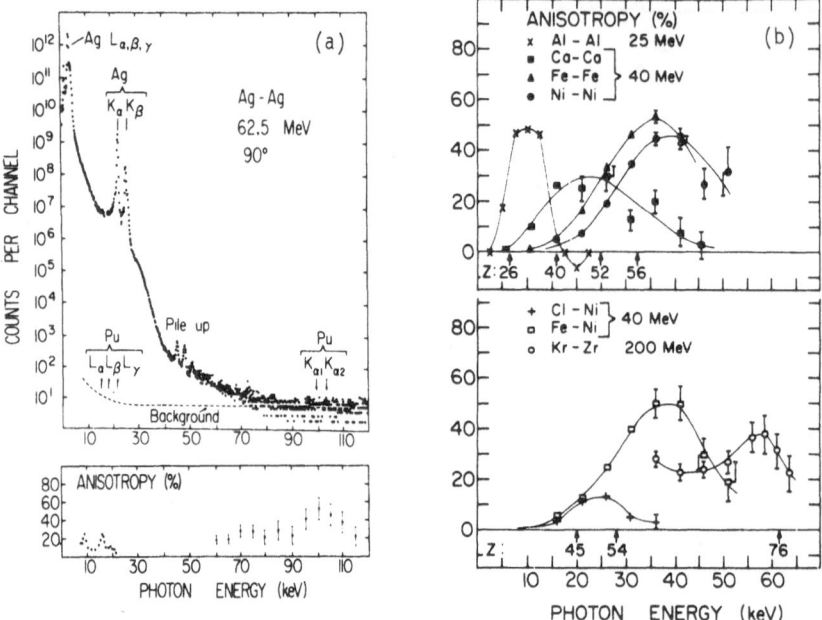

Fig. 6.40a and b. Anisotropy measurements for the K-MO radiation (see [6.128]).
(a) For Ag-Ag collisions. (b) For different symmetric and asymmetric collision systems

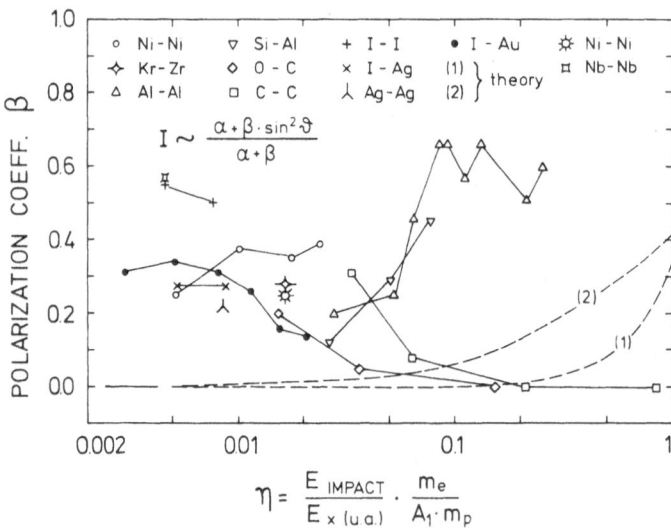

Fig. 6.41. The maximum anisotropy as a function of the adiabaticity parameter (see
[6.127]). The anisotropy is given by the polarization coefficient β, assuming a dipole emission pattern for the radiation

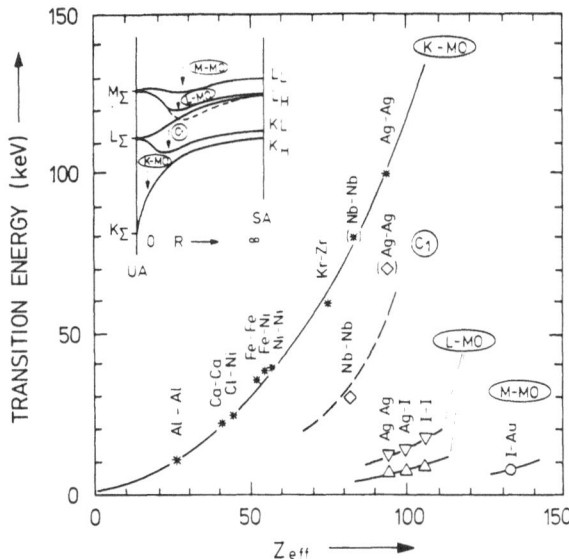

Fig. 6.42. Maximum position of the anisotropy distribution as a function of $Z_{eff}=Z_1+Z_2$

A nonisotropic emission pattern for the MO radiation can also be caused by alignment effects, i.e., by an unequal occupation probability of the various electronic substates contributing to a transition. For $\eta \ll 1$, i.e., for a rather adiabatic collision, high alignment might be expected. For higher collision velocities an increasingly statistical distribution of the electrons over the various substates may prevail, resulting in a reduction of the maximum anisotropy values, in accordance with the experimental findings (see Fig.6.41). But the true mechanism responsible for the anisotropy enhancement for small internuclear distances is not clear. Possibly, collision broadening effects via Doppler corrections may also contribute to anisotropy maxima around UA transition energies.

Independent of the true mechanism, there seems to be an experimental correlation between the maximum in the anisotropy distribution and the MO transition energies at small internuclear distances (see Fig.6.42). For I-Au collisions this correlation was tested for various low impact energies, i.e., for distances of closest approach (in head-on collisions) yielding different transition energies [6.131]. The dependence of the x-ray energy at the position of maximum anisotropy as a function of impact energy seems to reproduce the gross features of the corresponding MO level dependence. The general correlation between the anisotropy maximum and the transition energy at small internuclear distances gives some hope for a spectroscopy of quasi-atoms in the superheavy region.

In interpreting such measurements, important Doppler corrections must be applied. In order to make such corrections the velocity of the MO emitting system must be known. This velocity can be determined by comparing spectra for corresponding forward and backward emission angles (assuming a symmetric emission pattern). Within the experimental errors the emitter velocity is found to be equal to the center of mass velocity, giving further support to the quasi-molecular character of the radiation [6.127,131-133]. It should be mentioned that some continuum radiation, e.g., nuclear--nuclear bremsstrahlung, may have the same emitter velocity dependence. Nevertheless, many competing processes, as for example radiative electron capture, electron bremsstrahlung, radiation from activated materials and so on, have different velocity dependences.

Finally, it should be pointed out that refined anisotropy determinations show oscillations as a function of x-ray energy superimposed on the general anisotropy [6.128]. Whether this effect is caused by various rotational states or by rearrangement effects is an open question. Summarizing, anisotropy measurements may be a key consideration for a two-center level spectroscopy. Many questions, both experimentally and theoretically, must however be solved before experiments in the superheavy region can be suitably interpreted.

6.5 Bremsstrahlung and Radiative Electron Capture

The continuum x-ray emission produced by heavy ions includes, in addition to the quasi-molecular radiation, important contributions from processes in which electrons ejected during the collisions or present in the target interact with the projectile or with other target atoms. These processes which are classified in the following section both contribute to the measured spectra and give some information on the collision [6.134].

6.5.1 Electron Bremsstrahlung

The electron bremsstrahlung arises when electrons undergo strong accelerations in the nuclear electric field of an atom and are radiating part of their kinetic energy. The energy spectrum of the bremsstrahlung ranges from 0 to the full kinetic energy of the electron E_e, and the total cross section has only a weak dependence on the energy of the radiation E_r and on E_e, given in the classical limit by the simple Sommerfeld-Kramers result

$$\frac{d\sigma^B}{dE_r} = C_B \cdot Z^2/(E_r \cdot E_e) \quad \text{for} \quad 0 < E_r < E_e \tag{6.45}$$

with $C_B = 1.43 \cdot 10^{-24}$ cm^2 keV, which is discussed by LEE et al. [6.135] who also give more reliable results for 1 keV $\leq E_e < 500$ keV. The bremsstrahlung is peaked perpendicular to the incident electron direction, and due to the retardation for higher electron energies $E_e > 10$ keV, the maximum will be at somewhat forward angles in the laboratory system [6.136].

In ion-atom collisions, electrons are ejected from the collision system as part of the ionization process and these secondary electrons may then later make bremsstrahlung with a fairly high probability and produce a continuous radiation spectrum. This process is pronounced for solid targets, where the secondary electrons are stopped, but is also to be seen in many gas targets and thin solid targets, where the electrons are only somewhat slowed down [6.136]. For the direct ionization by a light ion incident on a heavy target (6.10), the energy distribution of secondary electrons in a single collision is well known and may be calculated in PWBA or BEA. The intensity of ejected electrons is generally decreasing with increasing electron energy E_e, roughly as E_e^{-2}, from the lowest kinetic energies to the maximum energy T_m transferred to a free electron in a head on collision with the ion:

$$T_m = \frac{4 \cdot m \cdot A_1 M}{(A_1 \cdot M + m)^2} \cdot E_1 \simeq 0.00218 \cdot E_1/A_1 \quad . \tag{6.46}$$

For energies higher than this value the intensity decreases rapidly, e.g., like E_e^{-10}. In this classically forbidden region, only the high momentum tail of the bound electron wave function can contribute to ejection of high energy electrons. The secondary electrons will by subsequent collisions be slowed down and can also emit bremsstrahlung according to (6.45). The bremsstrahlung spectrum is simply a folding of the electron distribution with the ratio of (6.45) to the electronic stopping power [6.137], which essentially preserves the shape of the spectral distribution (see Fig.6.43). The secondary electron bremsstrahlung is the dominant source of continuum x-ray emission for $0 < E_r < T_m$. The angular distribution of the radiation is peaked perpendicular to the beam or at slightly forward angles, because the electrons are preferentially ejected in the forward direction (for high E_e). The intensity may change as much as a factor 2 with angle, but is difficult to account for theoretically as the electrons may be severely deflected from their direction of ejection, before they radiate, due to the slowing down process [6.136,138].

For heavy-ion collisions the secondary electron bremsstrahlung is more complex, as electrons are ejected both from target and projectile and from the transiently formed collision molecule. The target will be excited in the normal way with the Z^2 scaling of (6.24), and the projectile will also emit electrons because it is struck by the target atom which relative to the ion is moving with velocity $-\underline{v}$. The projectile electrons are thus mainly ejected in the backward direction in the projectile frame. Addition of the beam velocity tends to cancel the resulting velocity in the laboratory system (for the higher electron energies). The formation of collision molecules is further expected to affect the energy spectrum of the secondary elec-

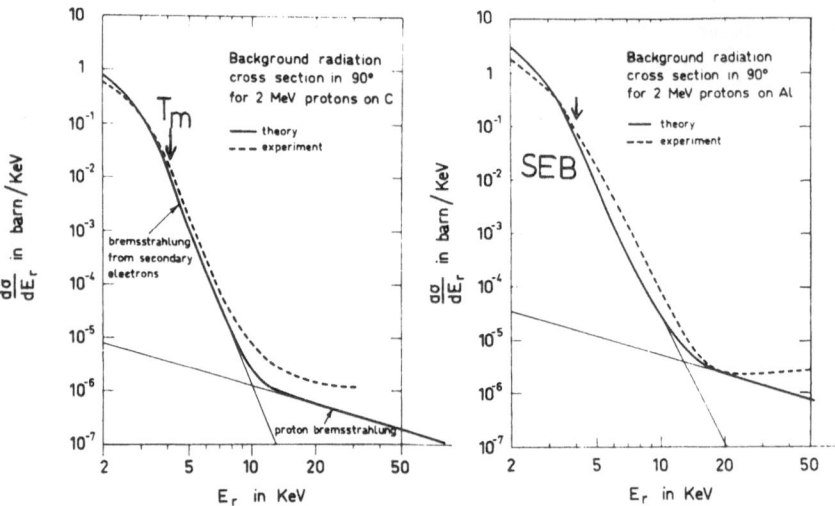

Fig. 6.43. Continuum x-ray cross sections for 2 MeV protons on solid carbon and aluminium as a function of the radiation energy E_r. Theoretical curves are given for secondary electron bremsstrahlung (SEB) using BEA ionization and for the proton-nucleus bremsstrahlung, (6.47) [6.137]

trons in a way similar to that for quasi-molecular excitation or radiation, so that the corresponding bremsstrahlung distribution is also influenced. This MO dependence of the secondary electron bremsstrahlung has not yet been experimentally investigated. Estimates can be given based on a simple change of the effective binding energy.

The interest in secondary electron bremsstrahlung is partly motivated by the desire to know how much it contributes, when other processes such as quasi-molecular radiation are studied or when the emisssion of characteristic x-rays from the target is used for trace element analysis of small samples [6.72,136,139-141]. It may contain information on the electron distribution in target and projectile and on the MO formation during the collision, just as does the primary energy distribution of the secondary electrons.

Inner bremsstrahlung is the first step in the secondary electron bremsstrahlung, namely the radiation from an electron in the field of the colliding nuclei, from which it has been ejected. Generally its contribution is much less than that from many succeeding collisions, but it is a one-collision process and will survive for low pressure gas targets and may be important for heavy projectiles.

Inverse bremsstrahlung occurs for heavy-ion beams simply as the normal bremsstrahlung of target electrons in the field of the projectile, i.e., as (6.45), and has a fairly flat spectrum from zero up to an energy $\sim T_m/4$, because a free target electron has a velocity $-\underline{v}$ relative to the projectile and thus an energy $E_e = m \cdot v^2/2 = T_m/4$ [6.142].

6.5.2 Nucleus-Nucleus Bremsstrahlung

At high radiation energies the various electron bremsstrahlung continua are reduced
to very small values, and what remains is often the direct bremsstrahlung of the
projectile and target nuclei in their relative motion during the collision. This radiation extends up to the projectile energy E_1 and is normally dominated by the
electric dipole contribution to the multipole expansion, which in the Born approximation [6.142,143] has the total cross section

$$\frac{d\sigma_{NNB}^{E1}}{dE_r} = C_{E1} \cdot \frac{Z_1^2 Z_2^2}{E_1 E_r} \cdot A_1 \left(\frac{Z_1}{A_1} - \frac{Z_2}{A_2}\right)^2 \quad , \tag{6.47}$$

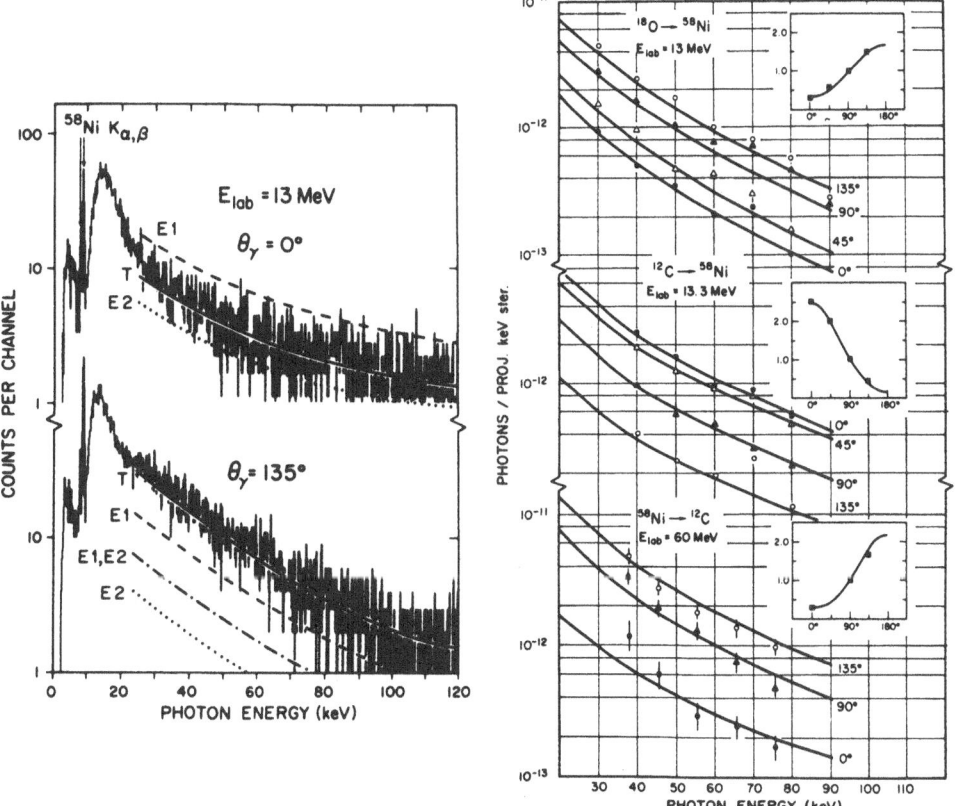

Fig. 6.44. Nucleus-nucleus bremsstrahlung in heavy-ion collisions, and its angular
distributions, which exhibit strong interferences between dipole and quadrupole radiation, as calculated classically by REINHARDT et al. [6.143]. The measurements
and the figures are from TRAUTVETTER et al. [6.144]

with $C_{E1}=\left\{\ln \frac{[1+(1-x)^{1/2}]^2}{x}\right\}\cdot 4.3\cdot 10^{-28}$ cm^{2}keV, where $x=E_r/E_{CM}=(A_1+A_2)\cdot E_r/(A_2\cdot E_1)$. The nucleus-nucleus bremsstrahlung (NNB) of incident protons is seen in Fig.6.43. For collision nuclei of similar charge to mass ratios, the dipole term vanishes or becomes small and the electric quadrupole bremsstrahlung

$$\frac{d\sigma_{NNB}^{E2}}{dE_r} = C_{E2}\frac{z_1^2 z_2^2}{E_r}\cdot\frac{A_1^2 A_2^2}{(A_1+A_2)^2}\cdot\left(\frac{z_1}{A_1^2}+\frac{z_2}{A_2^2}\right)^2 \qquad (6.48)$$

with $C_{E2}=\left\{(1-x)^{1/2}+0.3\cdot(2-x)\cdot\ln\frac{[1+(1-x)^{1/2}]^2}{x}\right\}\cdot 9.2\cdot 10^{-34}$ cm^2 will be dominating [6.143]. In these cases interferences between dipole and quadrupole radiation arise as shown in the classical calculations by REINHARDT et al. [6.143] and demonstrated very nicely in the experiment by TRAUTVETTER et al. [6.144], employing ^{18}O, ^{12}C, and ^{58}Ni beams. Fig.6.44 shows this correspondence between theory and experiment both in the absolute magnitude of the bremsstrahlung and for the angular distribution of the radiation. In Fig.6.44 the interference pattern between E1 and E2 radiation is clearly observed as well as the dramatic change from one collision system to another.

At high radiation energies another continuum contribution is often encountered in gas- and solid-state detectors, namely Compton scattering of high energy radiation, e.g., of γ-rays from excited nuclear states, but as it is not a primary target effect, it will not be further discussed.

6.5.3 Radiative Electron Capture

A target electron can be captured into a vacancy state of the projectile, and may directly radiate the energy gained by this transition as an x-ray photon. When the ion velocity is $\underline{v}$, the target electrons are moving with a velocity $-\underline{v}$ relative to the projectile, to which one has to add the internal orbital velocity $\underline{v}_e$ in the target atom. By a transition to a vacancy state the electron will gain the binding energy U_p in the projectile and the kinetic translational energy $m\cdot v^2/2$ and will lose its binding of energy U_t to the target. With the kinetic energy determined by the vector addition $m\cdot(\underline{v}+\underline{v}_e)^2/2$ the radiated energy is

$$E_{REC} = U_p + \frac{1}{2}\cdot m\cdot v^2 - U_t + m\cdot \underline{v}\cdot \underline{v}_e \qquad (6.49)$$

where the direction and magnitude of $\underline{v}_e$ enters [6.145]. The radiative electron capture (REC) thus results in a peak-like continuum centered $mv^2/2-U_t$ above the projectile binding energy. The REC width is proportional to the mean orbital velocity of the target electrons. The outer target electrons give normally the main contribution to the REC, which has then a width increasing with Z_2. In Fig.6.45 the REC to the Ar K-shell by bombardment with 56 MeV Ar is shown. The REC continuum accompanies the

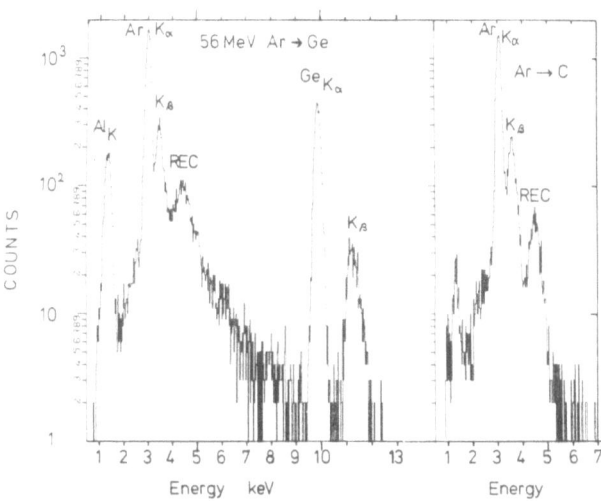

<u>Fig. 6.45.</u> Radiative electron capture (REC) peaks in x-ray spectra from 1.4 MeV/amu Ar^{12+} bombardment of solid Ge and C targets. The REC peak structure follows the characteristic projectile x-rays, at a somewhat higher energy given by (6.49), and with a width which increases with Z_2. An 18 μm Al absorber was used in front of the Si(Li) detector in this measurement

characteristic K lines of Ar as a somewhat broader peak at higher energy. The binding energy U_p in (6.49) is that of the excited vacancy state, to which capture takes place, and is higher than the characteristic projectile x-ray energies. This state may be produced in the first layers of a solid target, and the binding energy will be better approximated for the few electron systems of interest by hydrogenic wave function estimates than by the neutral binding energies. The angular distribution of the radiation is proportional to $sin^2\theta$, where θ is the emission angle relative to the beam direction. An expression for the REC cross section is given by KLEBER and JAKUBASSA [6.145], who treat the theoretical shape of the continuum and how it is related to the velocity distribution of the target electrons - through v_e in (6.49).

The shape of the REC peak is sensitive to the orbital velocities of mainly the outer electrons in the target atoms, and the detailed REC shape is best studied at high beam velocities, where the peak is well separated from the characteristic lines. Several approaches have been made to account for the observed shape. One is to employ hydrogenic wave functions [6.146]. Another more advanced one is to use Hartree--Fock wave functions [6.68], but still the width was not satisfactorily explained. The third approach, used by BETZ et al. [6.147], is to employ velocity distributions, experimentally determined by Compton profile measurements, and by this method an excellent agreement with the observed REC peak-shape for solid targets is achieved. Conversely, the study of the REC continuum can elucidate the electron distribution and may represent an alternative way of determining Compton profiles, at least for outer-shell electrons. Inner-shell electrons are also captured into the projectile, but no distinct radiation peak is observed, only a continuum ranging to high energies This tail extends into the region where the bremsstrahlung and quasi-molecular continua reside, and is itself affected by molecular orbital formation during collision. Thus radiative electron capture constitutes a last example of how the x radiation can be utilized for getting information on atomic structure and collision dynamics.

References

6.1 E. Merzbacher, H.W. Lewis: In *Handbuch der Physik*, Vol. 34, ed. by S. Flügge (Berlin-Göttingen-Heidelberg: Springer 1958) pp. 166-192
6.2 J.D. Garcia, R.J. Fortner, T.M. Kavanagh: Rev. Mod. Phys. 45, 111 (1973)
6.3 Q.C. Kessel, B. Fastrup: In *Case Studies in Atomic Physics*, Vol. 3, ed. by M.R.C. McDowell, E.W. Daniel (North Holland, Amsterdam 1973) pp. 137-213
6.4 B. Crasemann (ed.): *Atomic Inner Shell Processes* (Academic Press, New York 1975)
6.5 D.H. Madison, E. Merzbacher: Ref. 6.4, Vol. 1, pp. 1-72
6.6 P. Richard: Ref. 6.4, Vol. 1, pp. 73-158
6.7 W. Mehlhorn, R. Brenn (eds.): *Proceedins of the Second International Conference on Inner Shell Ionization Phenomena, Invited Papers* (Freiburg, 1976)
6.8 R.W. Fink, S.T. Manson, J.M. Palms, P.V. Rao (eds.): *Proceedings of the International Conference on Inner Shell Ionization Phenomena and Future Applications* (USAEC, Oak Ridge, Tennessee 1973)
6.9 J.S. Risley, R. Geballe (eds.): *The Physics of Electronic and Atomic Collisions* (University of Washington Press, Seattle 1976)
6.10 B.C. Čobić, M.V. Kurepa (eds.): *The Physics of Electronic and Atomic Collisions* (Institute of Physics, Beograd 1973)
6.11 F.W. Saris, W.F. van der Weg (eds.): Nucl. Instr. Methods 132 (1976)
6.12 R. Marrus, M. Prior, H. Shugart (eds.): *Atomic Physics 5* (Plenum Press, New York 1977)
6.13 G. zu Putliz, E.W. Weber, A. Winnacker (eds.): *Atomic Physics 4* (Plenum Press, New York 1975)
6.14 C.L. Cocke: In *Beam-Foil Spectroscopy*, Vol. 1, ed. by I.A. Sellin, D.J. Pegg (Plenum Press, New York 1976) pp. 283-298
6.15 E. Storm, H.I. Israel: Nuclear Data Tables A7, 565 (1970)
6.16 W. Bambynek, B. Crasemann, R.W. Fink, H.U. Freund, H. Mark, C.D. Swift, R.E. Price, P.V. Rao: Rev. Mod. Phys. 44, 716 (1972)
6.17 E.J. McGuire: Ref. 6.4, pp. 293-330
6.18 D. Burch, N. Stolterfoht, D. Schneider, H. Wieman, J.S. Risley: Phys. Rev. Lett. 32, 1151 (1974)
6.19 J.H. Scofield: Ref. 6.4, pp. 265-292
6.20 O. Keski - Rahkonen, M.O. Krause: Atomic Data and Nuclear Data Tables 14, 139 (1974)
6.21 H.-D. Betz, F. Bell, H. Panke, G. Kalkoffen, M. Welz, D. Evers: Phys. Rev. Lett. 33, 807 (1974)
6.22 R.L. Kauffman, P. Richard: In *Methods of Experimental Physics*, Vol. 13, Spectroscopy, ed. by D. Williams (Academic Press, New York 1976) pp. 148-203
6.23 F.S. Goulding: Nucl. Instr. Methods 142, 213 (1977)
6.24 G. Bertolini, G. Restelli: Ref. 6.4, Vol. 2, pp. 123-167
6.25 Y. Cauchois, C. Bonnelle: Ref. 6.4, Vol. 2, pp. 83-121
6.26 M.W. Charles, B.A. Cooke: Nucl. Instr. Methods 61, 31 (1968)
6.27 A.J.P.L. Policarpo, M.A.F. Alves, M.C.M. Dos Santos, M.J.T. Carvalho: Nucl. Instr. Methods 102, 337 (1972)
6.28 H.E. Palmer, L.A. Brady: Nucl. Instr. Methods 116, 587 (1974)

6.29 W. Koenig, F.W. Richter, U. Steiner, R. Stock, R. Thielmann, U. Wätjen: Nucl. Instr. Methods 142, 225 (1977)

6.30 R.L. Watson, F.E. Jenson, T. Chiao: Phys. Rev. A10, 1230 (1974)

6.31 J. McWherter, J. Bolger, C.F. Moore, P. Richard: Z. Physik 263, 283 (1973)

6.32 D. Burch, L. Wilets, W.E. Meyerhof: Phys. Rev. A9, 1007 (1974)

6.33 J.C. Slater: Phys. Rev. 36, 57 (1930)

6.34 F. Hopkins, A. Little, N. Cue, V. Dutkiewicz: Phys. Rev. Lett. 37, 1100 (1976)

6.35 R.L. Watson, A.K. Leeper, B.I. Sonobe, T. Chiao, F.E. Jenson: Phys. Rev. A15, 914 (1977)

6.36 P. Richard: In *Scientific and Industrial Applications of Small Accelerators* (Publ. no 76 CH 1175-9 NPS, IEEE, New York 1976) pp. 293-298

6.37 W. Wölfli, C. Stoller, G. Bonani, M. Stöckli, M. Suter: Ref. 6.7, pp. 272-278

6.38 G. Basbas, W. Brandt, R. Laubert: Phys. Rev. A7, 983 (1973)

6.39 B.H. Choi, E. Merzbacher, G.S. Khandelwal: Atomic Data 5, 291 (1973)

6.40 W. Brandt, G. Lapicki: Phys. Rev. A10, 474 (1974)

6.41 W. Brandt: Ref. 6.8, pp. 948-978

6.42 D.H. Madison: Ref. 6.7, pp. 321-338

6.43 J. Bang, J.M. Hansteen: Kgl. Danske Vidensk. Selsk., Mat.-Fys. Medd. 31, No. 13, 1 (1959)

6.44 J.M. Hansteen, O.M. Johnsen, L. Kocbach: Atomic Data and Nuclear Data Tables 15, 305 (1975)

6.45 L. Kocbach: J. Phys. B9, 2269 (1976)

6.46 J.M. Hansteen: Adv. At. Mol. Phys. 11, 299 (1975)

6.47 K. Taulbjerg: Ref. 6.7, pp. 130-144

6.48 P.A. Amundsen, L. Kocbach: J. Phys. B8, L 122 (1975)

6.49 D.L. Clark, T.K. Li, J.M. Moss, G.W. Greenlees, M.E. Cage, J.H. Broadhurst: J. Phys. B8, L 378 (1975)

6.50 G. Basbas, W. Brandt, R.H. Ritchie: Phys. Rev. A7, 1971 (1973)

6.51 R. Laubert, W. Losonsky: Phys. Rev. A14, 2043 (1976)

6.52 G. Basbas, W. Brandt, R. Laubert: Bull. Amer. Phys. Soc. 18, 103 (1973) and to be published

6.53 C.G. Soares, R.D. Lear, J.T. Sanders, H.A. van Rinsvelt: Phys. Rev. A13. 953 (1976)

6.54 T.J. Gray, P. Richard, R.L. Kauffman, T.C. Holloway, R.K. Gardner, G.M. Light, J. Guertin: Phys. Rev. A13, 1344 (1976)

6.55 J.U. Andersen, E. Laegsgaard, M. Lund, C.D. Moak: Nucl. Instr. Methods 132, 507 (1976)

6.56 J.S. Hansen: Phys. Rev. A8, 822 (1973)

6.57 C.L. Cocke, R. Randall, B. Curnutte: In *Abstracts of Papers VIII ICPEAC*, ed. by B.C. Cobič, M.V. Kurepa (Institute of Physics, Beograd 1973) pp. 714-715

6.58 R.R. Randall, J.A. Bednar, B. Curnutte, C.L. Cocke: Phys. Rev. A13, 204 (1976)

6.59 R. Schulē. K.E. Stiebing, H. Zekl, H. Schmidt-Böcking, I. Tserruya, K. Bethge: In *Abstracts of Papers IX IXPEAC*, ed. by J.S. Risley, R. Geballe (University of Washington Press, Seattle 1975) pp. 498-499, and to be published

6.60 D. Burch, W.B. Ingalls, H. Wiemann, R. Vandenbosch: Phys. Rev. A10, 1245 (1974)

6.61 W. Brandt, K.W. Jones, H.W. Kramer: Phys. Rev. Lett. 30, 351 (1973)

6.62 E. Laegsgaard, J.U. Andersen, L.C. Feldmann: Phys. Rev. Lett. 29, 1206 (1972)

6.63 H.O. Lutz: Ref. 6.9, pp. 432-446; and Ref. 6.7, pp. 104-129

6.64 K. Stiebing, H. Schmidt-Böcking, R. Schulē, K. Bethge: Phys. Rev. A14, 146 (1976)

6.65 H.J. Stein, H.O. Lutz, P.H. Mokler, K. Sistemich, P. Armbruster: Phys. Rev. A5, 2126 (1972); H.J. Stein, Thesis, K.F.A.-report Jül-NP-799 (1971)

6.66 J.R. Macdonald, M.D. Brown, S.J. Czuchlewski, L.M. Winters, R. Laubert, I.A. Sellin, J.R. Mowat: Phys. Rev. A14, 1997 (1976)

6.67 G. Lapicki, W. Losonsky: Phys. Rev. A15, 896 (1977)

6.68 H.-D. Betz, M. Kleber, E. Spindler, F. Bell, H. Panke, W. Stehling: Ref. 6.9, pp. 520-530

6.69 D. Belkic, A. Salin: J. Phys. B9, L 397 (1976)

6.70 K. Taulbjerg, P. Sigmund: Phys. Rev. A5, 1285 (1972)

6.71 K. Taulbjerg, B. Fastrup, E. Laegsgaard: Phys. Rev. A8, 1814 (1973)

6.72 F. Folkmann: In *Material Characterization Using Ion Beams*, ed. by A. Cachard, J.P. Thomas (Plenum Press, New York 1977) pp. 239-279

6.73 R.K. Gardner, T.J. Gray, P. Richard, C. Schmiedekamp, K.A. Jaminson, J.M. Hall: Phys. Rev. A15, 2202 (1977)

6.74 T.J. Gray, P. Richard, K.A. Jamison, J.M. Hall, R.K. Gardner: Phys. Rev. A14, 1333 (1976)

6.75 W. Brandt, K.W. Jones: Phys. Lett. 57A, 35 (1976)

6.76 J.F. Hofmann, H. Stöcker, W. Scheid, W. Greiner, V. Ceausescu, E. Badralexe: Z. Physik A280, 131 (1977)

6.77 H.J. Specht: Z. Physik 185, 301 (1965)

6.78 S. Datz, C.D. Moak, B.R. Appleton, T.A. Carlson: Phys. Rev. Lett. 27, 363 (1971); P.H. Mokler, H.J. Stein: In Annual Report 1971, Inst. für Kernphysik and Inst. für Neutronenphysik, KFA Jülich, p. 197; and [5.96]

6.79 F.W. Saris: Physica 52, 290 (1971)

6.80 F. Folkmann, P.H. Mokler: unpublished (1976)

6.81 D.K. Olsen, C.F. Moore, P. Richard: Phys. Rev. A7, 1244 (1973)

6.82 U. Fano, W. Lichten: Phys. Rev. Lett. 14, 627 (1965); W. Lichten: Phys. Rev. 164, 131 (1967); M. Barat, W. Lichten: Phys. Rev. A6, 211 (1972)

6.83 H. Hartmann, K. Helfrich: Theor. Chim. Acta 10, 406 (1968); K. Helfrich, H. Hartmann: Theor. Chim. Acta 16, 263 (1970)

6.84 B. Müller, W. Greiner: Z. Naturforsch. 31A, 1 (1976)

6.85 B. Fricke, T. Morović, W.-D. Sepp, A. Rosen, D.E. Ellis: Phys. Lett. 59A, 375 (1976)

6.86 J. Eichler, U. Wille: Phys. Rev. A11, 1973 (1975); J. Eichler, U. Wille, B. Fastrup, K. Taulbjerg: Phys. Rev. A14, 707 (1976)

6.87 B. Fastrup: Ref. 6.9, pp. 361-383

6.88 L. Landau: Phys. Z. Sowjet. 2, 40 (1932); C. Zener: Proc. Roy. Soc. Ser. A137, 696 (1932)

6.89 Yu. Demkov: Sov. Phys. JETP 18, 138 (1964)

6.90 A. Russek: Phys. Rev. A4, 1918 (1971)

6.91 J.S. Briggs, J. Macek: J. Phys. B5, 579 (1972); J.S. Briggs: Ref. 6.9, pp. 384-407; J.S. Briggs: Reports on Progress in Physics 39, 217 (1976)

6.92 J.S. Briggs, K. Taulbjerg: J. Phys. B8, 1909 (1975)

6.93 C. Foster, T. Hoogkamer, P. Woerlee, F. Saris: J. Phys. B9, 1943 (1976)

6.94 W.E. Meyerhof. R. Anholt, T. Saylor, P. Bond: Phys. Rev. A11, 1083 (1975)

6.95 W.E. Meyerhof: Phys. Rev. Lett. 31, 1341 (1973); W.E. Meyerhof, R. Anholt, T. Saylor, S. Lazarus, A. Little, L. Chase: Phys. Rev. A14, 1653 (1976)

6.96 S. Hagmann: (Thesis, University of Köln) GSI-report Pa-1-77 (1977)

6.97 N. Stolterfoht, P. Ziem, D. Ridder: J. Phys. B7, L 409 (1974); see also Chapter 5

6.98 K. Taulbjerg, J.S. Briggs, J. Vaaben: J. Phys. B8, 1351 (1976)

6.99 W. Betz, G. Heiligenthal, J. Reinhardt, R.K. Smith, W. Greiner: Ref. 6.9, pp. 531-559

6.100 D. Burch, K. Taulbjerg: Phys. Rev. A12, 508 (1975)

6.101 Q.C. Kessel: Bull. Amer. Phys. Soc. 14, 946 (1969); cf. also Ref. 6.79

6.102 F.W. Saris, D. Onderdelinden: Physica 49, 441 (1970); P.K. Cacak, Q.C. Kessel, M.E. Rudd: Phys. Rev. A2, 1327 (1970)

6.103 P. Armbruster, P.H. Mokler, H.-J. Stein: Phys. Rev. Lett. 27, 1623 (1971)

6.104 P. Armbruster, P.H. Mokler, H.-J. Stein: Ref. 6.8, pp. 396-413

6.105 J. Briggs: J. Phys. B7, 47 (1974)

6.106 W.E. Meyerhof, T.K. Saylor, S.M. Lazarus, A. Little, B.B. Triplett, L.F. Chase, R. Anholt: Phys. Rev. Lett. 32, 1279 (1974)

6.107 W. Lichten: Ref. 6.11, pp. 249-285

6.108 P. Armbruster: Ref. 6.7, pp. 21-41

6.109 F.W. Saris, W.F. van der Weg, H. Tawara, R. Laubert: Phys. Rev. Lett. 28, 717 (1972); F.W. Saris, I.V. Mitchell, D.C. Santry, J.A. Davies, R. Laubert: Ref. 6.8, pp. 1255-1282

6.110 F.W. Saris, F.J. de Heer: Ref. 6.13, pp. 287-300; and Refs. cited there

6.111 P.H. Mokler, S. Hagmann, P. Armbruster, G. Kraft, H.-J. Stein, K. Rashid, B. Fricke· Ref. 6.13, pp. 301-324

6.112 W. Wölfli, Ch. Stoller, G. Bonani, M. Suter, M. Stöckli: Lett. Nuovo Cimento 14, 577 (1975)

6.113 B. Fricke, T. Morović, W.-D. Sepp: Private communication and to be published in Phys. Lett. (1977)

6.114 P. Armbruster, G. Kraft, P.H. Mokler, B. Fricke, H.-J. Stein: Physica Scripta 10A, 175 (1974)

6.115 P.H. Mokler, H.-J. Stein, P. Armbruster: Phys. Rev. Lett. 29, 827 (1972); P.H. Mokler, H.-J. Stein, P. Armbruster: Ref. 6.8, pp. 1283-1296

6.116 F.C. Jundt, H. Kubo, H.E. Gove: Phys. Rev. A10, 1053 (1974)

6.117 Ch. Heitz, P. Armbruster, W. Enders, F. Folkmann, S. Hagmann, G. Kraft, P.H. Mokler: In *Abstracts of Contr. Papers to II Int. Conf. Inner Shell Ioniz. Phen.*, ed. by W. Mehlhorn and R. Brenn (Freiburg, 1976) pp. 15-17; and G. Kraft, P. Armbruster, W. Enders, F. Folkmann, S. Hagmann, Ch. Heitz, P.H. Mokler: Phys. Lett. 62A, 409 (1972)

6.118 J.R. Macdonald, M.D. Brown, T. Chiao: Phys. Rev. Lett. 30, 471 (1973)

6.119 B. Knaf, G. Presser: Phys. Lett. 49A, 89 (1974); K.O. Groeneveld, B. Knaf, G. Presser: Physica Fennica 9S, 36 (1974)

6.120 W.E. Meyerhof: Ref. 6.9, pp. 470-480

6.121 B. Müller: Ref. 6.9, pp. 481-500

6.122 H.-D. Betz, F. Bell, H. Panke, W. Stehling, E. Spindler, M. Kleber: Phys. Rev. Lett. 34, 1256 (1975); H.-D. Betz, F. Bell, E. Spindler, M. Kleber: In *Abstracts of Contr. Papers to II Int. Conf. Inner Shell Ioniz. Phen.*, ed. by W. Mehlhorn, R. Brenn (Freiburg, 1976) pp. 11-13

6.123 W.E. Meyerhof: Science 193, 839 (1976)

6.124 K.H. Kaun, W. Frank, P. Manfrass: Ref. 6.7, pp. 68-78

6.125 K.H. Heinig, H.U. Jäger, H. Richter, H. Woittennek: Phys. Lett. 60B, 249 (1976)

6.126 G. Kraft, P.H. Mokler, H.-J. Stein: Phys. Rev. Lett. 33, 476 (1974); G. Kraft, H.-J. Stein, P.H. Mokler, S. Hagmann: Verhandl. DPG (VI) 9, 77 (1974)

6.127 P.H. Mokler, P. Armbruster, F. Folkmann, S. Hagmann, G. Kraft, H.-J. Stein: Ref. 6.9, pp. 501-519

6.128 W. Wölfli, C. Stoller, G. Bonani, M. Stöckli, M. Suter, W. Däppen: Phys. Rev. Lett. 36, 309 (1976)

6.129 B. Müller, R.K. Smith, W. Greiner: Phys. Lett. 49B, 219 (1974)

6.130 K. Dettmann: Ref. 6.7, pp. 57-67

6.131 F. Folkmann, P. Armbruster, S. Hagmann, G. Kraft, P.H. Mokler, H.-J. Stein: Z. Physik A276, 15 (1976)

6.132 W.E. Meyerhof, T.K. Saylor, R. Anholt: Phys. Rev. A12, 2641 (1975)

6.133 W. Frank, K.H. Kaun, P. Manfrass, N.V. Pronin, Y.P. Tretyakov: Z. Physik A279, 213 (1976)

6.134 F.W. Saris, T.P. Hoogkamer: Ref. 6.12, pp. 509-536

6.135 C.M. Lee, L. Kissel, R.H. Pratt, H.K. Tseng: Phys. Rev. A13, 1714 (1976)

6.136 F. Folkmann: In *Ion Beam Surface Layer Analysis*, ed. by O. Meyer, G. Linker, F. Käppeler (Plenum Press, New York 1976) pp. 695-718

6.137 F. Folkmann, C. Gaarde, T. Huus, K. Kemp: Nucl. Instr. Methods 116, 487 (1974)

6.138 K. Ishii, S. Morita, H. Tawara: Phys. Rev. A13, 131 (1976)

6.139 S.A.E. Johansson (ed.): Proc. Int. Conf. Particle Induced X-Ray Emission and its Analytical Applications: Nucl. Instr. Methods 142, Nos. 1,2 (1977)

6.140 S.A.E. Johansson, T.B. Johansson: Nucl. Instr. Methods 137, 473 (1976)

6.141 F. Folkmann: J. Phys. E8, 429 (1975)

6.142 D.H. Jakubassa, M. Kleber: Z. Physik A273, 29 (1975)

6.143 J. Reinhardt, G. Soff, W. Greiner: Z. Physik A276, 285 (1976)

6.144 H.P. Trautvetter, J.S. Greenberg, P. Vincent: Phys. Rev. Lett. 37, 202 (1976)

6.145 M. Kleber, D.H. Jakubassa: Nucl. Phys. A252, 152 (1975)

6.146 A.R. Sohval, J.P. Delvaille, K. Kalata, K. Kirby-Docken, H.W. Schnopper: J. Phys. B9, L 25 (1976)

6.147 H.-D. Betz, F. Bell, E. Spindler: Ref. 6.12, pp.493-508; E. Spindler, H.-D. Betz, F. Bell: J. Phys. B9, L1 (1977)

7. Extensions of Beam Foil Spectroscopy

I. A. Sellin

With 22 Figures

The key advantage of beam-foil spectroscopy over alternative methods of studying related atomic properties is its simple, straightforward application to the direct determination of lifetimes of excited states of atoms and ions through measurements of the decay in flight of such particles in beams of well-defined and easily measured velocity. The linear relation between velocity (~ 1 cm ns^{-1}) and time after excitation in a thin target (thickness $\sim$ a few hundred Å, transit time $\sim 10^{-14}$ s) allows good measurements of macroscopic radiation decay lengths ~ 1 cm in the laboratory and equivalent fractional accuracy in the determination of lifetimes ~ 1 ns of excited ions undergoing decay in flight. Through such lifetime measurements, related quantities like transition probabilities, or equivalently, oscillator strengths for dominant decay modes, can frequently be established to accuracies limited by various systematic effects (e.g., target aging effects) to errors of a few percent or less. Fig.7.1 displays a typical decay in flight curve for the ~ 0.54 ns 1^1P_0-2^3P_1 inter-combination transition in heliumlike fluorine, measured in our laboratory by MOWAT et al. [7.1] a few years ago. This particular curve illustrates the principle of differential metastability, in that the fast decays of, for example, the unresolved resonance transition 1^1S_0-2^1P_1 had initially died out at the slight decay times after excitation corresponding to the leftmost data point shown. The longer-lived cascade feeding transitions which eventually dominate at points far downstream are sufficiently weak in the primary decay region of interest to permit an approximate background subtraction from the upstream portion of the curve. Establishment of the beam-velocity independence of the fitted lifetime (within experimental errors) for the upstream portion of the curve is an important test of the adequacy of this kind of background subtraction procedure. In this regard the measurement of metastable state decays, which arise through the violation of some selection rule whose magnitude is frequently the object of measurement, is often eased by the tendency of these decays to outlive all but very weak, long-lived cascade transitions which rarely repopulate the metastable state.

In the illustration cited, the $\Delta S=0$ selection rule is violated by spin-orbit couplings which admix n^1P_1 states into the 2^3P_1 wave function. The sum of these admixtures as manifested in the 2^3P_1 decay rate was the goal of our measurements, and is of interest because of the strong-Z scaling ($\sim Z^{10}$) of the 2^3P_1 decay rate in the low-Z region. Hence the traditional triplet-singlet classification scheme for helium-

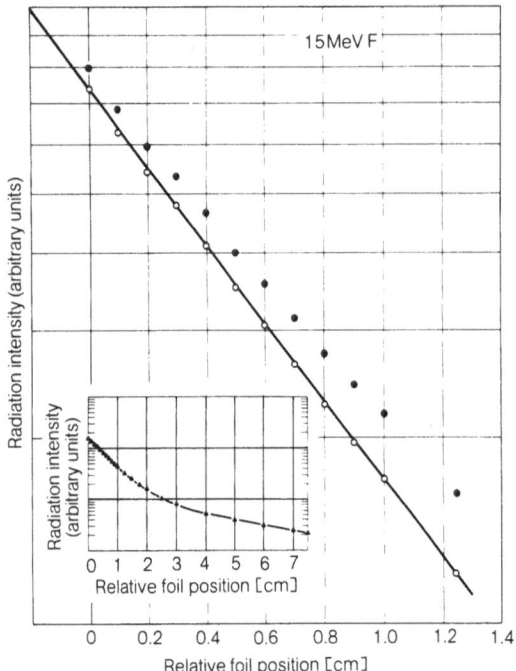

Fig. 7.1. Experimental decay curves for foil excited $1\ ^1S_0 - 2\ ^3P_1$ transitions in F^{7+}, obtained at 15 MeV beam energy. The inset shows the full curve fitted to a sum of two decaying exponentials (solid line). Closed circles are uncorrected data and open circles are data after subtraction of the apparent background, which has been represented to a useful degree of accuracy by a single exponential. The statistical counting errors are smaller than the circles. The solid line passing through the open circles is a least-squares fit to a single, decaying exponential

like atoms breaks down badly due to strong violation of L-S coupling for atoms as light as, for example, C^{4+}. The principle of differential metastability is especially useful under conditions of modest spectral resolution, for levels which are not necessarily resolved spectroscopically can often easily be resolved temporally owing to distinct portions of a decay in flight curve being dominated by one or another of otherwise unresolved transitions. Identification of lines of otherwise uncertain origin is frequently aided by qualitative knowledge of decay times, as the oscillator strength of a given transition is frequently determined by angular momentum quantum numbers which govern the relative forbiddenness of a transition.

A counterbalance to the assertion that the ability to make reasonably accurate lifetime measurements is a key advantage of the beam-foil method is the fact that accompanying Doppler shifts and spreads have often severely hindered spectroscopic resolution and transition energy measurements. Related line blending problems then hinder lifetime work as well. Typical values of the ratio of beam velocity to speed

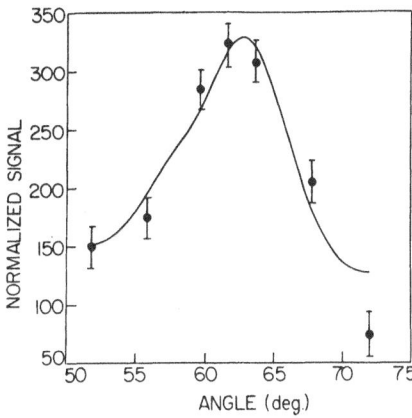

Fig. 7.2. Resonance curve used to obtain the line center for the ΔE-S transition in F^{8+}

of light $v_B/c \sim 0.05$ imply accompanying line shifts and widths which have been successfully combatted only by intensity limiting collimation, or by Doppler compensation techniques appropriate only to the visible region (see discussion by MARTINSON [7.2]). On a few occasions such Doppler shifts as opposed to spreads have been used to good advantage. One example is the use of a Doppler-tuned x-ray absorption edge spectrometer to study spectra and lifetimes of foil-excited heavy ions as discussed recently by VARGHESE et al. [7.3]. Another is the use of Doppler tuning of an HF laser to measure the $2s_{1/2}$-$2p_{3/2}$ fine structure transition in hydrogenic fluorine by KUGEL et al. [7.4]. While the success of the latter experiment was notable in view of signal strength and signal to noise limitations in addition to Doppler tuning problems, it is also clear that the accurary of measurement ($\sim 1\%$) was limited by inability to deconvolute the line shape of the observed resonance. The beam Doppler profile overlapped two strong laser lines. Fig.7.2 shows an example from the work of KUGEL et al. [7.4].

It is clear that many experiments of a related character where the precision of measurement is critical would greatly profit from the use of techniques which in some way reduce the severity of the Doppler spread problem. One of the extensions of beam-foil spectroscopy (BFS) to be discussed concerns a three to four order of magnitude reduction in Doppler shifts and spreads for xuv/soft x-ray spectra through the use of target ion recoil spectroscopy (TIRS), as has been recently investigated in preliminary experiments by SELLIN et al. [7.5]. These experiments will be discussed in Section 7.2.

In the years of early BFS experiments, it was widely believed that the accessible lifetime range was practically limited to a range of about two orders of magnitude on either side of the 10^{-8} to 10^{-9} s range characterized by $\lesssim 1$ cm ns^{-1} beam velocities. As we shall note, in recent years it has been demonstrated through the use of a variety of techniques that the range available to fast beam measurements is now at least as large as $\lesssim 10^{-14}$ to $\gtrsim 10^{-5}$ s, representing a ten order of magnitude dynamic range in lifetime measurement capability! The recent widening of this dynamic range is a principal subject of Section 7.1.

Progress has also been made in dealing with cascade correction problems, as has for example been discussed recently by CURTIS [7.6]. In some cases, as for example in a number of the $\Delta n=0$ transitions characterizing the important resonance transitions of various ions in the lithiumlike, berylliumlike, boronlike, sodiumlike, magnesiumlike, etc., sequences, a number of important cascade feed transitions can be studied simultaneously with transitions fed by them, as the feeding and fed transitions often lie in the same wavelength and decay length regimes. A program to study such $\Delta n=0$ resonance transition lifetimes is underway in our laboratory under the leadership of D.J. Pegg. As decay rates for such transitions scale fairly slowly with Z and tend to be long compared to $\Delta n \neq 0$ transitions for which photon energies are larger (PEGG et al. [7.7]), the dominant transitions are often the feeding and fed transitions within the same principal quantum state. An example for such a decay scheme for carbonlike S is provided in Fig.7.3. The subtraction of measured cascade feeds from decay curves for fed transitions aids greatly in the determination of oscillator strengths for the fed transitions.

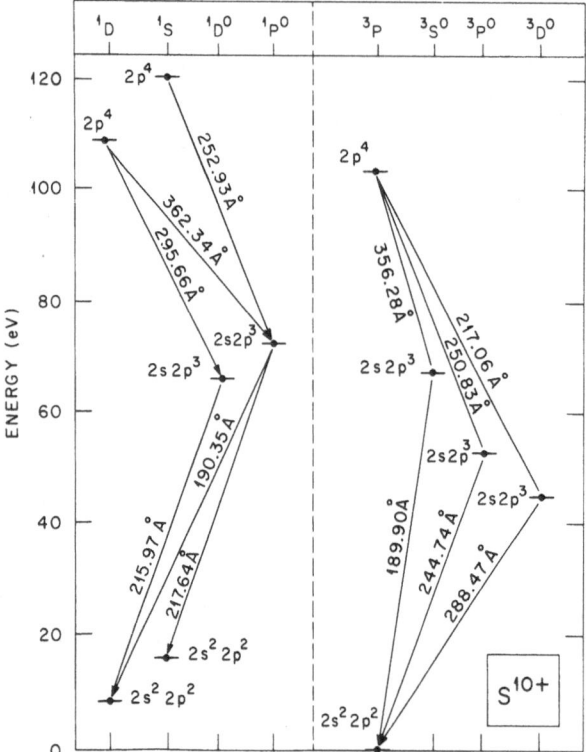

Fig. 7.3. Decay scheme for $\Delta n=0$ transitions in C-like S ions

A beam-foil lifetime measurement apparatus which operates in the xuv/soft x-ray region and has therefore been useful for such $\Delta n=0$ lifetime measurements in, for example, lithium, beryllium, and boron sequence S ions and sodiumlike and magnesium-like Fe is depicted in Fig.7.4. The same diagram is archetypical of similar apparatus for lifetime measurements in nearly any spectral region, as generally only the type and design of the spectrometer differs significantly with desired wavelength region. Grazing incidence is used in the xuv/soft x-ray region owing to the poor reflectivity of grating coatings at such wavelengths for more nearly normal angles of incidence. For harder radiations, crystal spectrometers of similar optical design, semiconductor Si(Li) or Ge(Li) detectors, or Doppler tuned x-ray absorption edge spectrometers would normally be employed.

A typical arrangement for the use of semiconductor radiation detectors is shown in Fig.7.5. This arrangement has served us well for studying excitation by foil-stripped ions in thin gas targets. Both projectile and target x-ray production for magnetically selected charge states can be studied as a function of gas pressure in this apparatus. Multiple electron rearrangement phenomena produced by impact of foil-stripped, highly ionized heavy ions in lighter gas targets, and characterized by extensive target ionization and excitation, are interesting extensions of BFS. Colli-sional aspects of the recent data obtained by SELLIN et al. [7.5] in connection with exploring the TIRS technique are also included in Section 7.2. A foil assembly (not shown) is also installed within the gas cell from time to time to permit comparison of collisions in solid and gaseous media containing the same target element, and to permit conventional beam-foil lifetime measurements. For example, the data of Fig.7.1

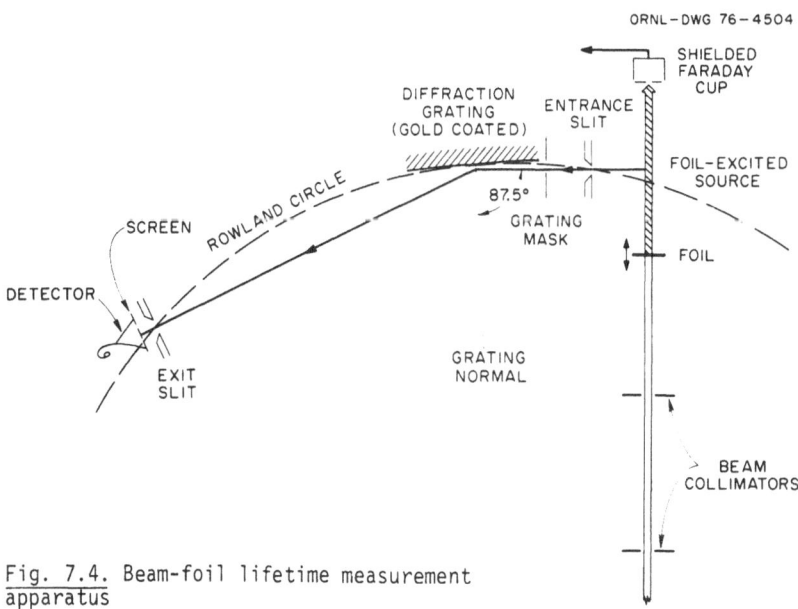

Fig. 7.4. Beam-foil lifetime measurement apparatus

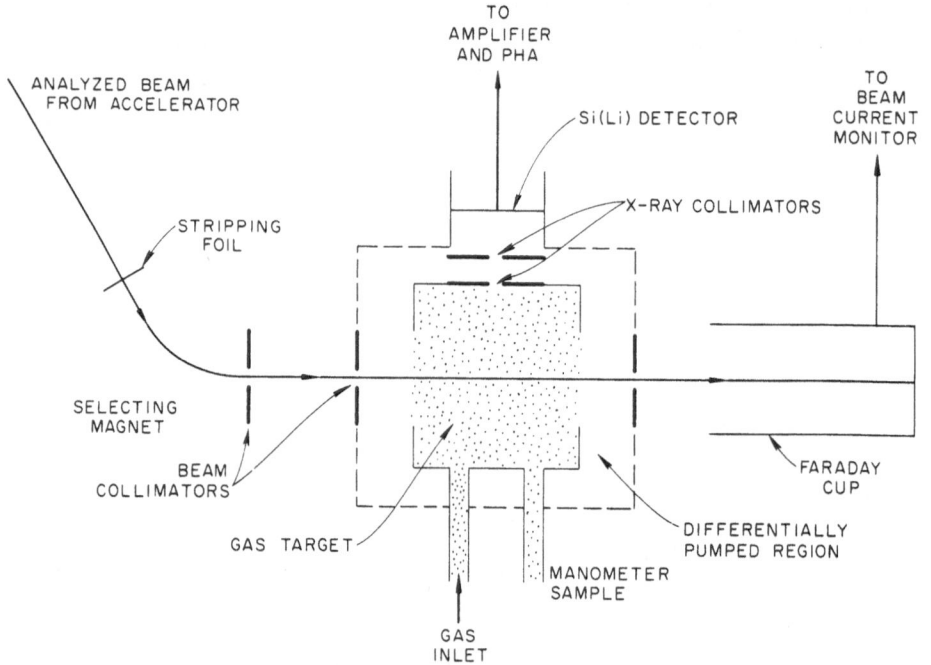

Fig. 7.5. Schematic diagram for apparatus useful for making excitation distribution and cross section measurements for single charge state component projectiles incident on gas target atoms under single collision conditions

were obtained with this apparatus. The study of comparative gas or solid target differences is a subject discussed by DATZ in Chapter 8, Section 8.4, and will not receive further attention here.

The Doppler-tuned x-ray absorption edge spectrometer arrangement used by VARGHESE et al. [7.3] is shown in Fig.7.6. Further discussion of its use and sample lifetime results obtained with it will be deferred to Section 7.1.

Section 7.3 concerns the extension of BFS to the study of Auger electrons emitted by excited ions as they undergo decay in flight. Because the preferred decay mode of most atoms in most ionization-excitation states for at least the first third of the periodic table is decay by Auger electron emission as opposed to photon emission, this particular extension may be argued to be more important to future understanding of how excited atomic systems relax than the more familiar optical branch of beam--foil spectroscopy. For example, typical K-shell vacancy fluorescence yields lie in the range 10^{-1} to 10^{-2}, and typical L-shell vacancy fluorescence yields are some two orders of magnitude smaller. Since fluorescence yields express the probability of photon decay as a percentage of total decay rate, it can be seen how comparatively rare photon emission is whenever Auger deexcitation channels are open. Standard ref-

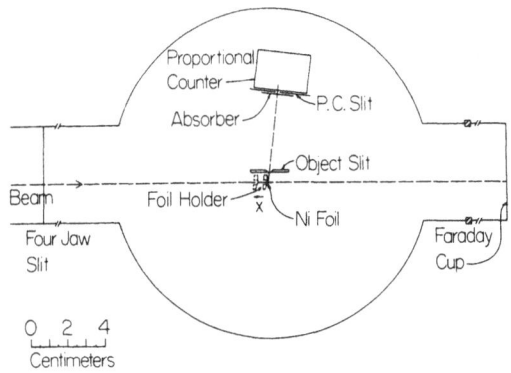

Fig. 7.6. Schematic diagram of Doppler-tuned x-ray absorption edge spectrometer

erence sources concerning atomic energy levels give quite the opposite impression, since the data entries therein normally are mostly derived from photon data on only optically allowed, single-particle, valence-shell excitations in low states of ionization. For such single-particle, valence-shell excited states, Auger emission is forbidden, but for the more numerous core-excited states which are copiously produced in energetic ion-atom collisions and in hot plasmas of either terrestrial or stellar origin, Auger emission and its inverse are very common processes. The inverse process is known to plasma and astrophysicists as dielectronic recombination, in which initially free electrons are attached to heavy particles through simultaneous excitation of a bound electron to a doubly excited state, followed by relaxation by photon emission. The first stage of this process can easily be thought of as an inverse Auger process, and treated as such by use of the principle of detailed balance.

The discussion of electron emission accompanying ion-atom excitation is the subject of an entire chapter of this book (see Chap.5 by STOLTERFOHT). STOLTERFOHT does not, however, discuss the BFS aspects of Auger electron emission measurements. Section 7.3 therefore complements Chapter 5 in this sense. Because work on autoionizing ion levels and lifetimes by fast projectile electron spectroscopy has been reviewed fairly recently by SELLIN [7.8], the emphasis in Section 7.3 is on a basic description of the technique and its limitations, and on more recent related BFS work in other laboratories.

As in the case of the optical branch of beam-foil spectroscopy, a key advantage of fast projectile electron spectroscopy is its single, straightforward application to the direct determination of lifetimes of excited states by the decay in flight method. A corresponding Doppler broadening limitation again occurs, since the important parameter is not the ratio of projectile velocity to the velocity of light, but to the velocity of the emitted electron. The possible use of TIRS to achieve resolution improvement will be even more desirable in this case, except for high

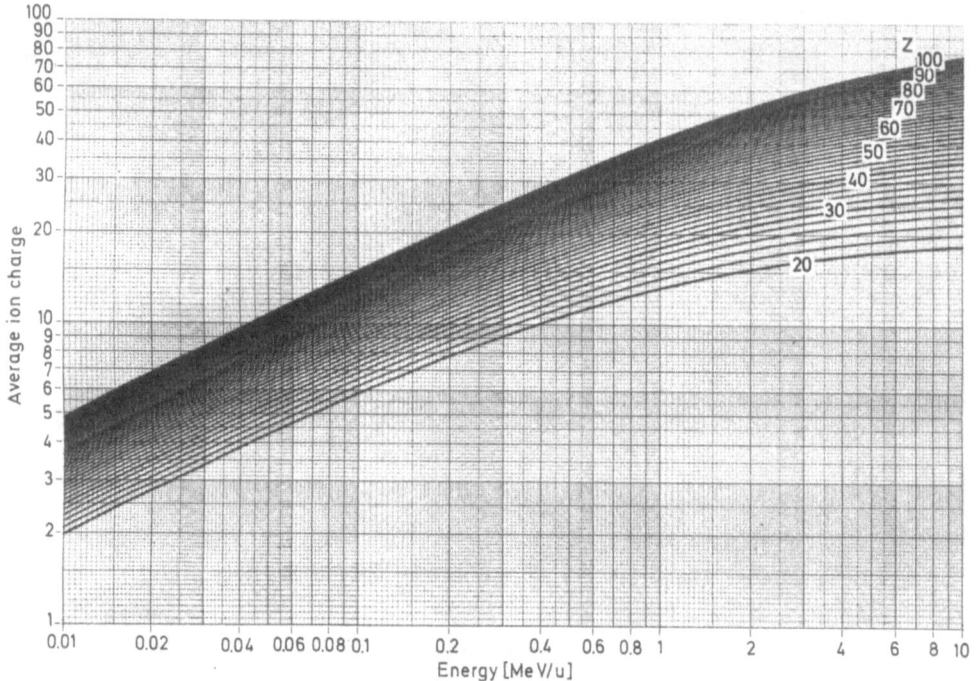

<u>Fig. 7.7.</u> Mean charge state of ions emergent from solids as a function of energy up to 10 MeV/nucleon as estimated by MARTIN [7.11] using semi-empirical formulas of DMITRIEV and NIKOLAEV [7.9]

velocity beams whose post-foil multiple scattering angular half-widths are small, and whose associated Auger electron emission energies are large ($\gtrsim 1/2$ keV) owing to the high equilibrium charge of the emergent ions.

We include here mention of three useful guides to qualitative estimates of the degree of ionization achieved for various foil-transmitted species as a function of their speed in typical solid media. They are fundamentally based on the Bohr criterion, whose basic assertion is that those electrons whose internal atomic velocities are less than the velocity of passage through the medium tend to be stripped off, whereas those whose internal atomic velocities are higher tend to remain attached. The semi-empirical model calculations of DMITRIEV and NIKOLAEV [7.9] have been incorporated in useful graphical form by MARION and YOUNG [7.10] for very low Z, by MARTIN [7.11] for higher Z and higher velocities (neglecting shell effects), and by STELSON [7.12] for still higher velocities (in [7.12], modifications suggested by BETZ [7.13] are also incorporated, and shell effects are approximately considered). Figs.7.7 and 7.8 are taken from the work of MARTIN [7.11] and STELSON [7.12], respectively. Together these graphical data serve as a useful guide to the mean equili-

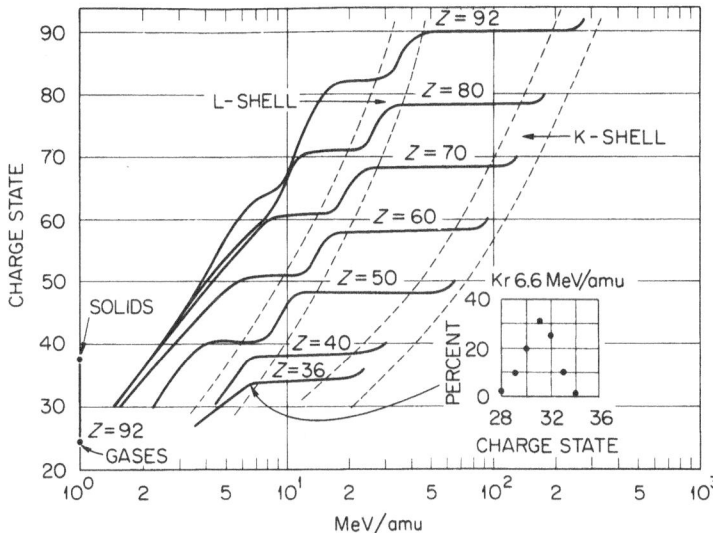

Fig. 7.8. Mean charge states of ions as a function of ion energy above 1 MeV per nucleon, as estimated by STELSON [7.12] using the Bohr criterion as modified by BETZ [7.13]

brium charge states to be expected for ions of various Z transversing solid targets as a function of their velocity. The distribution of charge states about the mean value has been discussed by a number of authors (cf. the review by BETZ [7.13]). For low Z ($\lesssim$ 10) the distribution of charge states for which appreciable post-foil populations exist is narrow ($\sim$ ±2 charge states), but for heavy elements (e.g., Br) in intermediate stages of ionization, a dozen of more emergent charge states may be important. As the situation of nearly complete stripping of a shell is approached, strong shell effects and a consequent narrowing of the charge state distributions occur, as is suggested by the zones marked "L" and "K" in Fig.7.8 which refer to those regions in which successive depletion of L-shells and K-shells occurs. For medium to heavy Z and intermediate as opposed to very high states of ionizations, differences of $\gtrsim$ 10 units in equilibrium charge distributions between foil and thick gas targets are often found to occur. This subject is discussed by DATZ in Chapter 8 of this volume and will not receive further discussion here.

The general plan for the subsequent sections of this chapter is as follows. In Section 7.1, progress in extending the dynamic range of lifetime measurements to some ten orders of magnitude is discussed. In Section 7.2, the Doppler spread advantages of extending BFS techniques to include target ion recoil spectroscopy (TIRS) are discussed. The collision physics of highly charged ions with both symmetric and lighter collision systems under single collision conditions is crucial to TIRS experiments. This subject is also discussed in Section 7.2. BFS lifetime measurements utilizing fast projectile Auger electron spectroscopy is the principal subject of Section 7.3. All of these areas of activity borrow heavily from not only beam-foil spectroscopy techniques but diverse other experimental techniques as well. The unifying theme is the use of charged particle accelerators to produce monoenergetic beams of ions in high ionization-excitation states for use in atomic structure and ion-atom collision experiments.

7.1 Extension of the Dynamic Range of Lifetime Measurements Using the Beam-Foil Technique

As noted in the introductory material at the beginning of this chapter, beam velocities ~ 1 cm ns^{-1} tend to encourage lifetime measurements primarily in the range of 0.1 to 100 ns. Measuring a mean life usually involves establishing a linear variation on a semi-logarithmic plot of line intensity vs target position relative to the viewing region of a spectrometric device, for the purpose of finding a 1/e folding length (decay length). Finding the decay length and establishing its linearity requires subdivision of one decay length into $\gtrsim 10$ increments, as in Fig.7.1. If the longitudinal dimensions of the viewing region are of the order of or larger than the decay length itself, this subdivision is impractical. Hence extending the lifetime measurement on the short lifetime end requires limiting the viewing region to as small a length as possible. In the apparatus depicted in Fig.7.4, some progress toward this goal can be made by appropriate collimation and adaption of a target chamber which vacuum couples to the entrance slit housing so that the beam trajectory passes as close as possible to the plane of the entrance slit. Typical viewing lengths of $\gtrsim 100$ μm are realized this way, for solid angles subtended of $\gtrsim 3 \times 10^{-4}$ sr. Further length reductions by collimation alone lead to unacceptable intensity loss due to solid angle loss. Such solid angle loss can and has been compensated by BARRETTE [7.14] as reported by KNYSTAUTAS and DROUIN [7.15], by displacing the effective entrance slit position along the Rowland circle to near intersection with the beam. Fig.7.9 is taken from the work of KNYSTAUTAS and DROUIN [7.15] and shows a decay curve for the resonance line in heliumlike carbon. These authors appear to have been the first to realize the measurement of ~ 1 ps lifetimes by this technique. A very similar technique has been applied by VARGHESE et al. [7.3] in the x-ray region. A sketch of this apparatus appears in Fig.7.6. Here, the spectroscopic technique is not dispersive, but relies on the kinematic variation of Doppler shift with angle to tune across x-ray absorption edges in wavelength matched absorbers interposed in front of an x-ray detector. The resulting integral spectra can be numerically differentiated to produce spectra like the one shown in Fig.7.10. A combination of col-

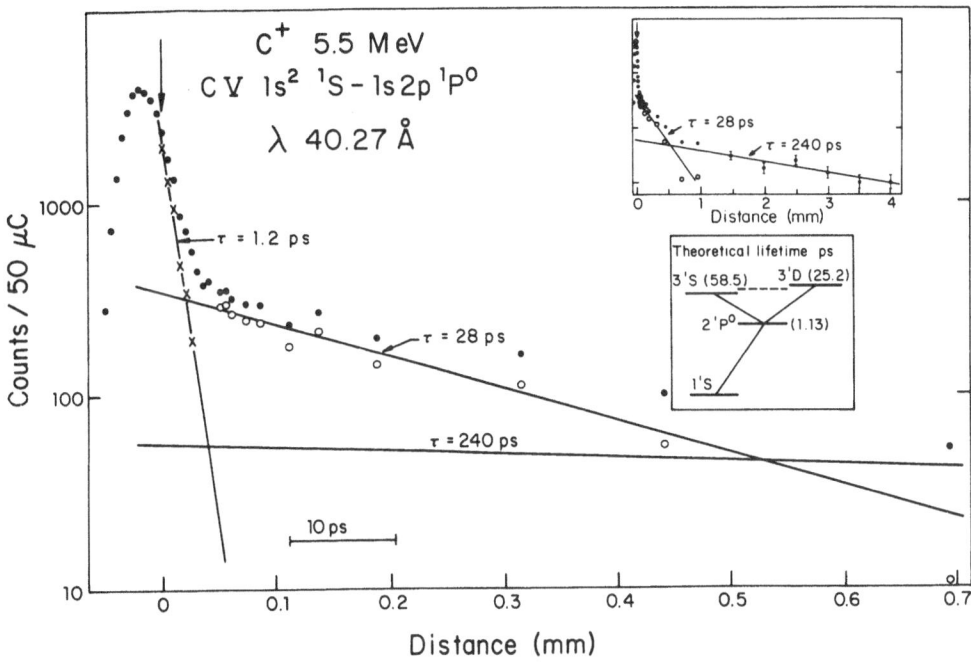

Fig. 7.9. Decay curve for the resonance line in heliumlike carbon

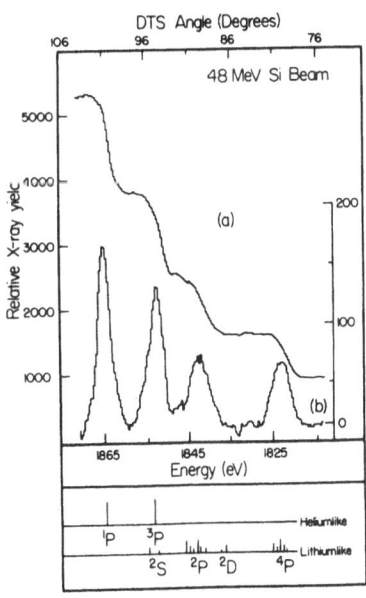

Fig. 7.10. Integral spectrum (a) observed using a Doppler-tuned spectrometer with a 48 MeV Si beam; differential spectrum (b) showing various lines identified at the bottom of the figure

limation together with proximity of a narrow defining slit to the excited beam produces the necessary spatial resolution. A precision micrometer drive is used to step
the target position with respect to the slit location in such a way that an otherwise standard beam-foil decay in flight curve is generated.

In such experiments, lifetime measurements at the $\lesssim$1ps level become increasingly
limited by considerations of uniformity of foil thickness, target flatness, and per-
pendicularity to the beam direction. For example, a deviation from flatness of
~0.04 mm produces a ~3 ps spread in the time scale subsequent to excitation. Techniques for producing precision motion of adequately flat and perpendicular targets
are well known from the Doppler shift recoil ion lifetime measurements method popular
in nuclear physics.

Substantially subpicosecond lifetime measurements clearly require techniques other
than straightforward studies of decay in flight. As the transit time of a ~1 MeV
per nucleon ion traversing a ~10 µg cm^{-2} target foil is $\lesssim 10^{-14}$ s, the decay time
can often be on the order of the transit time itself. BETZ et al. [7.16] and PANKE
et al. [7.17] have demonstrated a technique for making measurements of lifetimes
~10^{-14} s which is based on observing the growth in projectile x-ray radiation intensity with target thickness. For targets in a suitable thickness range (~100 µg
cm^{-2}), there is often an intensity component of a given x-ray line which is proportional to target thickness in addition to the post-foil beam-foil light, which is
independent of target thickness (for sufficiently thick targets). BETZ et al. [7.16]
and PANKE et al. [7.17] were able to show that within certain cross-section approximations, the thickness at which the contribution from the thickness-proportional part
of the total line intensity was equal to the thickness-independent contribution is
a reasonably good approximation to the decay length of a line radiated by the corresponding excited projectile ion. Their demonstration experiment concerned ~90 MeV
S ions ionized and excited (K-shell) in carbon and aluminium foils. The basic cross
section approximations are that the sum of the cross sections associated with K va-

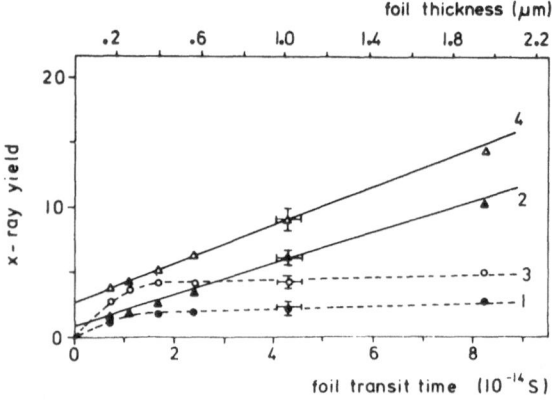

Fig. 7.11. Sulfur x-ray multiplet
yields produced by 92 MeV sulfur
ions are shown as a function of
target thickness

cancy production σ_v and destruction σ_c be larger than $(1/nv\tau)$, where n is the target number density, v the projectile velocity, and τ the excited state lifetime; and that the target thickness π in atoms per cm^2 be in the range $[\sigma_v + \sigma_c + (nv\tau)^{-1}]\pi \gg 1$.

Figure 7.11 presents lifetime data acquired when ~ 90 MeV S ions were passed through Al foils of various thickness in the range 30-400 μg cm^{-2}. Sulfur K_α x-rays were observed with a Bragg spectrometer fitted with an ADP crystal, which viewed radiation emitted perpendicular to the beam. Monitoring of the Rutherford scattering of projectile ions provided a suitable intensity scale normalization. The labels 1-4 correspond to $(1s2p^2)^4P \rightarrow (1s^22p)^2P$, $(1s2p^2)^2D,P \rightarrow (1s^22p)^2P$, $(1s2p)^3P \rightarrow (1s^2)^1S$, and $(1s2p)^1P \rightarrow (1s^2)^1S$ transitions, respectively. The intensities of the lines are presented as a function of transit time through Al foils of various thicknesses. For peaks 1 and 3, which are relatively metastable on a time scale of $\sim 10^{-14}$ s, the yield vs transit time curves are seen to have small slopes, indicating lifetimes long compared to the transit time through the target used. Multiple scattering, energy loss and straggling, and radiation damage tend to preclude the use of much thicker targets. Transitions 2 and 4, which concern transitions whose lifetimes fall in the accessible region, vary linearly in intensity with target thickness, with pronounced slopes and clearly positive intercepts. The foil thickness at which the intensity is twice its intercept value, expressed in terms of the ion transit time through the target, is a direct measure of the excited state lifetime. The corresponding experimental lifetimes are $(1.7\pm0.4)\times10^{-14}$ s (vs 1.49×10^{-14} s theoretically), and $(2.3\pm0.4)\times10^{-14}$ s (vs 2.44×10^{-14} s theoretically), respectively. The use of the target thickness dependence lifetime measurement technique has thus extended the range of BFS atomic lifetime measurements down to $\sim 1\times10^{-14}$ s.

For long lifetimes, the lifetime measurement range appropriate to fast beam experiments has also been extended appreciably into the ten to one hundred microsecond range through the use of delayed coincidence techniques. A review paper on this technique has been published recently by ERMAN [7.18]. In this method, the beam-excitation source is pulsed with as fast rise and fall times as can be arranged. The arrival of decay photons at the detector is registered as a function of time following onset of the excitation pulse, using a conventional time to amplitude converter. Here the excitation pulse could be used to derive a start pulse, and the photon pulse to derive a stop pulse. Good timing resolution at the nanosecond or subnanosecond level in both pulse trains is important, as the time resolution of a delayed coincidence spectrum will be limited by the jitter between timing pulses derived from the excitation and photon pulses. Owing to the usually smaller solid angle-efficiency product of the photon decay channel, it is often convenient to reverse the roles of the start and stop pulses by using a suitable delay in the excitation pulse channel, to minimize dead time in the electronics due to high excitation pulse repetition rates. Up to the present time, such delayed coincidence techniques have very frequently been used in nuclear physics, and have also often been used in atomic and molecular applications, usually in the context of pulsed electron or photon excitation of atoms and molecules. Such experiments are extensively discussed in the review of ERMAN [7.18]. Applications to heavy particle excitation have been comparatively rare. An exception is the pulsed proton excited, molecular lifetime work of DOTCHIN et al. [7.19].

Very recently SELLIN et al. 7.49 have applied the method to long-lived Auger emitting transitions in lithiumlike O^{5+}, Ne^{7+} in the 1s2s2p ^{4}P state. Discussion of this work is deferred until Section 7.2.

At the long lifetime end of the delayed coincidence lifetime measurement range (~ 1 μs to ~ 1 ms), serious corrections beyond those required by cascade repopulation phenomena, excitation transfer, and radiation trapping occur. Thermal and recoil drifts out of the field of view can often become very serious at the $\gtrsim 10$ μs lifetime level. Such effects have been analyzed in detail in the recent paper by CURTIS and ERMAN [7.20]. The recoil effects will of course be much more serious in the case of heavy particle impact at low collision energies. For the case of high collision energies, however, the heavy particle recoil effects become small (cf. the discussion of Ar^{18+}-Ne collisions at 2 MeV/nucleon in Section 7.2). On the other hand it is entirely conceivable that the apparent lifetime changes accompanying collision broadening will become an effective experimental tool for studying this broadening at somewhat lower collision energies. As pulsed, heavy particle beams are not generally available at pulse widths under one nanosecond, possibilities for short lifetime

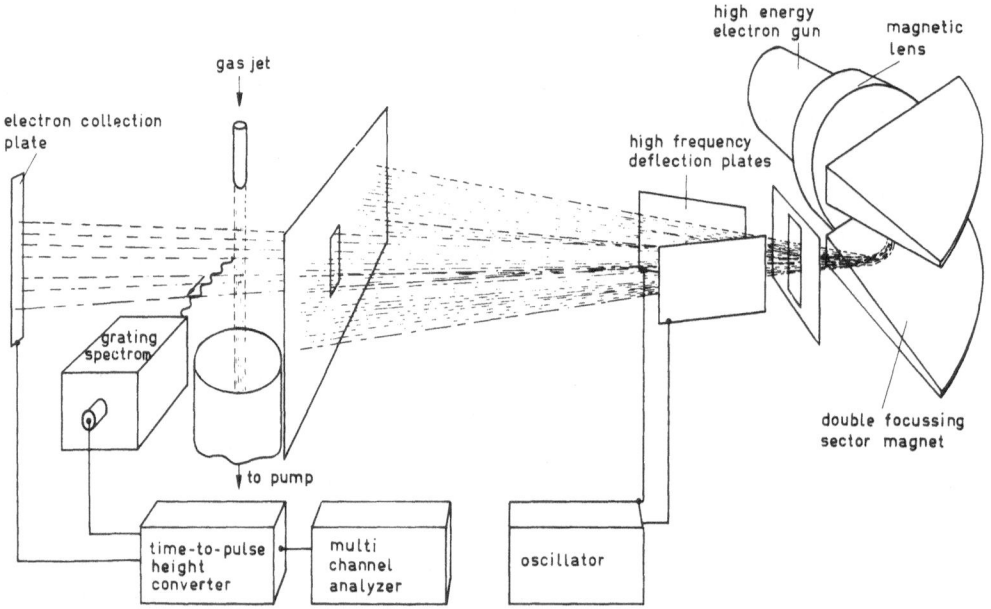

Fig. 7.12. Diagram of electron impact, delayed coincidence lifetime measurement apparatus

measurements are correspondingly limited. The apparatus used by ERMAN and collaborators is shown in Fig.7.12. The effects of thermal drift on measured lifetimes are illustrated in Fig.7.13, taken from [7.20]. The apparatus used by SELLIN et al. [7.49] is shown in Fig.7.14.

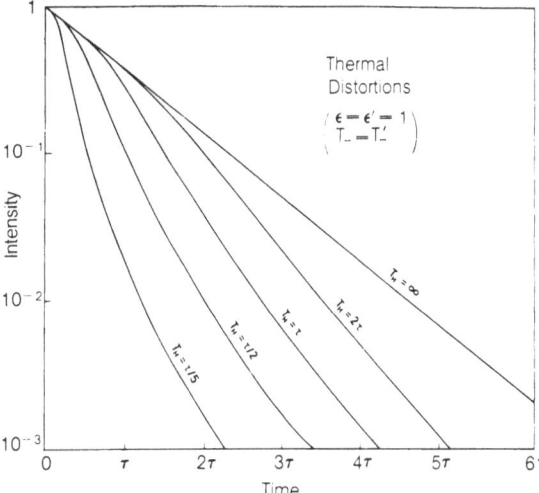

Fig.7.13. Decay curves for mean life τ, distorted by thermal escape from the field of view. T_H refers to a characteristic time for thermal drift across the viewing region. The relative dimensions of the excitation and viewing regions are assumed substantially different. For details see [7.20]

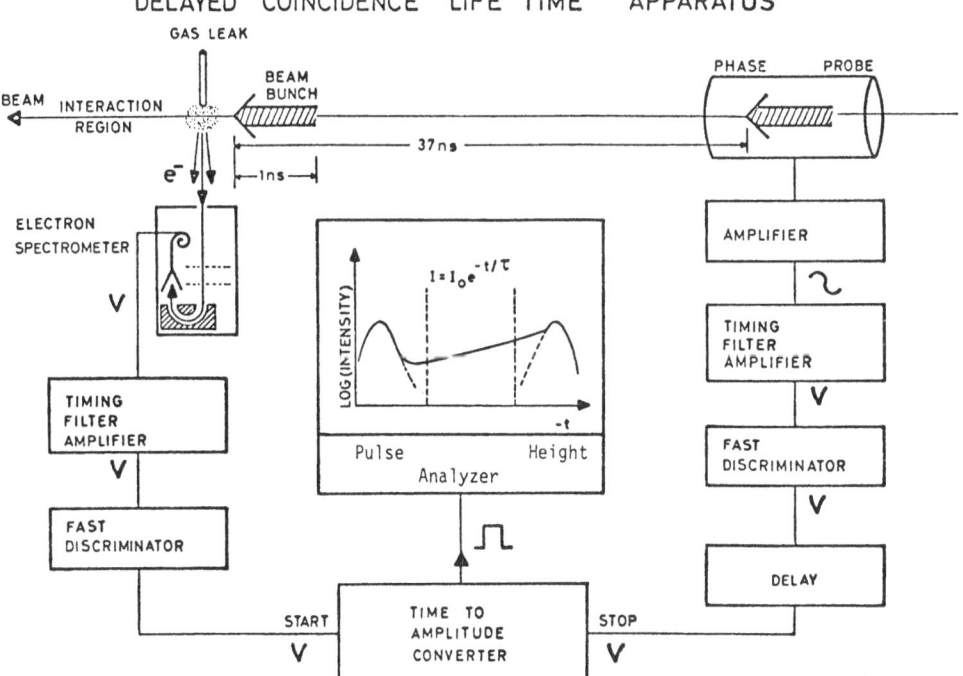

Fig. 7.14. Delayed coincidence apparatus for lifetime measurements on long-lived Auger transitions in target atoms and molecules. See [7.49]

It is therefore apparent that in recent years the dynamic range of lifetime measurements appropriate to fast beam techniques has grown to nearly ten decades in length—from $\sim 10^{-14}$ s to ~ 100 μs. Of the techniques discussed here, the delayed coincidence techniques are the least exploited.

7.2 Target Ion Recoil Spectroscopy

As described in the introductory paragraphs of this chapter, it is possible that precise spectroscopic measurements on multiply ionized and excited recoil ions would be feasible if target particles, ionized and excited by fast projectile ions in high charge states, had a more nearly thermal velocity distribution that fast beam ions, and if target ionization excitation cross sections were sufficiently high. Avoidance of the large Doppler shifts and spreads normally accompanying beam-foil measurements would be a great advantage. Recovery of lifetime measurement capabilities by means of the delayed coincidence technique discussed in Section 7.1 is a demonstrated attractive possibility. Study of very slow collisions of multiply charged target ions in subsequent ion-atom collisions might also be feasible. For example, the study of Coulomb explosion dissociation of Auger-emitting, multiply charged fragments of such molecules as CO following heavy-ion bombardment (e.g., by Ar^{12+}, Kr^{26+}) has recently been initiated by MANN et al. [7.21]. Experiments were recently undertaken in our laboratory to examine cross sections for L-shell excitation in neon by impact of highly charged heavy ions, to compare such target excitation spectra with the beam-foil spectra corresponding to comparable ionization states (BUCHET and DRUETTA [7.22]), to explore the projectile charge state effect in L-shell excitation (MOWAT et al. [7.23]), to study possible resolution advantages recoil ion spectroscopy has over beam-foil spectroscopy for equivalent collimation, to examine the question of possible spectator electron excitation accompanying inner-shell excitation, and to verify the absence of collisional quenching effects on excited levels of sufficiently short lifetime.

These experiments were in large part motivated by the discovery that impact of suitable, highly ionized, foil-transmitted heavy ions generate very high ionization-excitation states [7.23]. For example, substantial amounts of Lyman alpha from hydrogenic Ne^{9+} appeared in the target ion excitation spectrum when Ar^{q+} ions from the Oak Ridge Isochronous Cyclotron in the range $q \gtrsim 16$ were used as projectiles, *even under single collision conditions*. It is therefore plausible that precise spectroscopic measurements on multiply ionized and excited recoil ions which radiate in the soft x-ray to x-ray range would be feasible if the target ions has a more nearly thermal velocity distribution than fast beam ions, and if target ionization-excitation cross sections were sufficiently high. Avoidance of the need for refocusing or other Doppler compensation techniques [7.2] would be a key advantage over standard beam-foil methods, especially since optical compensating devices are largely unavailable in the grazing incidence and x-ray regions.

Estimates of recoil broadening are attractively low for two reasons. First, the long range, intense field of a highly ionized projectile permits electron removal and excitation processes for impact parameters larger than the shell radii of the affected target electrons [7.24]. Second, the high relative collision velocity ($\gtrsim 0.05$ c) provides a short interaction time, giving rise to slow recoil ions recoiling at almost exactly 90 deg to the beam direction. For example, a bare 2 MeV/A Ar nucleus passing by a Ne atom at an impact parameter of ~1/3 Å—a Ne L-shell radius—creates a Ne recoil ion traveling at about 89.99 deg to the beam direction at a recoil velocity $v_R \lesssim 2v_B(M_{RED}/M_{Ne})\cos\phi_R \cong 3\times10^5$ cm s$^{-1} \cong 10^{-5}$ c. Here v_B is the beam velocity, M_{RED} the reduced mass, and ϕ_R the laboratory recoil angle. Hence $v_R/c \ll v_B/c$ by almost four orders of magnitude. For viewing angles and collimation equivalent to those prevailing in beam-foil experiments, both Doppler shifts and spreads in recoil vs projectile lines—including Auger lines should be smaller by about the same factor. Moreover, v_R is only of the order of thermal velocities in a hollow cathode discharge, a standard source for xuv soft x-ray lines from single-particle valence shell excited states of ions in low states of ionization.

We have shown by direct measurement that L-shell excitation cross sections are attractively large for a wide range of ionization-excitation states, and that within the limitations of our present spectroscopic apparatus the low recoil estimates are fully justified. We have also found that the Ne target L-excitation spectrum is highly similar to that of a ~1 MeV foil-excited Ne beam [7.22], even though typical ion recoil velocities are a few thousand times smaller; that the L-shell excitation cross sections are of comparable magnitude ($\sim 10^{-18}$ cm^2) and only *weakly* dependent on charge state in the range of measurement (II-VIII); that there is a projectile charge state effect on the L-excitation cross sections of a factor ~5 for a corresponding projectile charge state change from 6^+ to $\sim 12^+$; that classification of K x-ray satellite spectra (KAUFFMAN et al. [7.25]) by L-shell vacancy labels (KL^0, KL^1, ...) is probably misleading and oversimplified because of extensive population of $n \geq 3$ spectator levels; that both the recoil ion and beam-foil spectra exhibit few lines corresponding to $n \geq 4$ for incompletely understood reasons; and that with the possible exception of metastable levels, excited state quenching effects due to collisions with other gas atoms in the target are negligible. The experimental procedures used have been discussed in detail by SELLIN et al. [7.26].

Figures 7.15 and 7.16 survey Cl on Ne target ion spectra obtained with a 1200 mm^{-1} grating between about 40 Å and 150 Å, and S on Ne spectra obtained with a 300 mm^{-1} grating between about 140 Å and 500 Å, respectively. The respective target pressures were 200 mTorr and 100 mTorr. These survey spectra were accumulated at modest resolution levels (200 μm slits) and took only about 4 hours apiece to complete. A number of selected regions of high intensity, blended lines were studied with significantly

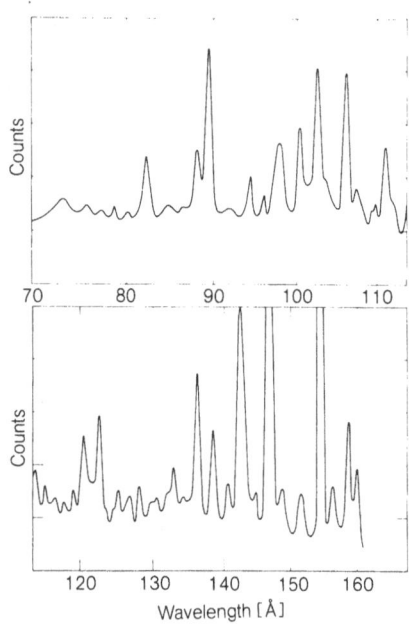

Fig. 7.15. Soft x-ray spectrum from bombardment of Ne gas at 200 μm pressure by foil-stripped Cl ions at 1.1 MeV A^{-1}. A 1200 mm^{-1} grating was used to obtain this spectrum

higher resolution, a few of which are also illustrated in Fig.7.16. The high intensities indicate large excited state production cross sections, which could be studied by taking account of measured beam currents, measured line intensities, solid angle subtended by the grating, target volume viewed, target number density, gold-coated grating reflectivity, astigmatism, and channel multiplier efficiency, as a function of wavelength.

An approximate excitation cross section based upon the intensity of the Ne II standard line at λ 460.7 and the experimental conditions noted above is $\gtrsim 3 \times 10^{-18}$ cm^2. While the estimate allows for grating reflectivity, astigmatism, and multiplier efficiency in addition to the more straightforward geometry factors, it does not allow for additional loss due to dispersion into higher diffraction orders nor for notoriously difficult estimates of loss due to possible time-dependent grating surface condition reflectivity losses. Hence this cross section estimate should be regarded as an approximate lower limit. Corresponding lower limits on excitation cross sections for other lines can be inferred from their relative intensities, provided account is taken of the grating blaze angle and wavelength dependent astigmatism. At the 87.5 deg angle of incidence used, the apparent relative intensity of lines in Figs.7.15 and 7.16 is estimated never to exceed a factor of two enhancement compared to the observed λ 460.7 intensity. A convenient cross-over between the spectra depicted in Figs.7.15 and 7.16 occurs in the intense feature near λ 154.5. While the cross sections for production of this feature by 1.4 MeV/A S$^{\sim 12+}$ impact and 1.1 MeV/A Cl$^{\sim 12+}$ impact are not *a priori* known to be the same, earlier experiments [7.24] on K ionizing collisions for Cl^{q+} in Ne indicated only weak dependence on beam energy,

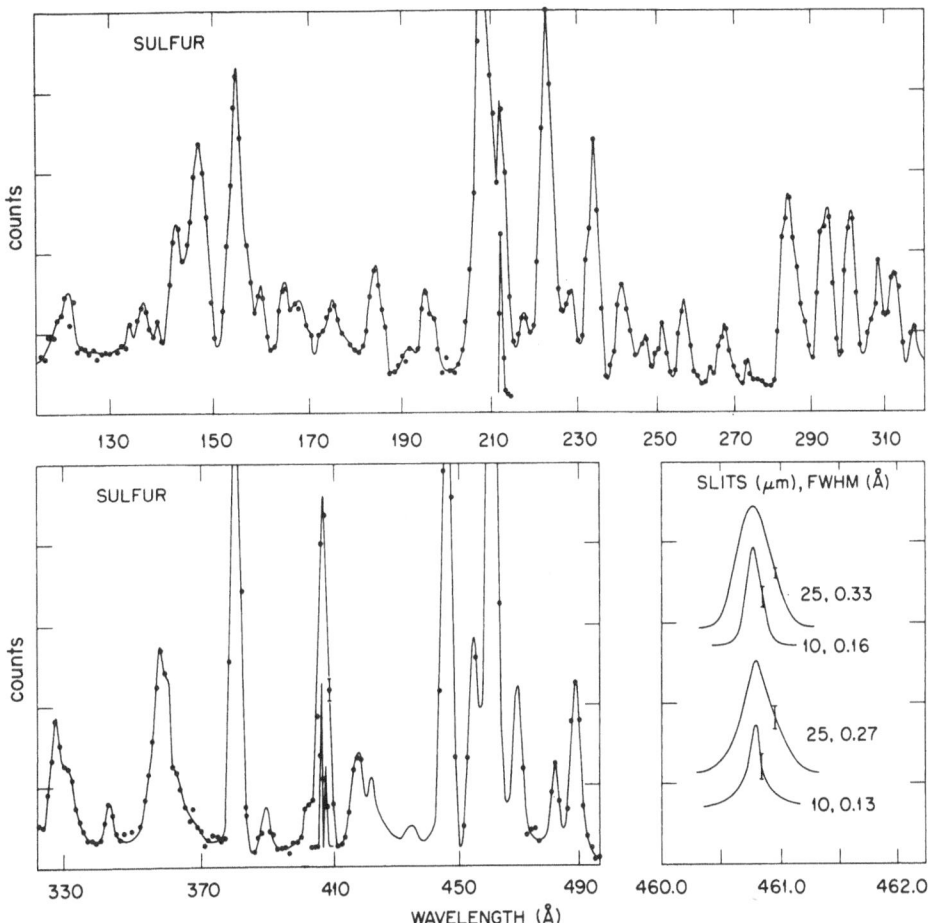

Fig. 7.16. Soft x-ray spectrum from bombardment of Ne gas at 100 μm pressure by foil-stripped S ions at 1.4 MeV A⁻¹. The main body of the spectrum was acquired using symmetric slit widths of 200 μm and a 300 mm⁻¹ grating. The inset spectra were acquired at 50 μm slit width. The standard Ne II line at λ 460.7 (lower right) was observed to have an instrument limited width of $\lesssim$ 7 meV at 10 μm, and is compared to a hollow cathode spectrum seen just above it

and a strong dependence on charge q. Indeed, independent evaluation of the cross section for production of the λ 154.5 feature by $S^{\sim 12+}$ and $Cl^{\sim 12+}$ ions indicated quantitative agreement within a factor of 2-3 (the $Cl^{\sim 12+}$ cross section being lower), despite the difference brought about by grating interchange. The exactness of this quantitative agreement is within uncertainties in relative grating efficiencies and dispersion into higher orders. In estimating approximate lower limits to cross sections, this factor will be omitted from further consideration.

Table 7.1 provides a list of intense lines observed;, their assigned configurations and terms where known from other reference sources; their intensities relative to λ 460.7 after adjustment for astigmatism, blaze, and electron multiplier efficiency;

Table 7.1. Intense Ne lines observed, associated transitions, relative intensities, and approximate lower limits on cross sections, adjusted for astigmatism, blaze, and electron multiplier efficiency for 1.4 MeV/nucleon S$^{\sim 12+}$ projectiles. Asterisks refer to transitions not noted in the beam-foil spectrum of BUCHET and DRUETTA [7.22]

[Å]	Spectrum	Transition		Relative intensity	Cross section (Mb) limit
82.2*	VII	2s2p $^1P^0$	-2s4d ^3D	60	0.2
88.1*	VIII	2s ^2S	-3p $^2P^0$	90	0.3
98.2	VIII	2p $^2P^0$	-3d ^2D	80	0.3
100.3	?	?	?	140	0.4
102.5*	?	?	?	190	0.6
106.2	VII	2s2p $^3P^0$	-2s3d ^3D	170	0.5
111.2	VII,VI	$2p^2, 2s^22p$	-2p3d, 2s2p($^3P^0$)3p	100	0.3
120.5	VII	$2p^2$ ^3P	-2p3s $^3P^0$	90	0.3
122.5	VI,V	$2p, 2p^2$	-3d, 2p4d	120	0.4
136.2	VI,V	$2s2p^2, 2s2p^3$	-2s2p3s, 2s2p^2(^4P)4s	120	0.4
138.5	VI	$2s2p^2$ ^2S	-2s2p($^3P^0$)3d $^2P^0$	90	0.3
142.5	V*	$2p^2$ ^3P	-2p3d $^3P^0$	170	0.5
147.1	V*	$2p^2$ ^1D	-2p3d $^1F^0$	320	1.0
154.5	V	$2p^2$ ^1S	-2p3d ^3D	270	0.8
184.7	V	$2p^2$ ^1S	-2p3s $^1P^0$	100	0.3
212.6	IV	$2p^3$ $^2D^0$	-$2p^2$(^1D)3s ^2D	280	0.8
222.6	IV	$2p^3$ $^2P^0$	-$2p^2$(^1D)3s ^2D	360	1.1
~234.3	IV	$2p^3$ $^2P^0$	-$2p^3$(^3P)3s ^2P	250	0.8
~283.7	III	$2p^4$ ^3P	-$2p^3$($^2D^0$)3s $^3D^0$	240	0.7
301.1	III	$2p^4$ ^1D	-$2p^3$($^2D^0$)3s $^1D^0$	230	0.7
379.3	III	$2s^22p^4$ ^1D	-$2s2p^5$ $^1P^0$	750	2.3
405.9	II	$2p^5$ $^2P^0$	-$2p^4$(^1D)3s ^2D	400	1.2
407.1	II	$2p^5$ $^2P^0$	-$2p^4$(^1D)3s ^2D	200	0.6
416.8	V	$2s^22p^2$ ^1S	-$2s2p^3$ $^1P^0$	170	0.5
455.3	II	$2p^5$ $^2P^0$	-$2p^4$(^3P)3s ^4P	400	1.2
460.7	II	$2s^22p^5$ $^2P^0$	-$2s2p^6$ ^2S	1000	3.0
462.4	II	$2s^22p^5$ $^2P^0$	-$2s2p^6$ ^2S	500	1.5

and the consequent approximate lower limits for excitation cross sections from 1.4 MeV/A $S^{\sim12+}$ ion impact. Asterisks denote a very few transitions which appear in the present spectrum which were not noted in the beam-foil spectrum of [7.22].

The attractively large sizes of the lower limits on excitation cross sections which are furthermore seen to be approximately independent of target charge state indicate a bright future for experiments dependent on such target ion spectra, particularly if the widths of target lines are sufficiently low. Because target x-ray spectra from similar experiments [7.24,25] involving impact of highly ionized projectiles have been characterized by linewidths in the range 0.3 to 3 eV, we sought to improve an experimental limit on the linewidths of a typical transition (the Ne II standard line λ 460.7) to $\lesssim 7$ meV, a factor of 50 to 500 improvement (about an order of magnitude on a $\lambda/\Delta\lambda$ basis). As the monochromator was operated at its limit of resolution for this measurement with the results shown in the lower right of Fig.7.16 for both the hollow cathode discharge (top) and $S^{\sim12+}$ on Ne source (bottom), the measured width was verified to be entirely instrumental. As noted above, the actual linewidth is suspected to be an order of magnitude smaller, because of the small estimated values of recoil velocity ($\sim 10^{-5}$ c). Additionally, a projectile charge state effect of a factor ~ 5 was noted in the intensity of many target lines in going from use of the primary accelerator beams (6^+) to foil-stripped beams ($\sim 12^+$).

Beyond establishing the magnitude of associated production cross sections and limits on Doppler broadening, it was noted above that the Ne L excitation spectrum is found to be remarkably similar to the beam-foil spectrum obtained at ~ 1 MeV energy [7.22], even though the typical recoil velocities are $\gtrsim 1000$ times smaller; that there is extensive population of initially unpopulated n=3,4 levels which if also present in K ionizing collisions constitute extensive population of spectator electron levels not previously noted [7.25]; that both the recoil-ion and beam-foil spectra exhibit few lines corresponding to upper states with $n \geq 4$; and that collisional quenching effects on excited levels having lifetimes typical of the high ionization-excitation states considered here are not only expected to be negligible at the target densities used but are found to be so for the typical transition at λ 147.1 for which this assertion was checked.

With respect to the classification problem for K x-ray satellites in terms of variable L-shell electron populations [7.25], the work strongly suggests that the classification in terms of vacancy states KL^n is at least ambiguous if not misleading. Owing to the large size of the L-shell excitation cross sections noted here it seems likely that the L-electrons are often present in excited states whose inner shielding affects the K x-ray satellite energy determination at about the 1 eV level. It was already a puzzle why K x-ray satellite distributions tend to form such distinct groups [7.25] classified by L-shell vacancy labels (KL^0, KL^1,...) in view of the tendency for a number of K satellite transition energies from a given L vacancy distribution to overlap that of another. For example, it is known from standard references that the energy of the $2s^2 2p^2\ ^3P-2s2p^3\ ^5S^0$ transition is shifted by about 1 eV when a 3d spectator electron is added, an energy comparable to the observed separation of various K x-ray satellite line [7.25]. The evidence for large cross sections for n=3 level population thus casts further doubt on the simple KL^n description of K x-ray satellite energies.

How to understand the relative rarity of n=4 upper levels and the nearly complete absence of intense transitions from $n \geq 5$ presents another problem. A plausible partial explanation is that $n \geq 4$ excitations are often accompanied by a second $n \geq 3$ excitation, resulting in intense deexcitation competition through the Auger effect, suppressing effective line intensities by a factor of $\sim 10^2$ to 10^5. The comparable production of ionization states II-VIII tends to support this suggestion, in that Auger cascades leading to increased production of the higher charge states could be initiated by intense multiple L-shell excitation followed by increasing ionization through successive Auger processes.

Another effect which could in principle account for the relative rarity of lines corresponding to $n \geq 4$ is the quenching of such states in collisions in the gas target in less than a radiative lifetime. Such quenching is however unexpected, since in typical state lifetimes ~ 10-1000 ps at velocities of $\sim 10^{-5}$ c, a mean free path of only $\lesssim 1000$ Å occurs. Hence for quenching cross sections even as large as 10^{-14} cm^2, negligible quenching should occur. Only for metastable states of mean life $\gtrsim 10$ ns would such quenching cross sections in gas targets at $\gtrsim 300$ mTorr pressure be likely to cause appreciable quenching. Indeed, a plot of the λ 147.1 intensity to pressure ratio vs pressure had only a slight (-8%±10%) slope consistent with zero quenching. The upper state in this transition is 2p3d $^1F^0$, a typical configuration among those found to be strongly populated.

Some additional valuable properties of the kind of target ion spectroscopy described here may be noted. First, the excited ions of interest are produced in an environment which is cold compared to both traditional hollow cathode discharges and to plasma sources. Collisional broadening and corresponding lifetime perturbation effects will be much reduced. Second, current technology easily permits recovery of one of the most valuable features of the beam-foil source, namely the ability to make lifetime measurements. Pulsed beams in the nanosecond regime are standard at numerous accelerator facilities. As discussed in Section 7.1, delayed coincidence measurements of the time interval between the beam pulse and photon emission are a well-known, standard technique which permits lifetime measurements on excited states of target ions in the nanosecond to millisecond range.

As noted in Section 7.1, SELLIN et al. [7.49] have applied the delayed coincidence method to long-lived Auger transitions in lithiumlike ions, where a start pulse generated by an energy selected Auger electron starts a time to amplitude converter. The apparatus for these measurements is sketched in Fig.7.14. Using this apparatus, we have measured the lifetime of the 1s2s2p $^4P_{5/2}$ state in O^{5+}, produced

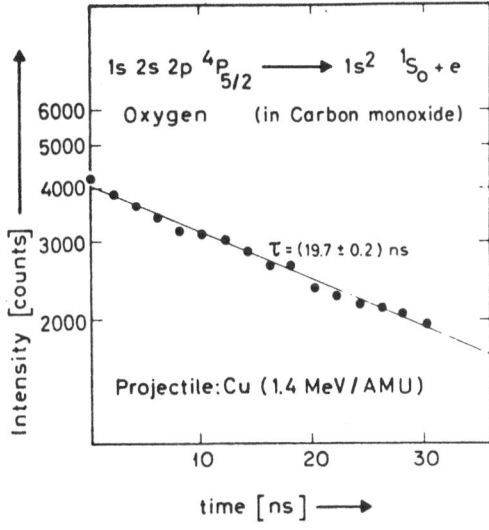

Fig. 7.17. Delayed coincidence decay curve for the Auger transition O^{5+} $1s2s2p$ $^4P_{5/2} \to O^{6+} + K\ ^2F_{5/2}$. The apparent lifetime is shortened by target quenching effects. See [7.49] for details

by ions like Ar^{12+}, Kr^{26+} from the GSI UNILAC accelerator at Darmstadt, in an O_2 target. Comparison with earlier beam-foil results shows good agreement, better intrinsic precision, and establishes that cascade phenomena cause negligible errors in either technique. Data on $^4P_{5/2}$ quenching in O_2 were also obtained.

An example of such lifetime data is shown in Fig.7.17. At the target pressure used, there is an appreciable collisional quenching effect on the $^4P_{5/2}$ state in O^{5+} because of the chance of collisions with other O_2 molecules in the target gas prior to decay. Data taken as a function of target gas pressure showed that a smooth, approximately linear extrapolation of the lifetime vs pressure curve to zero pressure permitted measurement of the unperturbed state lifetime. A number of such runs with a variety of projectile ions—e.g., Ar^{12+}, Cu^{20+}, Kr^{26+}—permitted extraction of a lifetime value of 26.0±1.5 ns, where the error bar is the range error of the various zero pressure extrapolations. This value can be compared to earlier DFS measurements of 25±3 ns in our laboratory at ORNL in which the decay in flight by Auger emission of MeV beams of O^{5+} ions was measured. A recent theoretical value of 23.1 ns (CHENG et al. [7.43]) is in fair agreement with both results.

The observed pressure dependence points the way to making excited state quenching cross section measurements for various slow (~ eV) ions in collisions with various gas targets. As the measurements at GSI have to date been carried out using gas jet targets, in which the pressure vs position profile is inadequately known, such measurements await future target designs which overcome this problem.

These lifetimes once again illustrate the tendency for all experimental values for the entire isoelectronic sequence from 8-18 to exceed those from theory by an average of 15%. Whether neglect of electron correlation in the calculations is responsible is still an unsettled question.

It is convenient to make Auger spectroscopy measurements on both the prompt and delayed transitions from both target and projectile systems using essentially the same apparatus, as has been done for the past few years by MANN et al. [7.21]. Auger lines of the core-excited lithiumlike charge states of light targets (e.g., Ne, O_2, CO) bombarded by high charge state ions (e.g., Ar^{12+}, Cu^{20+}, Kr^{26+}, Xe^{31+}) are particularly strongly excited. In fact the ^4P state of the lithiumlike ions is the most intense feature seen in most spectra.

When the decay time of a short-lived state—e.g., the lowest lying core-excited state $1s2s^2$—is comparable to the time required for the removal of the companion atom in a diatomic molecule to about one atomic unit, one can anticipate Auger line shift and broadening to occur in a post-collision interaction. This phenomenon, originally investigated in another context in the mid-sixties (BARKER and BERRY 7.50), is prominent in recent studies by a number of investigators in such areas as Auger decay following photoionization near threshold. Such post-collision phenomena are important again in the present context of line shifts and broadenings in Auger decay subsequent to excitation by highly ionized projectiles. In the GSI experiments, the Auger transitions $(1s2s^2)^2S_{1/2} \rightarrow (1s^2)^1S_0$ +e have been measured in light target elements after impact of Ar, Kr, and Xe ions from UNILAC. In monatomic targets the energy of this prompt transition agrees well with beam-foil data and with Hartree-Fock calculations. Transitions from target atoms in molecules, however, are strongly shifted to lower energy and broadened in line width. A simple model to account for this effect in terms of Stark perturbation of the emitted electron by the neighboring ion is being worked out.

7.3 Projectile Electron Spectroscopy and Lifetime Measurements

As noted in the introductory paragraphs in this chapter, the preferred decay mode of most atoms in most ionization-excitation states for at least the first third of the periodic table is decay by Auger electron emission as opposed to photon emission. For example, a neon atom lacking a single K electron is more than fifty times as likely to decay by electron emission as opposed to photon emission. Since the standard atomic data reference sources tabulate primarily energy level and lifetime data for low-lying, single-particle, valence shell excited states of neutral or neon-neutral atoms, it is clear that the study of atomic structure and lifetimes of atomic systems with arbitrary vacancy distributions has barely begun. The relative rarity of lifetime measurements on Auger-electron emitting states is striking. The experimentally inconvenient typical lifetime range ($\sim 10^{-12}$ to 10^{-16} s) is responsible for the lack of such data. The knowledge of autoionizing energy levels and transition rates that does exist concerns mainly neutral atoms and their singly charged positive and negative ions.

Data on Auger electron spectra and lifetimes have up to now been primarily acquired in experiments using photon and electrom impact to achieve the kinds of excitations that decay through Auger emission. RUDD and MACEK [7.27] give a number of examples of states which can autoionize: a) multiply excited states in which two or more electrons are simultaneously excited; b) inner-shell excitation states, in which an inner electron in an atom having more than one shell is excited to a higher energy orbit; c) single-particle excitation states with rearrangement of the core (same core configuration but different core angular momentum coupling); d) inner-shell vacancy states, in which an inner-shell electron in an atom with more than one shell is completely removed; e) single-particle excitation accompanied by some other internal energy change (e.g., in molecules, vibration-rotation energy may be present).

It is characteristic of electron excitation by photon and electrom impact that single-particle excitations are strongly dominant over multiple-electron excitation. Auger emissions from higher charge states of inner-shell excited atoms can often nonetheless be observed through the variety of electron shakeup and shakeoff processes which occur as a result of Auger cascade processes subsequent to the primary excitation process. A general discussion of such phenomena is provided in the descriptive volume of SEVIER [7.28]. A large amount of more recent data concerning such processes is found in the subsequent literature, particularly in recent issues of the Journal of Electron Spectroscopy [7.29]. In contrast, it is characteristic of heavy particle impact that violent multiple-electron excitation and rearrangement phenomena occur, particularly when the charge on a projectile incident on a lighter target atom is large. For example, MOWAT et al. [7.30] demonstrated simultaneous ten-electron excitation processes in single collisions of bare Ar^{18+} ions incident on Ne gas at energies ~ 2 MeV/nucleon. The subject of electron excitation in ion-atom collisions as manifested in photon and electron decay processes is a vast one. In particular, the discussion of electron emission accompanying ion-atom excitation is the primary subject of Chapter 5 of this book by STOLTERFOHT. Chapter 5 does not cover the BFS aspects of Auger electron emission processes, especially lifetime measurements on Auger electron emitting states. Because BFS work on autoionizing levels and lifetimes has been discussed fairly recently by SELLIN [7.8], the present discussion will be limited to a basic description of the technique as applied to BFS problems, and to more recent BFS work from various laboratories.

As indicated in the introductory material in this chapter, fast projectile electron spectroscopy (FPES) shares the advantages of differential metastability in making line assignments, lifetime measurements, and temporally resolving otherwise unresolved levels with its photon BFS counterpart. It also shares the corresponding Doppler shift and spread limitations. The ratio v/v_p of laboratory electron emission velocity v to projectile velocity v_p, and the angle θ between $\underline{v}$ and $\underline{v}_p$ determine the broadening. The velocity $\underline{v}_p$ is usually confined to a range of polar angles of ~ 0.1 to 1.0 deg about the projectile beam direction (z), and has a magnitude which is generally easily specified by accelerator analyzing-magnet systems to a precision which often renders the spread in v_p in the emergent beam negligible. Projectile passage through a thin foil is of course accompanied by an energy loss which reduces v_p. The post-foil mean value of v_p can either be directly measured, or reliably estimated through the use of energy loss tables like those of NORTHCLIFFE and SCHILLING [7.31] if the foil thickness is either known or measured. Thus $\underline{v}_p$ may often be regarded as having a transverse spread, but a negligible axial spread. Even the thinnest available foils contribute greatly to the transverse spread in $\underline{v}_p$ through multiple scattering processes. For a sharp Auger line, the energy of an electron emitted at a fixed polar angle $\bar{\theta}$ in the laboratory system will be spread because of the transverse spread in $\underline{v}_p$, but will be independent of the azimuthal angle of emission ϕ. The transmitted current of an electron energy analyzer observing electrons emitted in some range $\delta\bar{\theta}$ near $\bar{\theta}$ will exhibit both kinematic broadening due to the transformation between rest and laboratory frames and the transverse spread in $\underline{v}_p$. In the laboratory frame the narrowest projectile Auger lines and hence the best energy resolution will result from the smallest possible values for the analyzer polar angle acceptance range combined with the smallest possible spread in $\underline{v}_p$ consistent with needed line intensities Because of the consequent loss in solid angle subtended by an analyzer entrance slit at the site of an emitter, it is very convenient to take advantage of the cylindrical symmetry of ϕ to give intensity.

The electrostatic cylindrical mirror analyzer is a type whose collection geometry permits small values of $\delta\bar{\theta}$ and an azimuthal range $\geq \pi$. Such an analyzer, in use in our laboratory, is sketched in Fig.7.18. Its features are described in considerably greater detail in [7.8]. Electrons which are emitted from either the projectile or target ions at points in a roughly conical annular zone determined by the intersection of the beam with the field of view of the entrance slits are first retarded by a measured potential to an energy at which analysis is desired, deflected in the main cylindrical field region by a measured analyzer potential, and then detected by a channel electron multiplier detector placed at the axial location of the conjugate focus. The configuration shown in Fig.7.18 is appropriate to the study of projectile collision in gases. When foil excitation is used, the projectile beam-defining apertures are mounted behind a foil on a longitudinal drive rod which can adjust the separation of the foil and the spectrometer viewing region over the range 0 to ~50 cm. The mean angle of launch $\bar{\theta}$ into the analyzer is chosen to be 42.3°, for which angle there is for nonrelativistic particles a second order focus at the axial image point (ZASHKVARA et al. [7.32], DAHL et al. [7.33]). The dimensions of such analyzers scale to the inner cylinder radius, which is 5.71 cm for the analyzer pictured in Fig.7.18. ZASHKVARA et al. [7.32] give an approximate expression for the analyzer constant (ratio of outer cylinder potential to transmitted electron energy) as K=0.77 ln (b/a), where b and a are the outer and inner cylinder diameters, respectively. The calculated value K=0.586 is close to the measured one (0.582±0.002). If E_A is the energy at which analysis is made, the measured resolution (FWHM/E_A) for the analyzer depicted in Fig.7.18 is 1.3×10^{-3}. If R is the factor (typically $\lesssim$ 5) by which

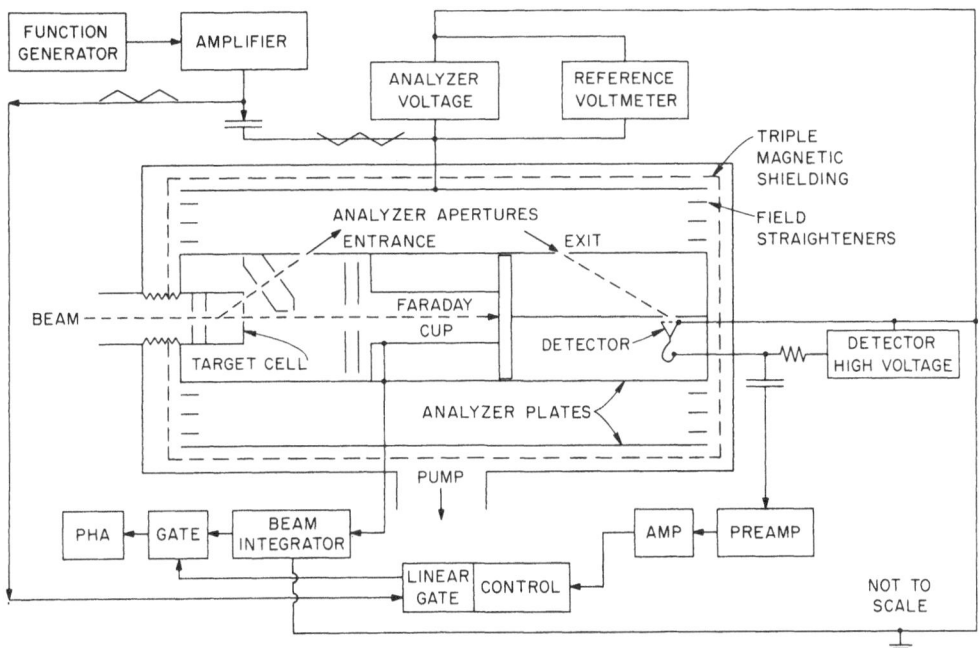

<u>Fig. 7.18.</u> Schematic diagram of an electrostatic cylindrical mirror analyzer suitable for use in projectile electron spectroscopy and lifetime experiments

electron energies are retarded to the level E_A, the overall resolution is then given by $1.3 \times 10^{-3}/R$. This resolution is critically dependent on the locations and widths of defining slits. Such details may be found in the more extensive description of the instrument provided in [7.8].

With foil targets, broadening arising from multiple scattering of beam particles in the target leads to linewidths much larger than such resolution figures would suggest, and often much larger than are found to occur in gas targets (in which recoil broadening is often an appreciable but not nearly so dominant source of resolution loss). As an illustration, the resolutions for Auger lines of core-excited states of Li projectiles excited in thin He gas targets as in the work of PEGG et al. [7.34] can be directly compared with that of BRUCH et al. [7.35], who used foil targets for excitation. Such line broadening problems for foil targets are discussed in [7.8], and for gas targets in the recent paper of DAHL et al. [7.33].

Since the key advantage of beam-foil spectroscopy is its applicability to making lifetime measurements, we concentrate here on such data. The most elementary system having an inner-shell (core) excited Auger decay node is the lithium atom and its three-electron, isoelectronic analogues. The lowest doublet state has the configuration $1s2s^2\ ^2S$, and decays through Coulomb repulsion among the electrons to a $1s^2\ ^1S$

$+^2kS$ final state. The lowest quartet state, in which all three electron spins are parallel, has the configuration $1s2s2p$ ^4P and cannot decay through Coulomb repulsion processes because the absolute conservation requirements on parity and total system angular momentum J are not consistent with the weaker conservation requirements on total electron orbital angular momentum and total electronic spin demanded by Coulomb repulsion forces. Here the final state is $1s^2$ $^1S_0+k^2L_J$. As discussed in [7.8], in which numerous references to earlier theoretical and experimental work on these meta-stable states can be found, the J=1/2 and J=3/2 states belonging to the ^4P term can decay through internal spin-orbit and spin-spin interactions among the electrons and the nucleus. The J=5/2 state can decay only through internal electronic spin-spin interactions, as then a spin change of one unit and an electronic orbital angular momentum change of two units is simultaneously required to produce a $k^2F_{5/2}$ state for the emitted electron, the only state permitted by the absolute conservation re-quirements on parity and total angular momentum.

The study of such metastable states provides a sensitive test of the spin-orbit and spin-spin interactions in three-electron systems, including the effects of elec-tron correlation. Comparison of experiments with theoretical predictions shows that accuracy in wave functions for the three-electron system is critical for predicting level lifetimes. The beam-foil method is ideally suited to such metastable state lifetime measurements over a rather wide range of Z, as picosecond to nanosecond lifetimes are fairly common. The allowed states generally decay faster; here line-width measurements can and have been used to estimate approximate lifetimes, as has for example been done by STOLTERFOHT et al. [7.36]. The accuracy of such estimates is, however, usually fairly limited by the need to deconvolute the intrinsic spectro-meter line shape function from the overall line shape, and compares unfavorably with respect to the accuracy of decay in flight lifetime measurements on the emitted electrons when these are possible.

Such measurements were first made in our laboratory. In particular, the $1s2s2p$ $^4P_{5/2}$ state lifetimes were investigated over the range Z=8-18 with the results de-picted in Fig.7.19, taken from HASELTON et al. [7.37]. A few other early measurements obtained by techniques involving measuring changes in ion charge states are also in-dicated, from the work of DMITRIEV et al. [7.38] and FELDMAN and NOVICK [7.39]. Esti-mates of decay rates for this state provided by LEVITT et al. [7.40], MANSON [7.41], and BALASHOV et al. [7.42] are also shown, together with the results of a more recent calculation involving Dirac-Fock-Slater wave functions by CHENG et al. [7.43]. It is clear that the recent calculations of CHENG et al. [7.43] represent a marked improve-ment over earlier estimates. The remaining $\sim15\%$ difference between theory and exper-iment may be due to the neglect of initial-state electron correlation and final state distortion by the out-going continuum electron. Though such correlation effects are conventionally thought to be most significant at low Z, it appears that significant correlation for inner electron effects may survive up to at least $Z\sim20$ at a level measurable to reasonable accuracy by the lifetime methods discussed here.

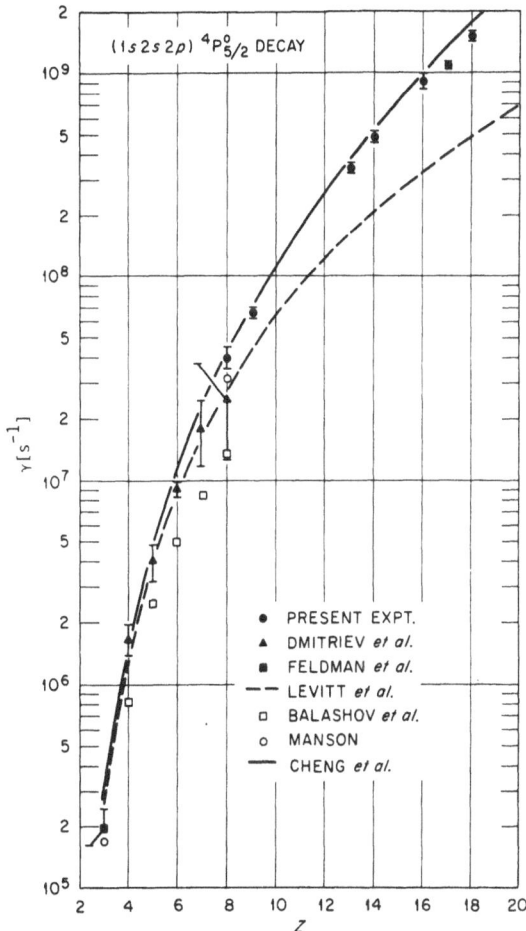

Fig. 7.19. Decay rate (inverse mean life) of the $(1s2s2p)^4P_{5/2}$ state of lithiumlike ions vs Z

More recent energy level and lifetime measurements on similarly metastable levels in lithiumlike, berylliumlike, and boronlike Ne ions, obtained by SCHUMANN and GROENE-VELD [7.44], are depicted in Table 7.2, together with theoretical calculations and estimates from the work of CHENG et al. [7.43] (using Dirac-Hartree-Fock wave functions) and of CHEN and CRASEMANN [7.45] and MATTHEWS et al. [7.46] (using Hartree-Fock wave functions).

A large number of Auger transition energies in the lithium, beryllium, and boron sequences, excited in C foils at beam energies ~ 200 keV, have been studied by BRUCH et al. [7.35]. Some corresponding lifetime information, e.g., for a few states of Be I and II, was obtained. For the $(1s2s2p)$ $^4P^0$, $(1s2p^2)$ ^4P, and $(1s2s3s)$ ^4S states of Be II lifetimes of ~ 164 ns, 3.1 ± 0.4 ns, and 5.2 ± 1 ns were obtained. A lifetime

302

Table 7.2. Transition energies and lifetimes in heliumlike, berylliumlike, and boron-like ions

Transition	Energy [eV] Theor.	Energy [eV] Exp.	Lifetime $(10^{-10}s)$ Theor.	Lifetime $(10^{-10}s)$ Exp.
$1s2s2p$ $^4P^0_{1/2} - {}^1S^e_0$	656.4	656±1	16.45^b	16.0± 2.4
$^4P^0_{3/2} - {}^1S^e_0$	656.2		6.38^b	4.0± 1.9
$^4P^0_{5/2} - {}^1S^e_0$	656.3		84.06^a	104.0±15.0
$1s2p^2$ $^4P^e_{1/2} - {}^1S^e_0$	674.4	675±1	0.94^b	
$^4P^e_{3/2} - {}^1S^e_0$	673.4		1.70^b	1.9± 0.9
$^4P^e_{5/2} - {}^1S^e_0$	673.5		0.28^b	
$1s2s2p^2$ $^5P^e_1 - {}^2P^0_{1/2}$	669.7	670±1	2.12^b	
$^5P^e_2 - {}^2P^0_{1/2}$	670.0		4.56^b	6.3± 1.8
$^5P^e_3 - {}^2P^0_{1/2}$	670.0		1.14^b	1.2± 0.6
$1s2s2p^2$ $^5P^e_1 - {}^2S^e_{1/2}$	685.7	686±1	2.12^b	
$^5P^e_2 - {}^2S^e_{1/2}$	686.0		4.56^b	6.0± 1.8
$^5P^e_3 - {}^2S^e_{1/2}$	686.0		1.14^b	0.9± 0.6
$1s2p^3$ $^5S^0_2 - {}^2P^0_{1/2}$	693.9	694±1	0.58^b	1.0± 0.6
$1s2s2p^3$ $^6S^0 - {}^1S^e$	671.7^c		1.09^b	
$^6S^0 - {}^1D^e$	678.4^c		1.09^b	
$^6S^0 - {}^3P^e$	682.8^c		1.09^b	
$^6S^0 - {}^1P^0$	690.6^c	690±1	1.09^b	
$^6S^0 - {}^3P^0$	697.7^c	698±1	1.09^b	

[a]CHENG et al. [7.43].
[b]CHEN and CRASEMANN [7.45].
[c]MATTHEWS et al. [7.46].

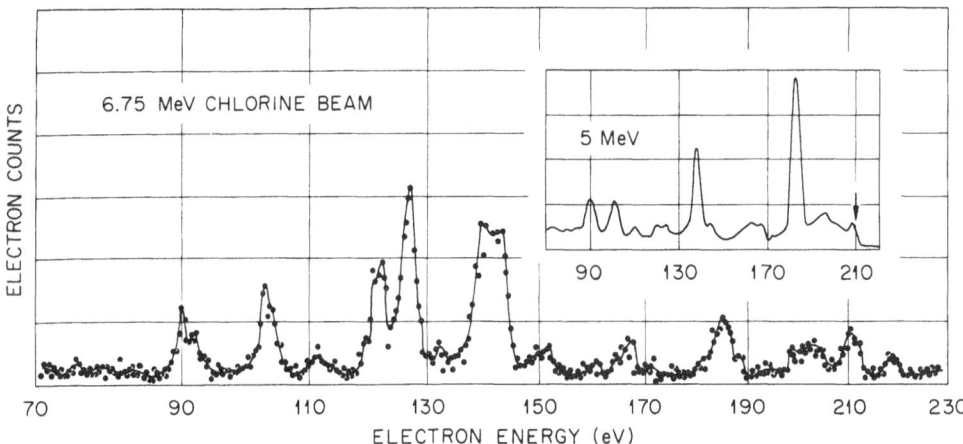

Fig. 7.20. Spectrum of autoionization electrons emitted by ($\sim$2 µg cm^{-2}) foil-excited 6.75 MeV chlorine ions undergoing decay in flight. The electron energies are expressed in the rest frame of the emitting atom. The inset spectrum contains lower-resolution data obtained at a relative delay of about 6 ns

for the metastable $(1s2s2p^2)$ ^{5}P state of Be I of 132±50 ns was also found. In addition, a number of interesting triply excited states in Li were characterized in this work.

Other alkalilike and alkaline earthlike states core-excited states have been studied in such systems as Na, Mg, Mg$^+$, and K by PEGG et al. [7.34], and in sodiumlike Cl by PEGG et al. [7.47]. The sodiumlike Cl states, excited in $\sim$2 µg cm^{-2} C foils, are depicted in a more recent spectrum shown in Fig.7.20, and are of particular interest because very many of the Auger-emitting states encountered are long-lived (or are perhaps fed by other, also metastable states). The interpretation of the spectra of these very long-lived ($\sim$50 ns) autoionizing levels of core-excited states of sodiumlike chlorine ions (with perhaps some contributions from adjacent charge states) is not yet clear, as needed theoretical calculations concerning these levels have apparently not yet been made. Fig.7.20 displays a spectrum for 6.75 MeV Cl ions in 2 µg cm^{-2} C foils, at a time delay following excitation of a few tenths of a nanosecond. It can be compared to the slightly lower resolution inset spectrum taken from [7.47], obtained at a time delay of about 6 ns. Some of the Auger lines are thought to arise from such configurations at $2p^5n\ell n'\ell'$. As in the case of core-excited, three-electron systems, such core-excited states will be metastable if the spins of the two valence electrons are parallel to the spin of the inner-shell hole. Taking the difference in resolution into account, the spectra are quite similar, despite the difference in time delay. One clear difference is that the series limit marked by the arrow in the inset does not seem to occur in the higher resolution spectrum. This

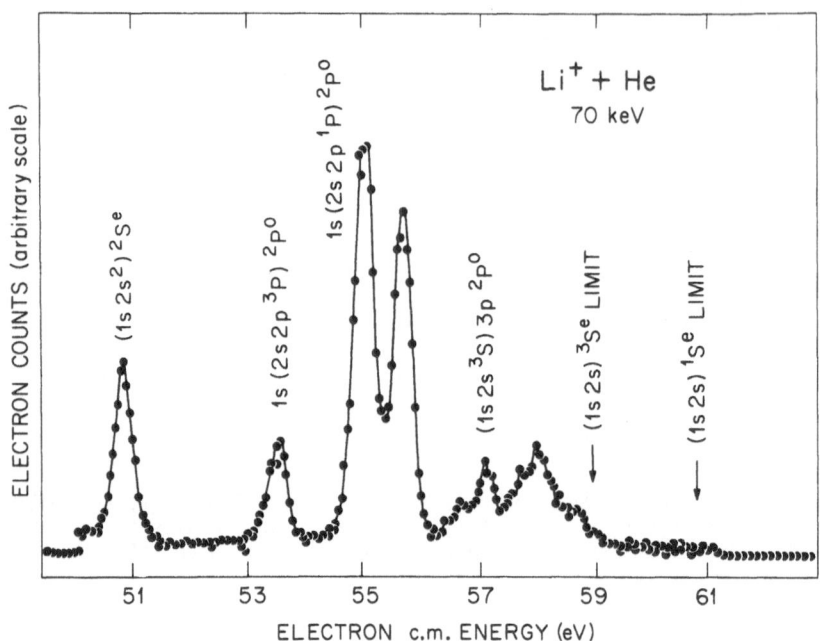

Fig. 7.21. Prompt spectrum of autoionization electrons emitted by 70 keV lithium atoms undergoing decay in flight after collisional excitation in a He gas target. The electron energies are expressed in the rest frame of the emitting atom. Excitation energies are obtained by adding the lithium ionization potential of 5.392 eV

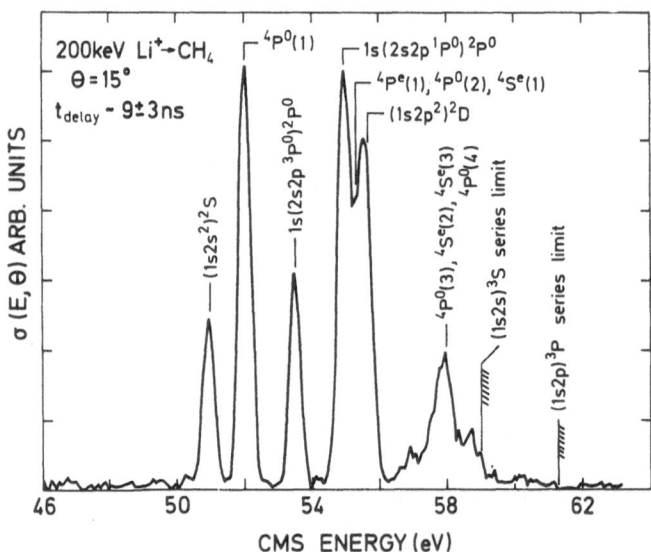

Fig. 7.22. Delayed spectrum of autoionization electrons emitted by 200 keV lithium ions undergoing decay in flight after collisional excitation in CH₄ gas. Obtained at a delay time of 9±3 ns

series limit corresponds to the limiting-energy Auger electrons that would arise from configurations like $2p^5$ 3sn'ℓ', with n'ℓ' very large. It thus appears that decays from states with parent terms like $2p^5$3p amd $2p^5$3d appear in the spectrum, but have mean lives too short to appear in the inset spectrum. What is surprising is the large number of very long-lived states (on an Auger time scale of $\sim 10^{-14}$ s). Once again, the principle of differential metastability is illustrated in the comparison of the relative intensities of the features between 120 and 130 eV in the two spectra with those of other features.

Even though the time resolution following excitation in a gas target ($\sim$ns) is relatively poor compared to that available from a foil target ($\lesssim 10^{-12}$ s), decay in flight measurements following excitation in such a gas target can certainly be made, at least for metastable states. The advantages of higher resolution due to reduced kinematic broadening is sometimes indispensable, particularly when fairly low energy electrons from closely spaced Auger levels are involved. By way of illustration, Fig.7.21 shows a prompt Auger spectrum resulting from excitation of core-excited Li levels in He gas at 70 keV beam energy from PEGG et al. [7.34]. Fig.7.22 shows a similar but delayed spectrum from excitation of Li core-excited states at 200 keV beam energy in CH_4, obtained by RØDBRO et al. [7.48]. A conspicuous difference is the occurrence of the strong line from the metastable 1s2s2p ^{4}P state [labeled $^4P^0$(1)], which is nearly absent from the Li-He spectrum. The occurrence of many allowed transitions in the delayed spectrum may be due either to residual gas excitation or alternatively to prompt allowed transitions following photon decay from higher states to the corresponding Auger emitting levels. As such higher levels would have opposite parity from the Auger emitting levels, they would be likely to be metastable against Coulomb autoionization. Further experiments would be needed to confirm this alternative suggestion.

References

7.1 J.R. Mowat, I.A. Sellin, R.S. Peterson, D.J. Pegg, M.D. Brown, J.R. Macdonald: Phys. Rev. A 8, 145 (1973)
7.2 I. Martinson: *Beam Foil Spectroscopy*, ed. by S. Bashkin, Topics in Current Physics, Vol. 1 (Springer, Berlin, Heidelberg, New York 1976) pp. 33-57
7.3 S.L. Varghese, C.L. Cocke, B. Curnutte: Phys. Rev. A 14, 1729 (1976)
7.4 H.W. Kugel, M. Leventhal, D.E. Murnick, C.K.N. Patel, O.R. Wood, II: *Beam Foil Spectroscopy*, Vol. 2, ed. by I.A. Sellin and D.J. Pegg (Plenum, New York 1976) pp. 815-827
7.5 I.A. Sellin, S.B. Elston, J.P. Forester, P.M. Griffin, D.J. Pegg, R.S. Peterson, R.S. Thoe, C.R. Vane, J.J. Wright, K.-O. Groeneveld, F. Chen, R. Laubert: Phys. Lett. 61A, 107 (1977)
7.6 L.J. Curtis: *Beam Foil Spectroscopy*, ed. by S. Bashkin, Topics in Current Phy-Physics, Vol. 1 (Springer, Berlin, Heidelberg, New York 1976) pp. 63-104
7.7 D.J. Pegg, S.B. Elston, P.M. Griffin, H.C. Hayden, J.P. Forester, R.S. Thoe, R.S. Peterson, I.A. Sellin: Phys. Rev. A 14, 1036 (1976)

7.8   I.A. Sellin: *Beam Foil Spectroscopy*, ed. by S. Bashkin, Topics in Current Phy-
      Physics, Vol. 1 (Springer, Berlin, Heidelberg, New York 1976) pp. 265-295
7.9   I.S. Dmitriev, V.S. Nikolaev: Zh. Eksper. I. Teor. Fiz. $\underline{47}$, 615 (1964) [Sov.
      Phys. - JETP $\underline{20}$, 409 (1964)]
7.10 J.B. Marion, F.C. Young: *Nuclear Reaction Analysis - Graphs and Tables* (North-
      -Holland, Amsterdam 1968)
7.11 J.A. Martin: unpublished (1970)
7.12 P.H. Stelson: *Beam Foil Spectroscopy*, Vol. 1, ed. by I.A. Sellin and D.J. Pegg
      (Plenum, New York 1976) pp. 401-417
7.13 H.-D. Betz: Rev. Mod. Phys. 44, 465 (1972)
7.14 L. Barrette: unpublished (1975)
7.15 E. Knystautas, R. Drouin: *Beam Foil Spectroscopy*, Vol. 1, ed. by I.A. Sellin and
      D.J. Pegg (Plenum, New York 1976) pp. 377-384
7.16 H.-D. Betz, F. Bell, H. Panke, G. Kalkolten, M. Welz, D. Evers: Phys. Rev. Lett.
      $\underline{33}$, 807 (1974)
7.17 H. Panke, F. Bell, H.-D. Betz, W. Stehling, E. Spinder, R. Laubert: unpublished
      (1975)
7.18 P. Erman: Physica Scripta $\underline{11}$, 65 (1975)
7.19 L.W. Dotchin, E.C. Chupp, D.J. Pegg: J. Chem. Phys. $\underline{59}$, 3960 (1973)
7.20 L.J. Curtis, P. Erman: private communication and to be published in J.O.S.A.
      (1977)
7.21 R. Mann: unpublished thesis; R. Mann, R. Peterson, F. Folkmann, G. Szabo, K.O.
      Groeneveld: Bull. Amer. Phys. Soc. $\underline{22}$, 610 (1977)
7.22 J.P. Buchet, M. Druetta: J. Opt. Soc. Am. $\underline{65}$, 991 (1975)
7.23 J.R. Mowat, I.A. Sellin, D.J. Pegg, R.S. Peterson, J.R. Macdonald, M.D. Brown:
      Phys. Rev. Lett. $\underline{20}$, 1289 (1974)
7.24 J.R. Mowat, I.A. Sellin, D.J. Pegg, R. Laubert, R.L. Kauffman, M.D. Brown, J.R.
      Macdonald, P. Richard: Phys. Rev. A $\underline{10}$, 1446 (1974), and references therein
7.25 R. Kauffman, C.W. Woods, K.A. Jamison, P. Richard: Phys. Lett. A $\underline{50}$, 117 (1974)
7.26 I.A. Sellin, C.R. Vane, S.B. Elston, J.P. Forester, P.M. Griffin, D.J. Pegg,
      R.S. Thoe, K.-O. Groeneveld, R. Laubert, F. Chen: Phys. Lett. $\underline{61A}$, 107 (1977)
7.27 M.E. Rudd, J. Macek: *Case Studies in Atomic Physics*, Vol. III, No. 2, ed. by
      M.A.C. McDowell and E.W. McDaniel (North-Holland, Amsterdam 1972) pp. 47-136
7.28 K.D. Sevier: *Low Energy Electron Spectroscopy* (Wiley, New York 1973)
7.29 J. Electron. Spectros. $\underline{1}$, 1 (1972 to present)
7.30 J.R. Mowat, I.A. Sellin, D.J. Pegg, R.S. Peterson, M.D. Brown, J.R. Macdonald:
      Phys. Rev. Lett. $\underline{30}$, 128 (1973)
7.31 L.C. Northcliffe, R.F. Schilling: Nuclear Data Tables A $\underline{7}$, No. 3-4 (1970)
7.32 V.V. Zashkvara, M.I. Korsunkskii, O.D. Komachev: Zh. Tekn. Fiz. $\underline{36}$, 132 (1966)
      [Sov. Phys. $\underline{11}$, 96 (1966)]
7.33 P. Dahl, M. Rødbro, B. Fastrup, M.E. Rudd: J. Phys. B $\underline{9}$, 1567 (1976)
7.34 D.J. Pegg, H.H. Haselton, R.S. Thoe, P.M. Griffin, M.D. Brown, I.A. Sellin:
      Phys. Rev. A $\underline{12}$, 1330 (1975)
7.35 R. Bruch, G. Paul, H.J. Andrä: Phys. Rev. A $\underline{12}$, 1908 (1975); R. Bruch: Thesis,
      unpublished (1976)
7.36 N. Stolterfoht: Abstracts of the Ninth International Conference on the Physics
      of Electronic and Atomic Collisions, Vol. 1, ed. by J.S. Risley and R. Geballe
      (University of Washington, Seattle 1975) p. 419, and references therein
7.37 H.H. Haselton, R.S. Thoe, J.R. Mowat, P.M. Griffin, D.J. Pegg, I.A. Sellin:
      Phys. Rev. A $\underline{11}$, 468 (1975)
7.38 I.S. Dmitriev, V.S. Nikolaev, Y.A. Teplova: Phys. Lett. A $\underline{26}$, 122 (1968)
7.39 P. Feldman, R. Novick: Phys. Rev. Lett. $\underline{11}$, 278 (1963); Phys. Rev. $\underline{160}$, 143
      (1967)
7.40 M. Levitt, R. Novick, P. Feldman: Phys. Rev. A $\underline{3}$, 130 (1971)
7.41 S.T. Manson: Phys. Lett. $\underline{23}$, 315 (1966)
7.42 V.V. Balashov, V.S. Senashanko, B. Tekou: Phys. Lett. A $\underline{25}$, 487 (1967)
7.43 K.T. Cheng, C.P. Lin, W.R. Johnson: Phys. Lett. A $\underline{48}$, 437 (1974)
7.44 R. Schumann: Thesis, unpublished (1976); R. Schumann, K.-O. Groeneveld: private
      communication
7.45 M.H. Chen, B. Crasemann: Phys. Rev. A $\underline{12}$, 959 (1975)
7.46 D.L. Matthews, B.M. Johnson, C.F. Moore: Atomic Data and Nuclear Data Tables
      $\underline{15}$, 41 (1975)

7.47 D.J. Pegg, I.A. Sellin, P.M. Griffin, W.W. Smith: Phys. Rev. Lett. 28, 1615 (1972)
7.48 M. Rødbro, R. Bruch, P. Bisgaard: private communication, and to be published (1977)
7.49 I.A. Sellin, R. Mann, H.J. Frischkorn, D. Rosich, S. Schumann, Gy. Szabo: Bull. Am. Phys. Soc. 22, 1320 (1977)
7.50 R.B. Barker, H.W. Berry: Phys. Rev. 151, 14 (1966)

8. Atomic Collisions in Solids

S. Datz

With 27 Figures

The concept of treating the penetration of energetic ions through solids in the context of the physics of binary atomic collisions requires some initial consideration.
Clearly atomic collision processes which occur at distances larger than interatomic
spaces in solids are difficult if not impossible to treat on a purely binary basis;
on the other hand as the velocity of the penetrating ion increases, many processes
involving these large impact parameters become less dominant and an atomistic treatment becomes more relevant. In the case of elastic scattering, for example, the large
impact parameter effects upon deflection by atomic cores and the energy transfer to
target atoms decrease with increasing projectile velocity. For collisions which lead
to inner-shell ionization, the impact parameters are so small that the presence of
the target atom in a solid medium will have hardly any effect except insofar as the
state of the projectile ion is altered by its previous collision history in the solid. The approach then is to apply our knowledge of single collision atomic processes
to the anticipated results of a sequence of binary processes.

Energetic ions penetrating matter undergo a series of collisions in which electrons are captured or lost by the ion and in which excitation and ionization of both
the ion and the atoms of the medium occur. These events are responsible for the
slowing of the ion (electronic stopping) and for the generation of free electrons
and radiation by the penetrating particle.

In a dilute gas, "charge state equilibrium" is in the main attained through a
series of electron capture and loss collisions involving the outer electrons of ground
state ions interacting with ground state target atoms. As the medium becomes more
dense, the time between collisions is too short to allow relaxation of excited states
so that one must think in terms of "excitation equilibrium" as well as "charge state
equilibrium". Ions channeled in single crystals have a restricted set of impact parameters with target atoms so that both their "charge state" and "excitation" equili-
bria will differ from either a dilute gas or a random solid.

Consider first in outline the separate processes involved for a heavy ion initially in a low charge state and with no inner-shell vacancies as it enters, penetrates,
and leaves a solid medium. The ion will rapidly lose outer-shell electrons with orbital velocities v_e lower than the ion velocity v_i; excitation of electrons with
$v_e \simeq v_i$ will occur with high probability. An electron in a high level of excitation
will be removed almost immediately in a dense medium, thus tending to increase the

charge state. The presence of even low lying excited states can reduce the effective capture cross section since capture into excited states can be destabilized by auto-ionization. This latter effect was postulated by BETZ and GRODZINS [8.1] and has been confirmed (at least for metastable states) by NICOLAEV et al. [8.2] for charge capture in gases. Since the time between collisions in solids is small compared to relaxation times, this effect could be important in the reduction of the capture cross section and would also tend to increase the charge state. In a heavy ion with many outer-shell electrons, BETZ [8.1] points out that the reduction in capture cross section is most important.

Inner-shell electrons (i.e., those lying below the shell in which rapid charge and excitation exchange is occurring) must, by definition, have velocities $v_e > v_i$ and will in general have low cross sections for ionization by Coulomb excitation. However, if promotion through quasi-molecular orbitals (MO) is possible, large cross sections for inner-shell vacancy formation may occur. In solids the ionization cross section can be affected by MO exit channel effects and by energy level shifts due to inner-shell ionization already present from previous collisions.

Solid-state effects can also occur in the relaxation of vacancies. For the moving ion the fluorescence yield can be changed by a reduction in the Auger lifetime through collisional mixing of states of the outer electrons which fill the vacancy; and an additional process, vacancy filling by electron capture (or hole transfer), can further reduce the number of x-rays generated per unit vacancy. The x-ray yield from the target atom can be affected by the presence of electrons upon adjacent atoms which, in some cases, can relax outer shells prior to inner-shell relaxation.

A true description of the state of the moving ion must also include its effect on its immediate environment. Conduction electrons scattered by the moving ion affect the electron density. For velocities $v_i \lesssim v_0$ the density is enhanced in the region of the nucleus of the ion and effects of this are seen, e.g., in transient field effects on hyperfine interaction studies; for velocities $v_i > v_0$, the density enhancement trails behind the ion, and troughs of lowered electron density as well follow in the wake. The details of the wake will depend on the ion charge and velocity and the plasmon spectrum of the material.

A state of equilibrium is said to be reached when all the processes of outer- and inner-shell electron capture and loss processes and ion wake formation have reached a steady state.

Upon emergence from the solid, the steady state excitation may be relaxed either radiatively or nonradiatively by autoionization in the outer-shell or by Auger processes in the inner-shells. The nonradiative processes will increase the ion charge but the charge state may be decreased by capture of electrons from surface states or from electrons trapped in the ion's wake.

Historically, the first observation of the net result of these solid-state effects was reported by LASSEN [8.3] who found that the charge states of fission fragments

emerging from solids were considerably higher than those emerging from gaseous targets. However, it is also found that the electronic stopping powers in solid media
are close to the same as those measured in gases (PIERCE and BLANN [8.4]) and the
"effective charge" derived from a comparison of stopping powers of heavy ions with
proton stopping powers in solids is that which is observed for ions in gases (BROWN
and MOAK [8.5]). The first of two possible solutions to this disparity suggests that
dynamical screening by electrons in the solid tends to neutralize the excess charge
on the ion moving in solids [8.6,7]. An alternative explanation [8.1] suggested that
the actual charge state in the solid is much like that in the gas but that many of
the bound electrons are in highly excited states which are lost by Auger events after
emergence.

Since the time of Lassen's experiments, a great deal of evidence has been accumulated which bears upon the description of collisions of heavy ions in solid media.
In addition to work on charge states and stopping powers, the measurements of x-ray
spectra and yields which describe the inner-shell excitation of the penetrating ions
have been carried out. The discovery of the phenomenon of "channeling" in 1963 [8.8]
and the subsequent quantitative elucidation of the mechanics of channeling have provided additional controllable variables to studies of the complex subject of interactions in solids.

8.1 Channeling

The complexity of multiple collision processes in solids is considerably reduced by
the process of "channeling" which occurs when a particle enters an ordered crystal
lattice within a certain critical angle with respect to a low index axial or planar
direction. In this case, because of the definite spacing of atoms in a lattice, the
penetrating ion undergoes a set of correlated collisions which affect its motion and
limit the set of interaction distances with the target atoms; i.e., in random motion
in a solid there is a maximum impact parameter of one-half the interatomic spacing;
channeling imposes a constraint on the minimum impact parameter.

The gross effect was actually first predicted in 1912 by STARK [8.9] and was re-
discovered in 1962 in the computer calculations of ROBINSON and OEN [8.8]. The first
analytical interpretation was given by LINDHARD [8.10], and a recent review (GEMMEL
[8.11]) lists over 700 references to study of the phenomenon and its application.

The general concept of channeling is illustrated in Fig.8.1. Consider a projectile atom aimed for a hard collision with an atom contained in a closely packed
atomic row in a crystal. The projectile will be slightly deflected by the repulsive
potential of the first atom, again by the second, and so forth. The final collision
will thereby be considerably softened, and, depending upon the angle with respect to

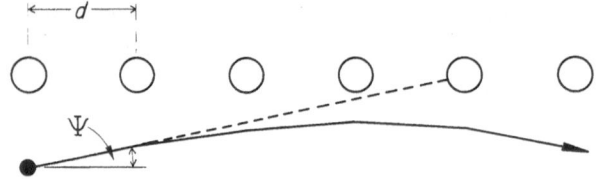

Fig. 8.1. Schematic showing the undeflected path of an ion in collision with a row atom compared with the deflected path due to a row repulsive potential

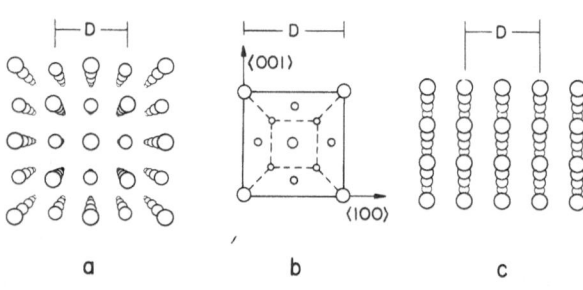

Fig.8.2a-c. Several views through a face-centered cubic crystal illustrating axial and planar channels

the atomic *row*, it may not collide at all with the original target atom. This is the basis of channeling; i.e., within an angle with respect to an atomic row or plane, determined by the interatomic potential and spacing of atoms, the particle will be deflected by a continuum potential made by properly summing the atomic potentials.

In Fig.8.2 we show some cases of low index axes and planes in the face-centered cubic crystal. For particles entering in a close to axial direction (e.g., Fig.8.2a) at high velocity, LINDHARD [8.10] showed that the close collisions with the resultant rows or "strings" of atoms are avoided for incidence angle ψ if

$$\psi \stackrel{\sim}{<} (2Z_1Z_2e^2/E \cdot d)^{1/2} \tag{8.1}$$

for a Coulomb potential where Z_1 and Z_2 are the incident and target atomic numbers, E is the ion's energy, and d is the lattice spacing between atoms (see Fig.8.1). This high energy condition is met when

$$E > 2Z_1Z_2e^2d/a_{TF}^2 \tag{8.2}$$

where a_{TF} is Thomas-Fermi screening length. If the incident ion is fully ionized

$$a_{TF} = 0.4685 \ Z_2^{-1/3} \ [\text{Å}] \tag{8.3}$$

and for a nonfully ionized ion

$$a_{TF} = 0.4685 \left(Z_1^{1/2}+Z_2^{1/2}\right)^{-2/3} \ [\text{Å}] \quad . \tag{8.4}$$

For planar channeling, i.e., confinement of the ion's motion between sheets of planar density n_p(atoms cm^{-2}) (see Fig.8.2c), the critical angle ψ_p is

$$\psi_p = (2\pi n_p Z_1 Z_2 e^2 a_{TF}/E)^{1/2} \tag{8.5a}$$

$$\simeq 0.545(n_p Z_1 Z_2 a_{TF}/E)^{1/2} \tag{8.5b}$$

The primary result of channeling is to prevent small impact parameter collisions from occurring. Hence, such effects as the elimination of Rutherford scattering and the reduction of x-ray yields are observed. The effect also leads to the elimination of nuclear stopping and restricts close electronic interactions to only valence and conduction electrons.

Here we shall limit our discussion to the determination of detailed trajectories of channeled ions from which interatomic scattering potentials and impact parameter dependent inelastic cross sections have been determined (DATZ et al. [8.12]). An understanding of these trajectories will also be important in the application of the concept to the following sections.

8.1.1 Atomic Interactions from Planar Channeling Measurements

The information obtained here arises mostly from measurements of the energy loss spectra of energetic ions which have been transmitted through thin single crystals. The setup consists simply of an ion beam, an orientable thin single crystal, and an energy-sensitive particle detector which subtends a small solid angle in the forward direction.

If we carry out this experiment in a planar channeling direction we observe spectra of the type shown in Fig.8.3. Here we see discrete peaks in energy loss. The discreteness of the energy loss can be understood from the following model.

An ion enters a planar potential, which has a minimum at the center and rises toward the atomic planes (see Fig.8.4). The particle will be deflected toward the center and will undergo oscillatory motion on its way through the crystal. Particles entering far from the midplane will have higher amplitude and, since the potential is anharmonic, will have a shorter wavelength than those entering close to the midplane. Further, these particles will suffer greater energy loss because they have penetrated regions of greater electron density and they do so more often. If all exiting particles were observed, a continuum of energy loss would be observed. But if we accept only particles in the forward direction we observe *only* particles with wavelengths λ, such that

$$n\lambda = L \tag{8.6a}$$

or

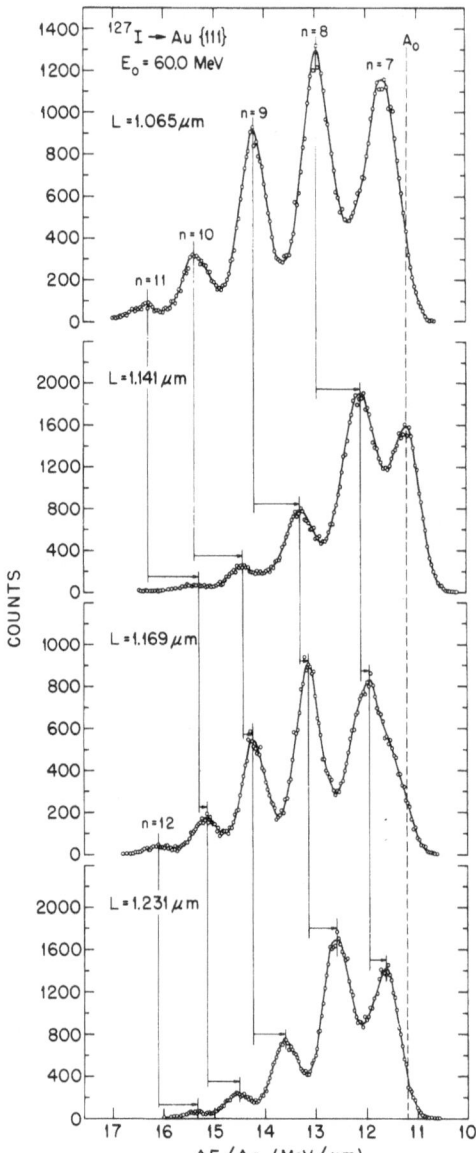

Fig. 8.3. Energy loss spectra observed in ^{127}I ions transmitted in the forward direction through {111} channels of an Au crystal as a function of path length

$$(n+1/2)\lambda = L \tag{8.6b}$$

where n is the number of oscillations and L is the pathlength. For reasons given in [8.12] the half-integral solutions do not lead to experimentally observable groups because of imperfections (mosaic spread) present in the crystal.

CHANNEL "WALL"

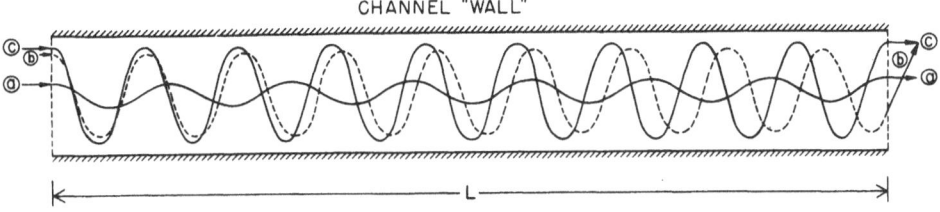

k——————————————————— L ———————————————————→|

Fig. 8.4. Trajectories for ions entering a planar channel at three different entrance points. Although all points are equally likely, only particles a and c will
reach a detector in line with the beam

The wavelength of the oscillations can be obtained from information of the type
shown in Fig.8.3. The position marked A_0 corresponds to the minimum-energy-loss group.
The particles in this group have been confined to trajectories near the channel midplane where the potential is almost harmonic. The peaks of the intermediate groups
move with changing pathlength as predicted.

A convenient measure has been found to be the oscillation "frequency"

$$\omega \equiv n(2m)^{1/2} \, v/L \quad .$$ (8.7)

When properly corrected, all values of n should yield the same relationship of initial stopping power E_0 to oscillation frequency. An example of this is shown for
60 MeV I ions in the {111} and {100} planes of Au in Fig.8.5. The linearity of the
curves obtained was at first surprising (and quite useful in easing data treatment).
Although we have used the case of 60 MeV I in Au, the linearity of the ω vs (dE/dz)
curves is by no means unique and has been found for all systems studied.

ROBINSON [8.13] developed a method for the extraction of atomic and planar potentials from the data obtained in these experiments.

$$V_0(\bar{x}) = V_1(\ell+x) + V_1(\ell-x) \quad , \quad -\ell \leq x \leq \ell$$ (8.8)

where ℓ is the half planar spacing and V_1 is the smoothed single planar potential.
The ion oscillates with a frequency

$$\omega^{-1} = 2 \int_0^{x_m} [V_2(x_m) - V_2(x)]^{-1/2} \, dx \quad ,$$ (8.9)

where V_2 is the potential between two adjacent planes and where x_m is the oscillation amplitude. The model stopping power is

$$S(x,E) = s_0 + s_1[\sigma(x) - 1]$$ (8.10)

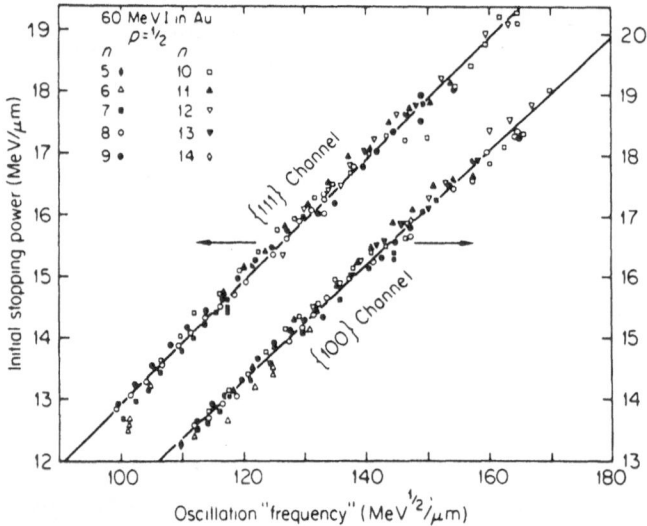

Fig. 8.5. The dependence of the stopping power of 60 MeV I ions on their transverse oscillation frequencies in two planar channels of Au

where s_0 and s_1 depend on the ion energy. The spatial dependence of the stopping power is described by the function $\sigma(x)$, normalized so that $\sigma(0)=1$. The value of s_0, the stopping power at the channel center, may be taken directly from the experiments by measuring the minimum energy loss (leading edge) of the spectrum. The stopping power of the channeled ions is found by averaging (8.10) over the oscillatory motion, weighting each portion of the channel cross section by the time which the ion spends there.

Empirically, it has been found that the data are well represented by the straight line

$$(-dE/dz)_{E=E_0} = \alpha + \beta\omega \tag{8.11}$$

as illustrated by Fig.8.5. Because of the apparent generality of this observation, it has been adopted as an empirical first principle, the consequences of which are to be explored. When (8.10) is averaged over the motion of the ion and the result is compared with (8.11), it is found that as long as the number of half-oscillations executed by the ion is integral, the intercept and slope of the straight line are given by

$$\alpha = s_0 - s_1 \quad , \tag{8.12}$$

$$\beta = 2s_0 \int_0^{x_m} \sigma(x)[V_2(x_m) - V_2(x)]^{1/2} \, dx \quad . \tag{8.13}$$

From (8.13) it is found that, as long as β is constant,

$$\sigma(x) - d/dx\left\{\left[2/V_2''(0)\right]\left[V_2(x) - V_2(0)2\right]\right\}^{1/2} \quad , \tag{8.14}$$

where the primes represent differentiation with respect to x.

The experimental data may be used to evaluate the curvature parameter,

$$y \equiv 2\pi^2(s_0-\alpha)^2/\beta^2$$

$$= V_2''(0)/\ell = 2V_1''(\ell)/\ell$$

$$= -8\pi\rho\kappa[V(\ell) + \ell V'(\ell)] \tag{8.15}$$

where the empirical form is given on the first line and two equivalent theoretical forms are given on the following lines. Given data for a sufficient number of channels with distinct values of ℓ, the potential itself can be evaluated by integration of the last of (8.15). For more limited data, the parameters of an assumed potential function can be evaluated.

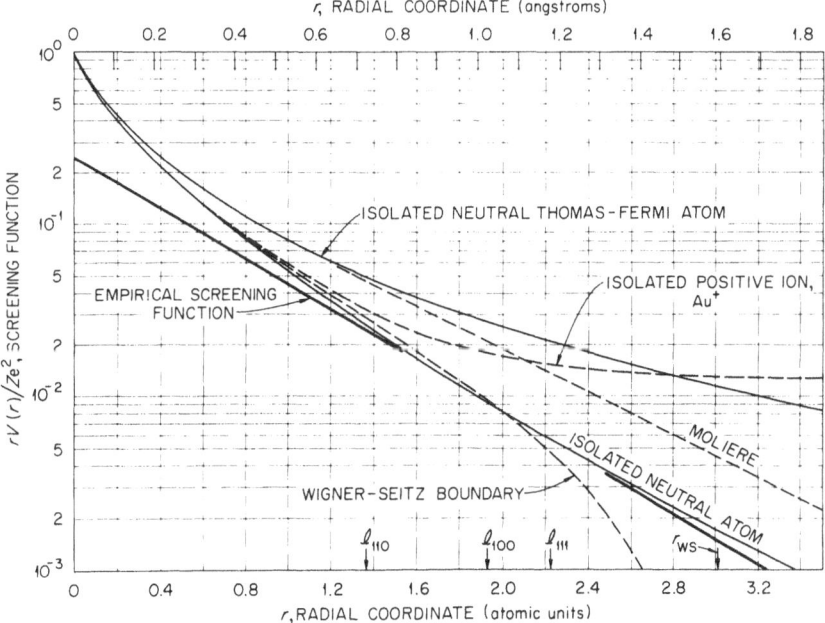

Fig. 8.6. Comparison of the screening function deduced empirically from planar channeling experiments in Au with a Thomas-Fermi, Moliere-Thomas-Fermi and relativistic Hartree potential ("isolated neutral atom")

Experimental curvature parameters have, for example, been determined for 3 MeV
He, 10 MeV O, and 15-60 MeV I in Au. The values are in no way dependent on assumptions about the potential function. The values of y are strongly dependent upon the
ion and its energy, showing that the planar-channel potential must also depend on
these variables. However, the ratio of the values for two channels appears to be
essentially independent of the ion and its energy within experimental error ±10%.
If the interaction potential between the channeled ion and the individual lattice
atoms is of the screened Coulomb type, the curvature parameter is

$$y = -8\pi\circ Z_1 Z_2 e^2 b \; \circ'(b\ell) \quad , \tag{8.16}$$

where $Z_1 e$ and $Z_2 e$ are the (effective) nuclear charges of the two particles, $\phi(b\ell)$
is the screening function, and b is a constant. The ratio of the values for two channels determines the value of the screening constant b, which is thus shown to be a
property of the gold crystal. Physically, the implication is that the nuclear charge
of the ion is always screened by the same number of electrons, whatever its position
in the channel. As it approaches a target atom, it never gets so close that this
screening is reduced significantly. Thus, it acts always as a simple charged particle
which does not much perturb the medium. That is, all three ions may be considered
as test charges. The strong variation of the experimental curvature parameters with
the ion and its energy results merely from changes in the magnitude of the ionic
charge with these variables.

After consideration of numerous forms of the potential energy it was shown by
ROBINSON (Fig.8.6) that the Hartree potential for the lattice atom describes the
interaction extremely well in the region of impact parameters sampled by planar channeling measurements [8.13].

8.1.2 Hyperchanneling

From the interatomic potentials determined by this method, the effective potentials
of the atomic rows ("strings") in the lattice may be developed. Such a potential
surface is shown in Fig.8.7. Since solid-state effects on the ion-atom interaction
potentials occur at large distances from the atoms, it is desirable to investigate
the largest impact parameter events possible. In the planar channeling experiments
discussed so far, the maximum impact parameters are determined by the planar spacings
It is possible by doing experiments in an axial channeling geometry to probe even
larger impact parameter interactions provided one can study ions confined to only
one particular axial channel. In planar channeling, a channeled ion travels in a
single channel between adjacent atomic planes, whereas in ordinary axial channeling
the majority of channeled ions wander from one axial channel to another. Axially
channeled ions of a limited class, however, are expected to have transverse energies
less than the potential barrier between adjacent rows, and accordingly should be

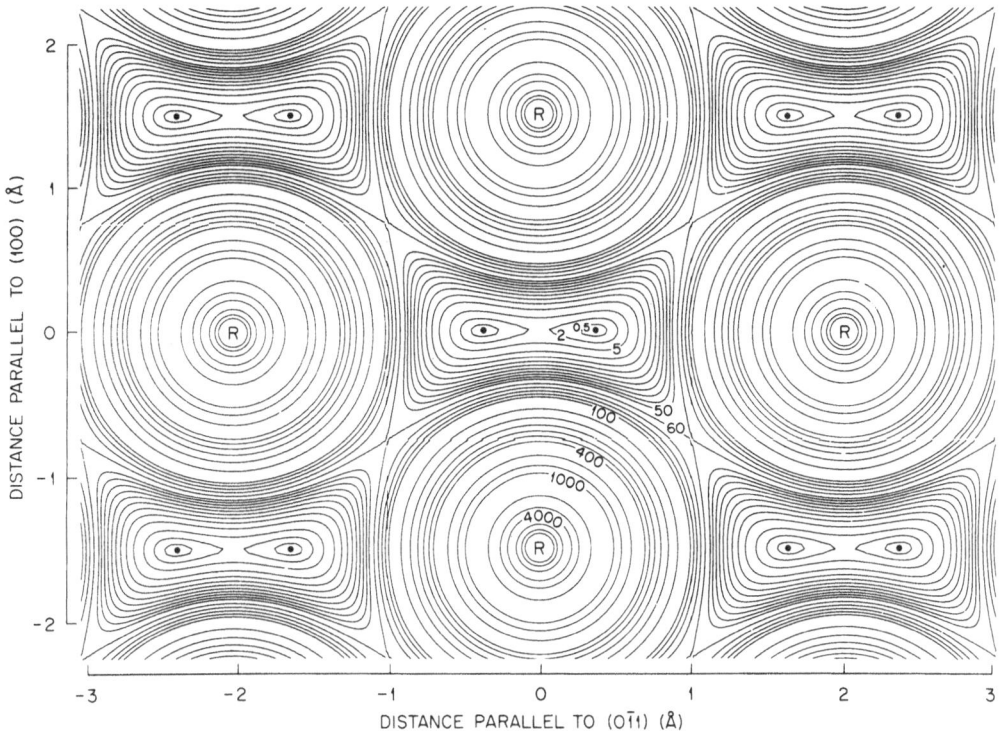

Fig. 8.7. Potential energy contours for I ions in the continuum potentials of (011) rows (R) of Ag atoms. All values are in eV and are relative to the minimum (•), which has an absolute value of 72 eV

constrained to travel through the crystal in a single axial channel. This phenomenon, called hyperchanneling or proper axial channeling, was recognized in the early computer calculations of ROBINSON and OEN [8.8] and was observed by EISEN [8.14] as a low-energy loss "tail" in axial energy loss distributions. In recent experiments (BARRETT et al. [8.15]), hyperchanneling was observed as a distant well-separated energy loss group, and systematic investigations of the effect indicate that solid--state effects may in fact be present.

Hyperchanneled ions are expected to have at least two distinctive characteristics; a smaller acceptance angle than ordinary axial channeling and lower energy losses because they sample larger average impact parameters than ions which wander from channel to channel. This effect was studied by APPLETON et al. [8.16] using a highly collimated beam of 21.6 MeV I ions onto a 0.60 μm thick <011> Ag crystal. Energy spectra measured with the beam incident at various angles to <011> in {111} are shown in Fig.8.8. The prominent group on the right, lying within a narrow angular range and centered in the energy loss distributions in Fig.8.8 taken at ±0.45° rela-

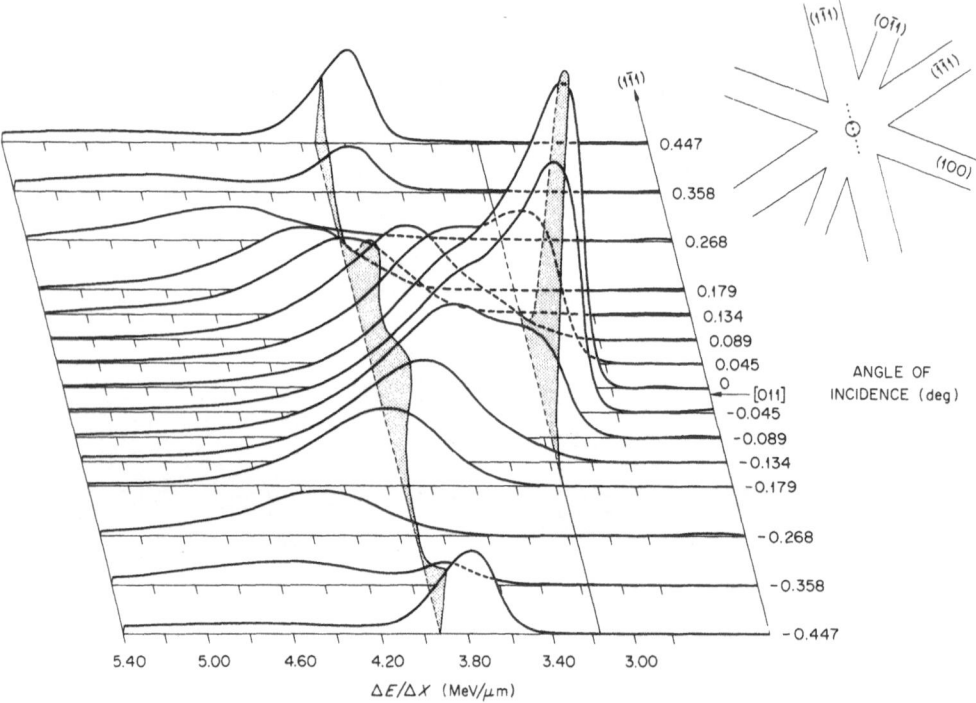

Fig. 8.8. Normalized energy loss distributions at various incident angles to (011) for 21.6 MeV I ions transmitted through 0.85 μm of Ag. The insert shows the angular extent of the measurements (dotted line) and the hyperchanneled region (shaded central spot)

tive to <011> are nearly identical to those obtained at angles 5 to 10° from <011> in {111}, i.e., normal planar channeling. As the incidence angle with respect to <011> is decreased, the energy loss distributions in Fig.8.8 lose the characteristic {111} shape at about 0.30°, and a new group of ions with significantly lower energy losses is dominant within ~0.12° of <011>. This low loss group has the two characteristics expected of hyperchanneled ions, i.e., those ions have been contained in the closed equipotentials of Fig.8.7.

The additional group seen at intermediate angles arises from particles which have spent their time in the region of the potential maximum between rows. From an analysis of the transverse energies required to remove particles from the hyperchanneled group, it has been shown that the free atom Hartree potential predicts too low a saddle point between rows and that, in fact, a correction such as that given in the Wigner-Seitz model is necessary to describe the data.

8.2 Charge State Distributions

As outlined in the introduction, an energetic ion beam emerging from a solid will, in general, be characterized by a distribution of charge states which depends upon electron capture, loss, and excitation processes occurring in the solid and which can be altered by exit surface effects and relaxation processes following emergence. As a simple example, consider starting with an energetic neutral atom and passing it into a medium. In the first collision that it makes it will in all probability lose an outer electron, in the second collision it may lose a second one, and so on. The cross section for electron loss σ_ℓ will, in general, decrease with increasing charge state q. In the meantime the probability of electron capture σ_c by the penetrating ion increases with increasing charge state and ultimately a steady state between capture and loss is attained where the most probable "equilibrium" charge is attained. This charge is unaffected by additional collisions, i.e., after traversing a given target thickness x, the fractional population ϕ_q of charge state q depends on relative cross sections for production and loss. Considering only single electron capture and loss

$$d\phi_q/dx = \phi_{q-1}\phi_\ell(q-1) + \phi_{q+1}\sigma_c(q+1) - \phi_q[\sigma_\ell(q) + \sigma_c(q)] \quad . \tag{8.17}$$

At equilibrium the derivative is zero, and if σ_c increases with q at about the same rate that σ_ℓ decreases, the most probable charge state q is that for which $\sigma_c \sim \sigma_\ell$.

In the case of a dilute gas target where all collisions which lead to excitation are relaxed prior to the next collision, this description is perfectly adequate and an equilibrium charge state distribution (CSD) can be calculated accurately from a knowledge of all of the capture and loss cross sections.

For heavy ions (Z_1) moving in a gaseous medium of heavy atoms (Z_2), BOHR and LINDHARD [8.6] showed that the most loosely bound electron remaining on the ion at equilibrium had an orbital velocity v_e about equal to the ion velocity v_i. Utilizing the Fermi-Thomas atom model, the average charge of the ion $\bar{q}$ was shown to be

$$\bar{q}/Z_1 = Z_1^{-2/3}(v/v_0) \quad , \quad (1 < v/v_0 < Z_1^{2/3}) \tag{8.18}$$

where v_0 is the Bohr velocity $(v_0 = 2.18 \times 10^8 \text{ cm s}^{-1})$.

This simple relationship has been shown to be remarkably accurate for gaseous targets $(Z_2 \gtrsim 10)$. A number of semi-empirical expressions have been developed which predict the average charges of emergent ions as a function of velocity. BETZ [8.17] gives

$$\bar{q}/Z_1 = 1 - C(0.71 \, Z^a)^{v/v_0} \quad , \quad (Z_1 \geq 10; \, v \gtrsim v_0) \tag{8.19}$$

where a=0.067 for air and a=0.053 for formvar foils and $C \simeq 1$.

For solid strippers, NIKOLAEV and DMITRIEV [8.18] give

$$\bar{q}/Z_1 = [1 + (Z_1^{-\alpha}v/v')^{-1/k}]^k \quad , \quad (Z_1 > 16) \tag{8.20}$$

where $v'=3.6\times10^8$ cm s^{-1}, $\alpha=0.45$, and k=0.6.

Note that the above expressions differentiate between solid and gaseous targets and that for high Z_1 significant differences in $\bar{q}/Z_1$ accrue. This is illustrated in Fig.8.9 which shows data reported by MOAK et al. [8.19] for Br ions in various solid and gaseous targets. These differences are attributable to the role of excited states in contributing to the cross sections of (8.1). Here we would expect that the presence of unrelaxed excited states arising either from capture or from a collision which excited but did not ionize a bound electron would lead, in the subsequent collision, to an effectively lower capture cross section and an effectively larger loss cross section. The result of these is an increase in the mean equilibrium charge in the solid. The effect on charge states of reducing the mean free time between collisions in gaseous targets by increasing pressure was noted by LASSEN [8.3] who found almost three units of charge out of a total of ~ 20 increase for heavy fission fragments recoiling through a gas at 500 Torr as compared to a dilute gas.

On the basis of the almost equal stopping powers for ions in solids and in gas targets [see (8.3)], BETZ and GRODZINS [8.1] argue that the charge state of high-Z_1 ions in solids is almost the same as those in gases. They point out that although the ionization cross section for an excited electron is indeed higher by a factor of ~ 2, the number of unexcited electrons in the outer shell of a high-Z_1 ion (perhaps as many as 10) militates against a great increase in the cross section for loss of an electron in collision. While the presence of an excited electron on the ion can effectively reduce the capture cross section through the formation of an autoionizing state, the increase in capture cross section into any state due to the higher ionic charge strongly damps any tendency toward higher charge states within the solid. Instead they postulate that the ion is pumped to a high degree of internal excitation of a large number of bound electrons which is relaxed by autoionizing and Auger events after the ion leaves the solid surface.

A numerical example of the above effect has been presented by BETZ [8.20]. Here he uses the case of 20 MeV Br ions passing through O_2 or N_2 gas with a solid carbon target. The mean emergent charges are $\bar{q}_g=7.2$ and $\bar{q}_s \simeq 12.5$, respectively. The outer--shell relaxed configuration for Br^{7+} is $3s^2\ 3p^6\ 3d^{10}$. For the ion in the solid Betz proposes, e.g., $3s^2\ 3p^4\ 3d^2\ 4s^2\ 4p^4\ 4d^3\ 4f^1$ (i.e., still Br^{7+}) and from calculations of the relative binding energies, Born approximations to the ionization cross sections, and Brinkman-Kramers calculations of the capture cross sections, arives at the conclusion that this postulated configuration is reasonable. The difference $\bar{q}_s - \bar{q}_g \simeq 5$ is then attained by relaxation to a $3s^2\ 3p^6\ 3d^5$ ground state by autoionizing

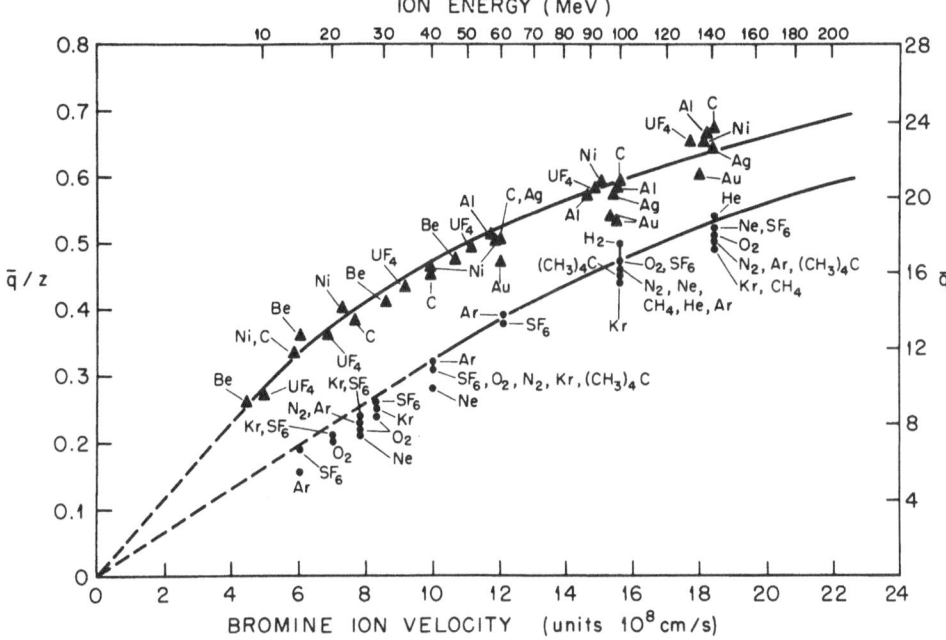

Fig. 8.9. Mean charge states of Br ions emerging from various solid and gaseous targets as a function of ion energy

events after emergence from the solid. Although the explanation is attractive, the inability of experiments to detect these autoionizing electrons (DATZ [8.21]) remains a puzzle.

Also missing from this picture is the effect of steady-state inner-shell excitation which, as we shall see in Section 8.4, is a significant contributor to total ion excitation in solids.

8.2.1 Effects of Channeling

As we have seen, it is difficult to isolate the effects of single charge changing collisions in solid media. Aside from the higher collision frequency which makes it difficult to consider the effects of isolated collisions, the range of impact para-meters is restricted, i.e., impact parameters greater than a lattice spacing are clearly forbidden. Thus, that part of the impact parameter dependence of a charge changing collision which extends past this distance is excluded from the physical picture. Channeled ions in crystal axes or planes on the other hand also restrict lower limit of impact parameters. Thus, if we study charge distributions for chan-neled ions we shall observe the results of collisions in a fixed range of relatively large impact parameter. Since these impact parameters are large, the ion can only

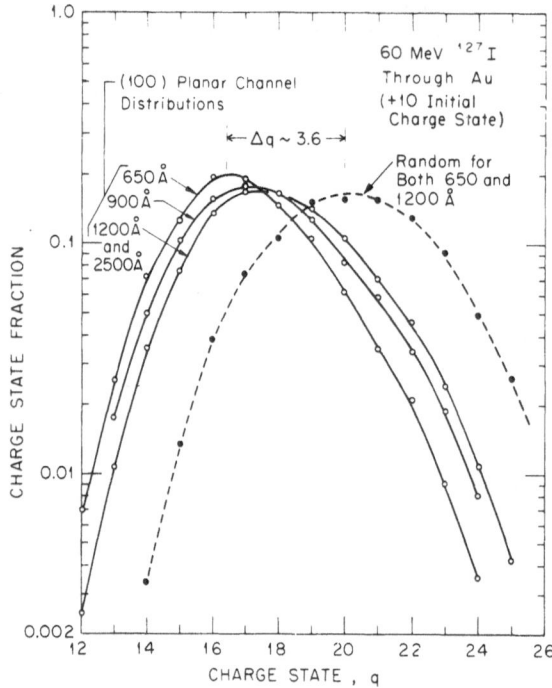

Fig. 8.10. Charge state distributions of 60 MeV I ions emerging
from Au in a random crystal direction and from a channeled direction as a function of crystal thickness

directly interact with valence and/or conduction electrons, all of which have velocities lower than the penetrating ion. In this case for ions with $v_i > v_0$ we have
simulated a high density, almost static electron gas.

A relatively straightforward example of the effect of channeling is seen in
Fig.8.10 for 60 MeV iodine ions channeled in Au (LUTZ et al. [8.22]). First, it can
be noted that the equilibrium charge state distribution for channeled ions is lower
here than for random lg directed particles. This is due to the suppression of multiply
ionizing collisions which are small impact parameter events. Second, it can be seen
that equilibrium for channeled ions does not take place until the target thickness
is ~1000 Å because of the reduced cross sections for both capture and loss events.
The suppression of the capture cross section occurs because these ions are traveling
at velocities equivalent to 250 eV electrons and can interact most directly with the
5d electrons of Au which are bound with only ~10 eV.

The effect is more startling for higher velocity almost totally stripped ions as
can be seen in Fig.8.11 (DATZ et al. [8.23]). Here we show the emergent charge state
distributions for 40 MeV oxygen ions channeled in the [110] axis of an Au crystal
0.33 μm (3300 Å) thick as a function of the input charge state of the oxygen ion
together with the CSD obtained from injection in a "random" direction. The random
distribution (dashed curve) is representative of charge-state equilibrium since it

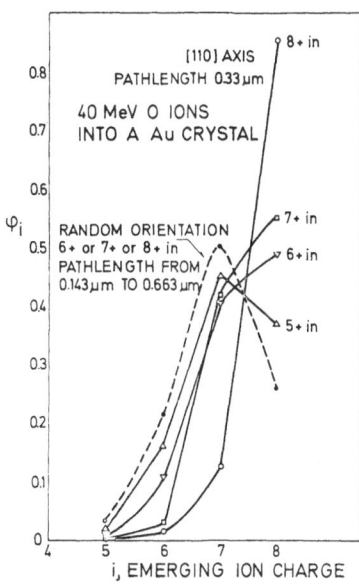

Fig. 8.11. Emergent charge state distributions for 40 MeV oxygen ions from Au in a random orientation (independent of incident charge or thickness) and from a [110] axial channel as a function of incident charge state

is independent of a pathlength (0.143 to 0.663 μm) in the crystal and independent of input charge state (6+, 7+, or 8+). The channeled-ion charge state distributions are clearly nonequilibrium cases. Since the 8+ fraction is enhanced for all initial charge states, it is immediately obvious that an equilibrium charge state distribution for channeled ions will contain a higher fraction of higher charge states than the random case. Moreover, it is clear that some of the charge-changing cross sections for the channeled ions in attaining equilibrium must be quite small, i.e., if we assume that equilibrium is attained in approximately 10 mean-free paths for charge exchange, then charge-changing cross sections of $\ll \sim 5 \times 10^{-18}$ cm^2 must be involved if equilibrium is not achieved in 0.5 μm with target densities of $\sim 5 \times 10^{22}$ atoms cm^{-3}.

These nonequilibrium effects persist at lower energies and are observed in planar as well as axial directions. The tendency toward higher charge states diminishes with decreased channel dimension and approaches the random distribution.

Since the emergent charge state distributions for these channeled ions are markedly far removed from equilibrium, it should be possible to adduce the relevant charge-changing cross sections from variations in charge state distribution with change in pathlength and, in some cases, from the additional information obtained from different input charge states. Thus, for example, a capture cross section from 8+ to 7+, $\sigma_{8,7}$ was determined to be 2.5×10^{-19} cm^2 for 40 MeV O^{8+} channeled in an Au {100} planar channel and the loss cross section from 7+ to 8+ $\sigma_{7,8}$ was 4.2×10^{-19} cm^2. The ratio $\sigma_{7,8}/\sigma_{8,7}$ causes the observed predominance of 8+ in the channel and the small magnitude of the cross section leads to the large distances required for equilibrium.

An interesting analogy can be made between the channeled and random charge-exchange cross sections and those for 40-MeV O ions in He and in Ar. In He $\sigma_{7,8}=2.5\times10^{-19}$ and $\sigma_{8,7}=9\times10^{-20}$ while in Ar $\sigma_{7,8}=3\times10^{-18}$ and $\sigma_{8,7}=1.5\times10^{-17}$. In the case of He, there are no electrons with sufficient orbital velocity to meet the matching criterion, and the capture cross section is a factor of 100 lower than that in Ar, where such electrons are available for transfer.

Collisions in He may thus be compared with collisions in channels both in magnitude and in capture to loss ratio—$\sigma_{7,8}/\sigma_{8,7}=2.8$ for He and 3.2 for Au {100}. When the fraction of O^{6+} is neglected, these ratios also represent the equilibrium charge state fraction ϕ_8/ϕ_7. Although, because of their larger magnitude, it is not possible to measure these cross sections in random Au, it can be noted that the ratio ϕ_8/ϕ_7 emerging from random Au is 0.5 which is comparable to the value 0.2 obtained in Ar gas targets.

8.3 Stopping Power

As the term implies, the "stopping power" of a medium represents its relative ability to stop or slow down an energetic penetrating particle. From the point of view of an atomic collisions physicist, it represents the sum of all processes, elastic and inelastic, by which the particle loses its energy; and although it is generally defined as $S=-dE/dx$, the energy loss per unit pathlength, it can equally well be defined per unit target atom. As we have seen in Section 8.1, information concerning single atom collitions is retrievable from properly designed experiments; and conversely, information obtained from isolated atomic collision experiments should aid us in an understanding of stopping phenomena. In this regard it should be noted that some theories of stopping were, by necessity, highly developed some time ago (e.g., BOHR [8.24]) and that more recent developments in atomic scattering are just now in the process of being included. An excellent introduction and review of the subject has been given by SIGMUND [8.25]. We shall concentrate here on the relation of stopping power to the collision process occurring in solids.

Stopping power is generally divided into two parts; "nuclear stopping", i.e., elastic energy transfer to atoms of the medium, and "electronic stopping", energy loss due to ionization or excitation of electrons of the medium or the penetrating ion. Our principal concern here will be with electronic stopping.

At very low velocities $v \ll v_0$, nuclear stopping predominates. As the velocity increases it reaches a maximum and then slowly decreases. The electronic stopping rises with increasing velocity, and soon becomes predominant. It levels off in the region $v_0 Z_1^{2/3}$ and then declines roughly as $(\ln E)/E$. (The even higher region of relativistic velocities will not be discussed here).

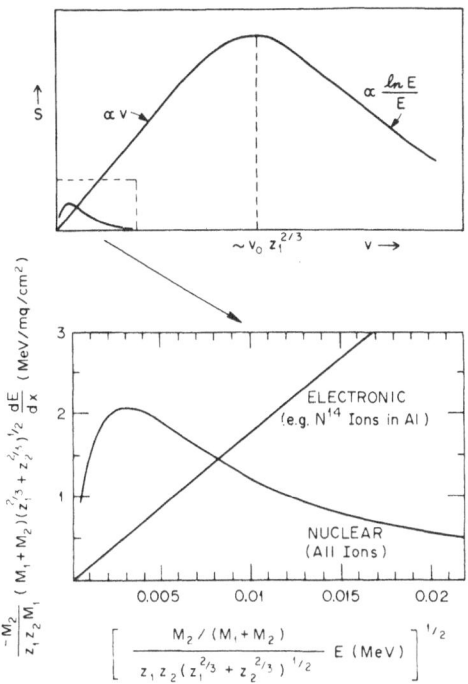

Fig. 8.12. Universal nuclear stopping power curve for heavy ions at low velocities and a typical electronic stopping power curve (*not* scaled) shown for comparison

Using a Thomas-Fermi model to determine the effects of screening, LINDHARD et al. [8.26] derived a universal curve for the nuclear stopping power shown in Fig.8.12, where it is compared with a typical (not scaled) electronic stopping power curve.

In the very high velocity region ($v \gg v_e Z^{2/3}$) where the ion is totally stripped of its electrons, the electronic stopping power may be formulated as

$$-dE/dx = NZ_2 \frac{4\pi Z_1^2 e^4}{mv^2} \cdot L \tag{8.21}$$

where N is the density of target atoms Z_2, and L is a dimensionless multiplier. Using first order perturbation theory, BETHE [8.27] showed that at high velocities L is independent of Z_1

$$L \simeq \ln 2m_e v^2/I \tag{8.22}$$

where $I \simeq I_0 Z_2$ is the average excitation potential of the target atom ($I_0 \approx 10$ eV for high Z_2). Experiments which showed differences in the stopping of negative and positive pions (BARKAS et al. [8.28]) and deviations from a pure Z_1^2 dependence in proton and helium ion stopping (ANDERSEN et al. [8.29]) indicated the need for higher order terms in L. This problem has been addressed in [8.30-32].

Electronic stopping at low velocities ($v \ll Z^{2/3} v_0$) decreases with decreasing velocity. Using a Thomas-Fermi model of an electron gas, LINDHARD et al. [8.33] obtained for the electronic stopping

$$-dE/dx = \kappa 8 - e^2 N a_0 \frac{Z_1 Z_2}{(Z_1^{2/3} + Z_2^{2/3})^{1/2}} \frac{v}{v_0}$$ (8.23)

where κ is a numerical factor in the order $Z_1^{1/6}$. An alternative treatment by FIRSOV [8.34], in which the ion's deceleration is attributed to the exchange of charge between projectile atom and atoms of the medium, leads to a similar linear velocity dependence.

Qualitatively the stopping power in the velocity region between the two mentioned above is often described in terms of a partially stripped ion with an effective charge which is in turn derived from a comparison of the ion's stopping power with that of a proton at the same velocity through the use of a parameter γ:

$$\left(\frac{dE}{dx}\right)_{Z_1 Z_2 v} \Big/ \left(\frac{dE}{dx}\right)_{p, Z_2 v} = \frac{\gamma^2 Z^2}{\gamma_p^2}$$ (8.24)

where

$$Z_{eff} = \gamma Z$$ (8.25)

and Z_{eff} is the effective charge for stopping. These methods have been used to classify the data taken in Oak Ridge for Br, I, and U stopping powers (BROWN and MOAK [8.35]). Over most of the energy range of interest here $\gamma_p = 1$. At the lower energies an empirical function due to BOOTH and GRANT [8.36] is used.

$$\gamma_p^2 = (1 - e^{-150 E_p}) \exp(-0.835 e^{-14.5 E_p})$$ (8.26)

where E_p is proton energy in MeV. Values of proton stopping power were taken from the tabulations of NORTHCLIFFE and SCHILLING [8.37]. More than 500 data points were used to obtain a least-squares fit to the function shown as a solid line in Fig.8.13. Although the individual points cannot be identified in this figure size, the conclusions are the same for each of the three ions. The analytical function which was derived is

$$\gamma = 1 - 1.034 \exp[-(v/v_0) Z^{0.688}] \quad .$$ (8.27)

The empirical exponent 0.688 lies remarkably close to the Thomas-Fermi value of 0.666. If either number (1.034 or 0.688) is varied by as much as 0.002, then no val-

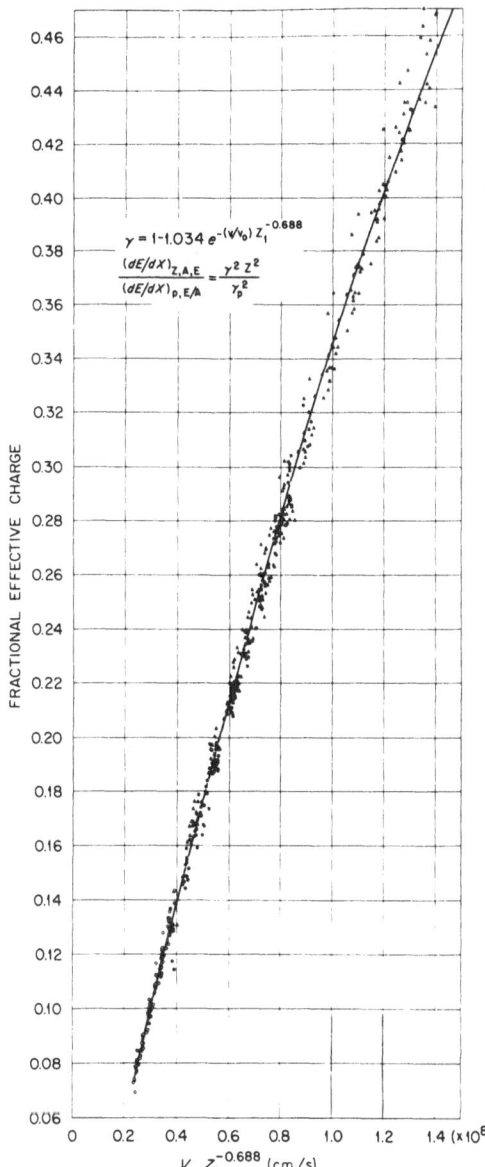

FRACTIONAL EFFECTIVE CHARGE

$$\gamma = 1 - 1.034\, e^{-(V/V_0)}\, Z_1^{-0.688}$$

$$\frac{(dE/dX)_{Z,A,E}}{(dE/dX)_{p,E/A}} = \frac{\gamma^2\, Z^2}{\gamma_p^2}$$

$V\, Z_1^{-0.688}$ (cm/s)

Fig. 8.13. Fractional effective charge (q_{eff}/Z_1) derived from stopping power measurements for Br, I, and U ions in various solid media as a function of reduced velocity

ue of the other number will produce an equally good fit to the data. The data for U, I, and Br, treated separately, produced the same numbers as given in (8.27). The function derived in (8.27) can be used to obtain estimates of electronic stopping power for heavy ions with $Z > 35$. To these values a small contribution due to nuclear stopping should be added as derived from [8.26] to obtain the total stopping power.

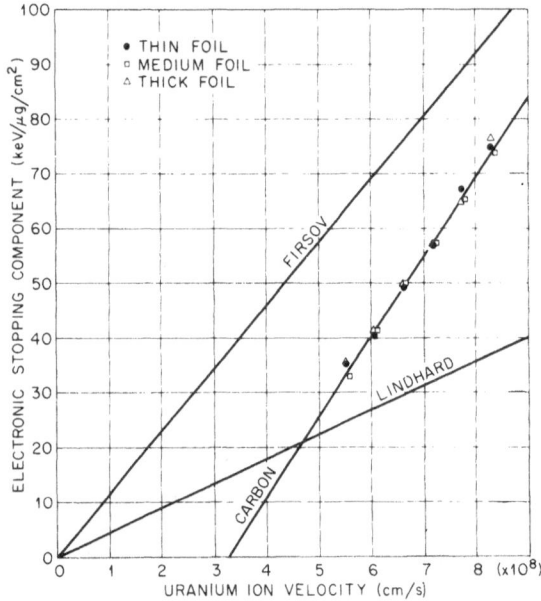

Fig. 8.14. Comparison of measured electronic stopping component for uranium in carbon compared with theoretical predictions

Additional data on the stopping powers of lighter ions, F, Mg, Al, S, and Cl, at velocities from 0.2 to 3.5 MeV/nucleon in Ti, Fe, Ni, Cu, Ag, and Au have been reported by FORSTER et al. [8.38] and have been shown to fit the same universal curve. This extends the range of γ up to 0.94.

It is interesting to observe that the values of the effective charge for stopping happen to agree closely with the charge states emerging from gas targets (see Fig.8.9) and not from solid targets. This would tend to reinforce the arguments of BETZ and GRODZINS [8.1], but for the reasons stated in Section 8.2 the connection remains uncertain.

In contrast to the monolithic description of the stopping process described above, recent experiments have shown marked deviations which require more subtle theoretical description.

Figure 8.14 shows some measurements of stopping power for U ions in C (MOAK et al. [8.39]) with a comparison to the LINDHARD et al. [8.33] and FIRSOV [8.34] theories. The linear dependence on velocity is evident but the intercept is not at zero. These measurements have been corrected for the anticipated nuclear stopping. However, to extend the measurements to lower velocities, where the nuclear stopping correction becomes more important, the channeling technique has been used. BØTTIGER and BASSON [8.40] showed that for a number of elements (17), low energy electronic stopping powers were not proportional to v but instead to E^p where p varied from .0.2 to 0.9. The results of MOAK et al. [8.39] are shown in Fig.8.15 for channeled I ions

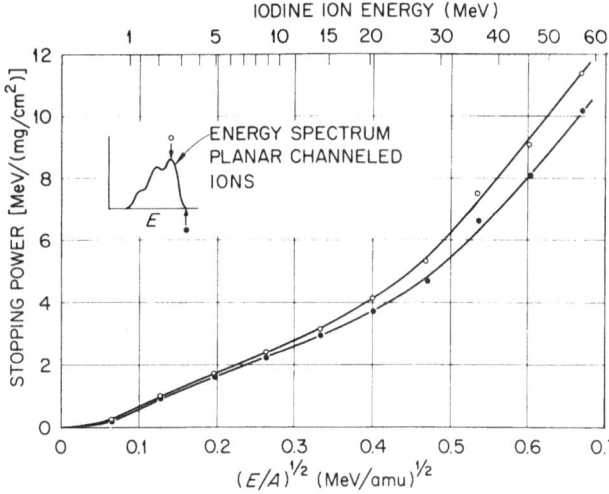

Fig. 8.15. Electronic stopping power for 0.6 to 60 MeV I ions in Ag (111) channel. Solid points which represent leading edge of the energy loss spectrum are for ions which follow the smallest channeling amplitude

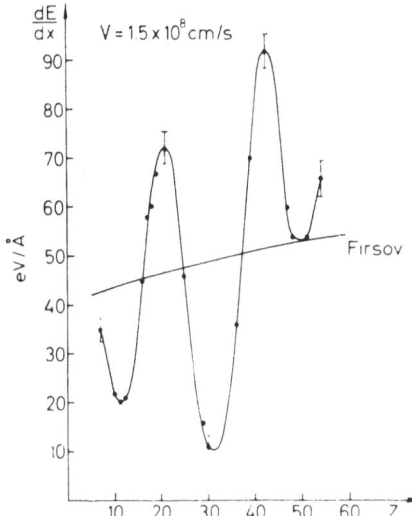

Fig. 8.16. Electronic stopping power for ions of varying Z_1 at a velocity of 1.5×10 cm s^{-1} channeled in Au

from 0.6 to 60 MeV. The behavior is clearly more complex than simple linearity with velocity, and reflects the velocity dependence of the ion's charge and excitation state in the medium.

At low velocities ($v \sim v_0$) the dependence of electronic stopping on projectile atomic number Z_1 is far from the monotonic behavior predicted by earlier theories as shown in Fig.8.16 (BØTTIGER and BASSON [8.40]). Here again the channeling effect is used to eliminate nuclear stopping and inner-shell electron interactions. (The oscillations shown in Fig.8.16 are also observed in nonchanneled experiments but are not as pronounced). Several explanations of the observed oscillations have been forwarded. In some the Firsov model is modified using more realistic atomic models (see, e.g., CHESHIRE et al. [8.41]) and agreement is qualitatively good but the peak to valley ratio predicted is not large enough. Another approach proposed by LINDHARD and FINNEMAN attributes the oscillations to a variation in the scattering cross section of target electrons near the Fermi level from the field of the incident ion. This corresponds to a variation in the phase shift induced in the electron wave function by the presence of the field of the moving ion. The stopping power in this description is proportional to electron momentum transfer cross section which can display transparencies (deep minima) similar to those observed in the Ramsauer-Townsend effect. Calculations on this basis (BRIGGS and PATHEK [8.42]) predict the observed effect rather well.

For high velocity ions we expect the stopping power to vary, in first order as the square of the charge on the ion. Measurement of this relationship for any but the lightest ions (e.g., H, He, and Li) is generally obscured by two effects. One of these effects is due to screening of the ion charge by electrons of the media and the other is that the number of bound electrons is constantly fluctuating because of exchange with electrons of the medium. These two effects are indeed difficult to separate but it has been done by measuring the energy loss for channeled ions which have not undergone charge exchange on their way through the crystal (see Sec.8.2.1). In this work DATZ et al. [8.43] injected beams of B^{5+}, $C^{5+,6+}$, $N^{5+,6+,7+}$, $O^{6+,7+,8+}$, and $F^{7+,8+,9+}$ ions at v=2 MeV/amu into a {111} planar channel of a 5000 Å Au single crystal and measured the energy loss spectra for particles emerging with the same charge as the input charge. The best defined trajectory here is for those particles whose paths were closest to the channel center (see Sec.8.1). The signature for this path is the minimum energy loss in the spectrum. The stopping powers for the bare nuclei when divided by the square of their respective charges [i.e., $(dE/dx)/q^2$] showed only a slight increase of 2% from B^{5+} to F^{9+}. Comparison $(dE/dx)/q^2$ for zero-, one-, and two-electron systems of the same q (e.g., C^{6+}, N^{6+}, O^{6+}) showed the effects of imperfect screening of the nuclear charge by the bound K electrons. By normalizing to the stopping power of the bare nucleus they observed that each of the electrons screened with $\sim$0.9 unit of charge. This is in accord with a picture that takes into account that electron collisions occurring outside the K-shell radius are perfectly screened while those with smaller impact parameters are not.

8.4 Inner-Shell Ionization

Because the time between collisions in solid media is shorter than the relaxation time of inner-shell vacancies, the penetrating ion often attains a steady state of inner-shell excitation which drastically affects its behavior in subsequent collisions. Considering first the projectile ion, ionization of low lying inner shells can occur because of steady state ionization of higher lying inner-shells; high degrees of outer-shell excitation can cause shifts of inner-shell levels which qualitatively change level correlations in the colliding pair; collisional mixing of outer levels can alter Auger lifetimes; the availability of electrons for capture to a projectile from target atoms can effectively reduce inner-shell vacancy lifetimes. The creation of inner-shell vacancies in a target atom is affected by the state of the projectile, and the relaxation of target atom vacancies can likewise be affected by its presence in a solid where valence and conduction electrons are always present.

In this section we shall illustrate each of these points and discuss the correlation between inner-shell vacancies and emergent charge state distribution.

8.4.1 Exit Channel Effects

A demonstration of exit channel effects in quasi-molecular promotion was given in a study of x-ray yields from collisions of 22-48 MeV I in Se, Br_2, and Kr targets (LUTZ et al. [8.44]). These are adjacent elements in the periodic table and the principal difference is that Se is a solid while Br and Kr are gases. From observations of the I L x-ray spectra (energy shift, etc.) it could be determined that the state of the I ion following an L-shell event was the same for all targets. This is not unexpected because an atypically violent collision is necessary to create such a vacancy. Nonetheless, the cross section for L-shell ionization was shown to be larger by a factor of 10 in Se as compared with Kr as shown in Fig.8.17 which shows

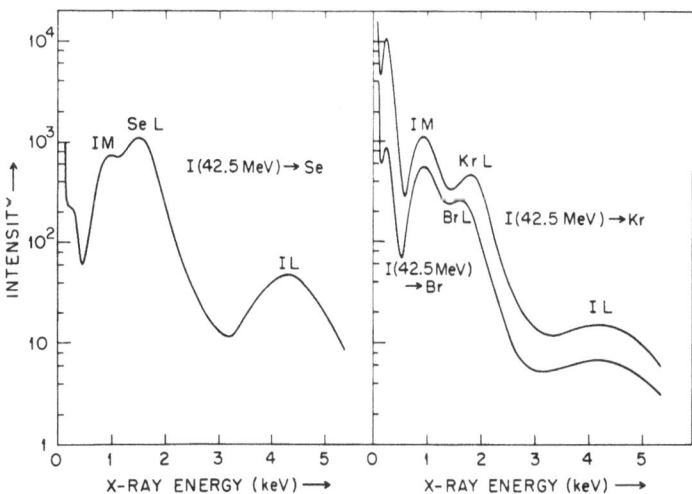

Fig. 8.17. X-ray spectra excited by 42 MeV I ions on solid Se and gaseous Kr and Br targets

334

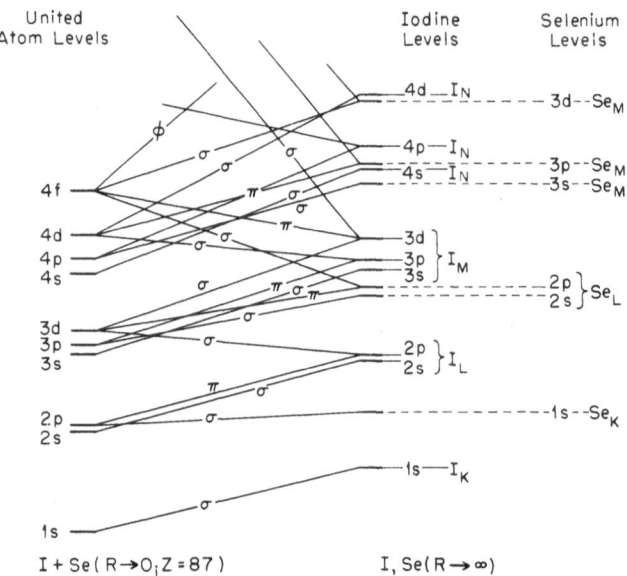

Fig. 8.18. Level correlation diagram for the I-Se system

absolute yields of x-rays observed with a proportional counter. The principal difference in the x-ray production in solid and gaseous targets arises from the state
of the ion prior to the collision. The outermost shell containing electrons here is
the N-shell; however, the high collision frequency in a solid can lead to considerable steady state ionic excitation in the M-shell of iodine. For example, in the
ion energy range investigated, the I M vacancy production cross section in the system I-Se is approximately 10^{-17} cm^2. The lifetime of an M vacancy in the free ion
will be about 5×10^{-14} s. (Changes in vacancy lifetime because of solid-state effects
will be discussed below). Thus, in a solid the I ion would be expected to have, on
the average, several M-shell vacancies. This should render possible an I L electron
promotion, e.g., via the $3d\sigma$ molecular level, to an unoccupied bound state in the I
M-shell (Fig.8.18). In a gast target, the I ion has time to deexcite M-shell excitations between collisions (principally by Auger processes). In this case, a colli-
sionally excited I L electron may have to be transferred high up to a weakly bound
state or directly into the continuum, resulting in a small cross section.

8.4.2 Level Shifting Effects

Changes in level correlations due to steady state excitation were first demonstrated
in the experiments of DER et al. [8.45] and FORTNER [8.46] who studied x-ray production in collisions of 30-200 keV Ar with carbon (graphite) and CH_4 gas. With CH_4

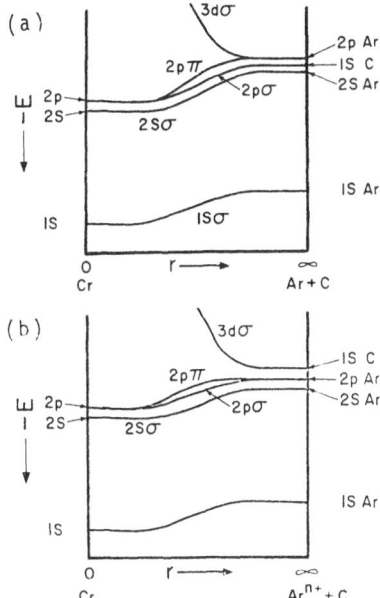

Fig. 8.19a and b. Level correlation diagram for the Ar-C system. The upper diagram is for Ar^0-C and the lower represents the Ar^{n+}-C system with a single L vacancy in the Ar

targets, only Ar L x-rays were observed. This is anticipated from the relevant correlation diagram [Fig.8.19(a)] which shows the Ar L level above the carbon K level and connecting with the $3d\sigma$ molecular orbital which leads to ionization. However, once the Ar ion loses a single L electron, the binding energies of the remaining L electrons increase sufficiently to change their position relative to carbon K [Fig. 8.19(b)]. This effect is observed when Ar beams are directed into a solid C target where carbon K radiation is predominant.

8.4.3 Solid-State Effects on Vacancy Lifetimes

When an inner-shell vacancy is created in an ion moving in a solid, the effects of subsequent collisions which occur while the vacancy is present cannot be neglected. For example, the mean free time for state changing processes with cross sections of 10^{-17} cm^2 is 10^{-14} s for ions moving in solids at $\sim v_0$. Thus states of the free ion with lifetimes 10^{-14} s will be significantly perturbed. These processes can involve the vacancy itself or the states of electrons in adjacent outer-shells which relax the vacancy.

One such effect has been noted by FORTNER and GARCIA [8.47] in their work on the Ar C system. They point out that collisions mix the outer-shell levels of long-lived Auger states with shorter-lived states and thereby reduce the fluorescence yield. This same effect can be seen in the results of LUTZ et al. [8.44] on I in Se and Kr.

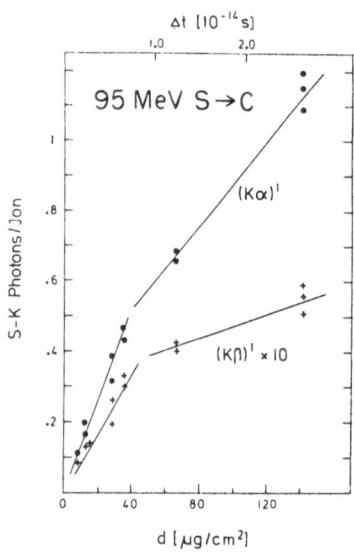

Fig. 8.20. Absolute yield of K x-ray lines from 95 MeV S ions (He-like) in C as a function of target thickness

Here, although the yield of I L x-rays is higher in solids by a factor of ~10 (exit channel effect), the yield of I M x-rays is lower in solids by a factor of ~3 (the outermost shell containing electrons being the N-shell).

An effective decrease in "fluorescence yield" in solids for high velocity S ions has recently been reported by BETZ et al. [8.48]. They observed that the yields of, for example, K_α x-rays from 90 MeV S^{14+} (He-like ions) penetrating carbon were not proportional to target thickness (Fig.8.20). They reasoned that for very thin targets most of the radiation occurred after emergence from the foil, and that the initial slope was thus characterized by a vacancy production cross section in the solid and the fluorescence yield characteristic of the ion in vacuum; but as the thickness is increased, decay inside the solid became predominant. The more gradual slope of the yield curve in the latter portion corresponds to a lower effective fluorescence yield inside the solid. To explain this phenomenon they point out that electron capture from the atoms in a solid constitutes an additional radiationless process for filling the vacancies within the solid.

They define effective "cross sections" for Auger relaxation $\sigma_A \equiv (nv\tau_A)^{-1}$ and radiative relaxation $\sigma_R \equiv (nv\tau_R)^{-1}$ where n is the target atom number density, v is the ion velocity, and τ_A and τ_R are the Auger and radiative lifetimes of the vacancy state, respectively. These are reasonable cross section definitions since, for example, the number of radiative decays taking place in a beam with atoms having a vacancy fraction Y within a time Δt for traversing a distance Δx is $\Delta Y = Y\sigma_R \Delta x = Y\Delta t/\tau_R$. The rate equation for Y(x) is

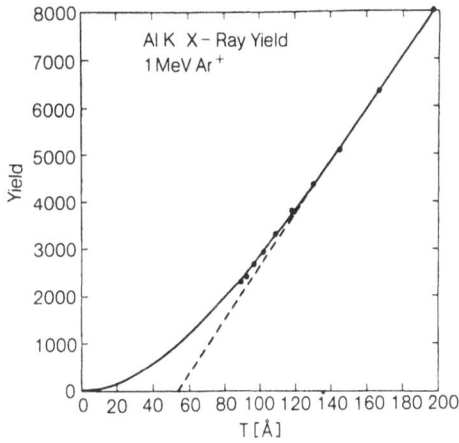

Fig. 8.21. Yield of Al K x-rays versus thickness for 1 MeV Ar^+ on A. The solid line is obtained from a fit to a vacancy transfer model. The dashed curve is an extrapolation from the linear portion of the curve

$$dY/dx = (1-Y)\sigma_V - (\sigma_A + \sigma_R + \sigma_C)Y \qquad (8.28)$$

and

$$Y(x) = (\sigma_V/\sigma_T)[1 - \exp(-\sigma_T x)] \qquad (8.29)$$

where σ_V is the vacancy creation cross section, σ_C is the electron capture cross section into the vacancy, and $\sigma_T \equiv (\sigma_A + \sigma_R + \sigma_V + \sigma_C)$. When the ion emerges from the target it will have a vacancy concentration $Y(x_0)$ and will emit, in vacuum, where $\sigma_C = 0$, x-rays in the amount ωY_0. The differing "fluorescence yield" in the solid occurs because of the additional term σ_C.

If the captured electron comes from an inner-shell of the target atom, then enhanced target atom x-ray emission will result from the movement of an ion containing a vacancy through a solid. FELDMAN et al. [8.49] measured the Al K x-ray yield produced by 0.3 to 2 MeV Ar ion bombardment of Al foils of varying thickness. In this work they demonstrated that the mechanism for the production of Al K vacancies required the prior production of Ar L vacancies which then transferred on a second collision to the Al K (equivalent to electron capture from Al K to Ar L). The slow initial rise in the Al K yield shown in Fig.8.21 as a function of target thickness is then due to distance required to attain steady state Ar L excitation. In this work the authors could derive the steady state vacancy fraction in Ar as a function of energy, and they showed that collisional electron transfer σ_C is the dominant process for quenching Ar L-shell vacancies in solid Al.

An alternative method for determining inner-shell vacancy populations in ions penetrating solids is based upon the observation (MOWAT et al. [8.50]) that the inner-shell vacancy formation cross section in a target atom can be strongly depen-

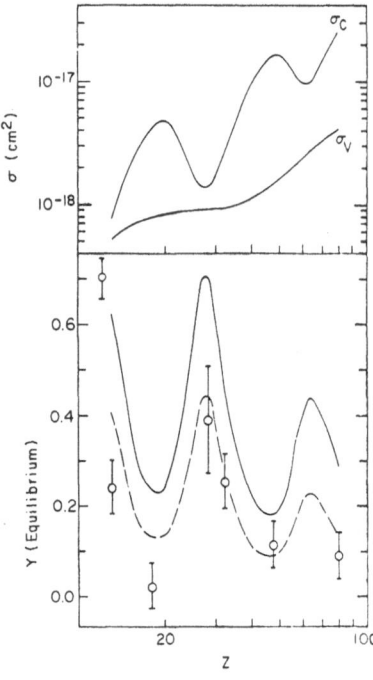

Fig. 8.22. Equilibrium K vacancy fraction of
2 MeV/amu Cl ions penetrating solid targets of
varying Z (lower portion). Upper portion shows
estimated K vacancy creation cross section σ_v
and calculated (Brinkmann-Kramers) capture
cross section σ_c

dent upon the ionization state of the projectile. DATZ et al. [8.51] reasoned that
if a knowledge of the charge state dependence of the target x-ray production cross
section could be obtained in a gaseous state where the initial charge could be defined, a measurement of the projectile charge in a solid target of the same target
element could be obtained from target x-ray yields. Using SiH_4 as a gaseous analogue
of solid Si, they observed that the charge state of 40 MeV oxygen ions determined
from the Si K x-ray yield in the solid did indeed correlate well with the anticipated
value. However, with 80 MeV Ar they observed x-ray yields which gave significantly
lower Si K x-ray yields from solids than were anticipated from the emergent charge
state of the penetrating Ar ion. However, it has since been shown (KAUFMAN et al.
[8.52], see Sec.8.4.) that the decay of Si K vacancies is significantly different
in SiH_4 and solid Si targets so that the effect may be due to a change in fluorescence yield.

This difficulty has been overcome by a technique developed by HOPKINS et al.
[8.53]. Using a very thin Cu target (~ 1 $\mu g/cm^2$) evaporated into the front of a carbon foil, they measured the cross section for Cu K x-rays produced by bombardment of
60 MeV Cl ions in varying charge states. They found the charge state dependence cross section to increase only slightly with increasing Cl ion charge until a single K vacancy (Cl^{16+})
was introduced whence the cross section increased by a factor of ~ 3. This increase
occurs because of the opening of another channel for inner-shell ionization of the

Cu, i.e., a transfer of a Cu K electron to the Cl K vacancy. Reasoning from other similar experiments that 2 K vacancies in the Cl ion would yield another factor of two increase in the Cu K cross section, they could devise a scale for Cl K ionization state from the Cu K x-ray yield. Now, by reversing the position of the foil, i.e., by placing the carbon towards the beam and the Cu deposited film away from it, they could determine the steady state population of Cl K vacancies for ions traversing the carbon from a measurement of the Cu K x-ray yield. Finally, by depositing their thin Cu "indicator" on foils of different elements, they measured the Z_2 dependence of the Cl K vacancy population. This is shown in the lower portion of Fig.8.22. The vertical axis Y represents the equilibrium vacancy fraction of the Cl K-shell electrons in the various media. The authors explain the oscillatory behavior on the basis of the anticipated oscillations in the capture cross section (σ_c) into the Cl K-shell from the target atom, the vacancy production cross section having been shown to have a monotonic behavior with target Z at this velocity (2 MeV/amu).

8.4.4 Effects on the Relaxation of Target Atom Vacancies

In Section 8.4.3 we have discussed the relation of the state of the projectile ion to the cross section to the production of target atom vacancies, the principal effect being the possibility of electron transfer from a target atom inner-shell to a pro-

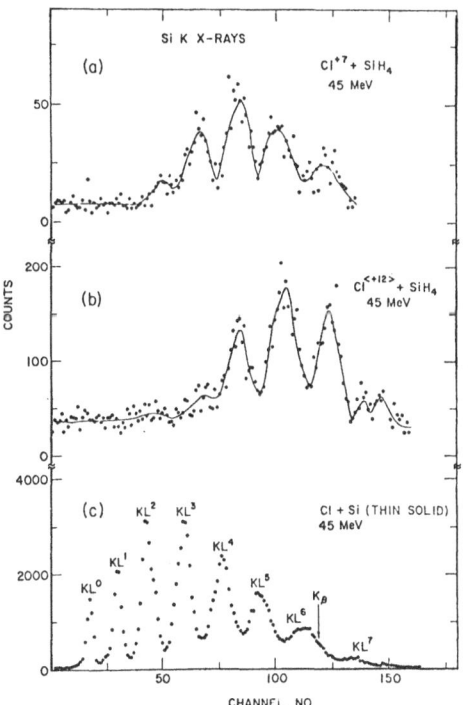

Fig. 8.23. High resolution Si K x-ray spectra from collision of 45 MeV Cl9+ on SiH₄ (gas) and Si (solid) targets. Additional shifts in the gas target spectra are attributed to the absence of M-shell electrons

jectile ion in a high state of ionization (excitation). The effect on vacancy relaxation arises from the multiple ionization of target atom by heavy ion projectiles.

KAUFMAN et al. [8.52] examined with high resolution the Si K_α x-ray spectra produced by 45 MeV Cl ion bombardment of thin targets of gaseous SiH_4 and solid Si with
the results shown in Fig.8.23. The spectrum obtained with the solid Si target displays satellite structure characteristic of multiple L-shell vacancies with a not
too strongly perturbed M-shell. On the other hand, the spectra from the SiH_4 gaseous
target can only be fit by assuming that essentially all of the Si M electrons (outer-
shell) have been removed in the initial K-shell ionizing collision. Within the solid target these M-shell vacancies can be readily filled by conduction or valance
electrons. Interestingly, the most probable L-shell configuration for the solid lies
between KL^2 and KL^3 (neglecting fluorescence yield differences) whereas in the gas
target (with Cl^{12+} projectiles) the peak lies at KL^5. This implies that L-shell relaxation is rapid compared to K-shell relaxation. This seems unusual but is in accord
with increased Auger lifetime of the K vacancy due to the highly stripped L-shell.
Aside from grossly changing the high resolution spectrum, the fluorescence yield for
the two systems should also be strongly affected. The authors estimate that the
fluorescence yield for Si in the SiH_4 target will be about a factor of two greater
than in the solid Si.

8.4.5 Relation of Inner-Shell Vacancies to Emergent Charge State Distributions

With the advent of extensive studies of heavy ion-induced x-rays in solid targets
a number of observations have been made which are relevant to the older problem of
relating measured charge state distributions to the states of ions penetrating solids.

In some work reported by KNUDSON et al. [8.54], outer-shell (M-shell) configurations have been deduced to some extent from K x-ray satellite energy shifts. High
resolution studies of K_α x-rays of Ar and Cl arising from 4 MeV Ar ion bombardment
of KCl are shown in Fig.8.24. The anticipated satellite structure due to the creation of L vacancies in the same collision which created the K vacancy are seen. The
positions of these satellite lines are calculated from a knowledge of the 2p-1s
energy differences with a given L vacancy configuration. The calculations based on
L vacancies only agree quite well with the observed spectrum of the target atom
(Cl); however, the lines from the moving projectile are both too broad and shifted
to higher energies. To fit the observations, M-shell (outer-shell) vacancies must
be added. Moreover, since the charge capture and loss cross sections for outer-shell
electrons are so high, it is anticipated that equilibrium has been attained here
prior to K-shell relaxation. The authors concluded that from five to six vacancies
are present in the M-shell which is in agreement with the charge state anticipated
for these ions after they emerge from solids. The experiment, however, cannot determine whether these electrons are totally removed from the ion or in excited states
since Hartree-Fock calculations give larger energy shifts for an electron bound in

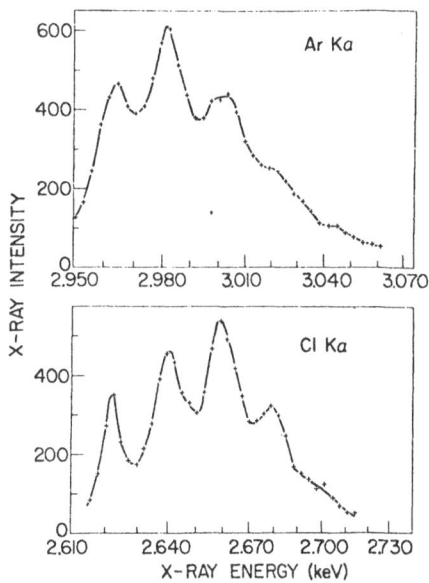

Fig. 8.24. High resolution K_α x-rays of Ar and Cl induced by 4 MeV Ar ion bombardment of KCl

3d level, for example, than for total removal of the electron. An additional note should be added here that the effective charge at the outer-shell will be enhanced by the presence of inner-shell vacancies (see below) which should decrease the loss and increase the capture over an ion with no inner-shell vacancies. Hence the M-shell of ions without these vacancies may in fact be even more highly ionized (excited).

FORTNER and GARCIA [8.47] considered the ion charge distribution in the solid deduced from x-ray measurements and compared them with measured emergent distributions for 90 keV Ar (TURKENBURG et al. [8.55]). The result is shown in Fig.8.25, and it is clear that the charge state inside the solid is higher than in the emergent beam. Moreover, when the ions emerge, loss of additional electrons by Auger processes would lead to an even higher charge state. For example, the authors showed that Ar ions moving in C would have 0.1 to 0.3 L vacancies at steady state at energies from 70 to 180 keV, respectively. Relaxation of these vacancies upon emergence would (almost entirely independent of outer-shell ionization) lead to the loss of another electron and would thus have a minimum charge state of 2+. Yet the mean charge measured on emergence is 0.5 and 1.25, respectively. For the case of Ar in C, no more than one vacancy can be sustained in the Ar L shell because the presence of this vacancy lowers the Ar L binding energy below the C K level (see Sec.8.4.2). This limitation is not present when Cl or S ions penetrate C, and multiple L vacancies have been observed in these collision systems (FORTNER and GARCIA [8.56]). TURKENBURG et al. [8.55], noting this observation, set out to measure the effect of inner-shell ionization on the emergent charge state distribution. They used beams

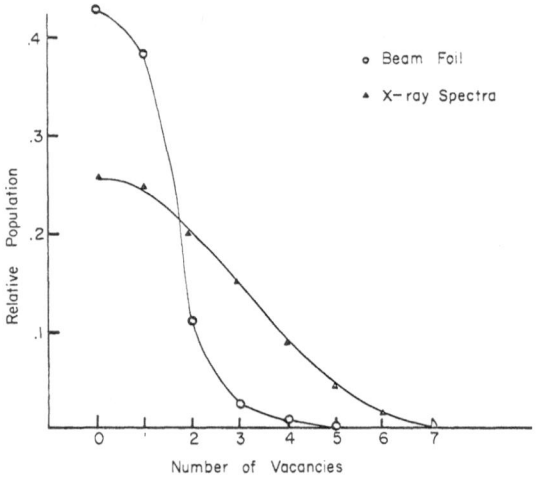

Fig. 8.25. Measured charge state
distribution for 90 keV Ar ions
for Ar ions emerging from a carbon foil compared with the number
of M and L vacancies for the same
ion in the solid as deduced from
x-ray measurements

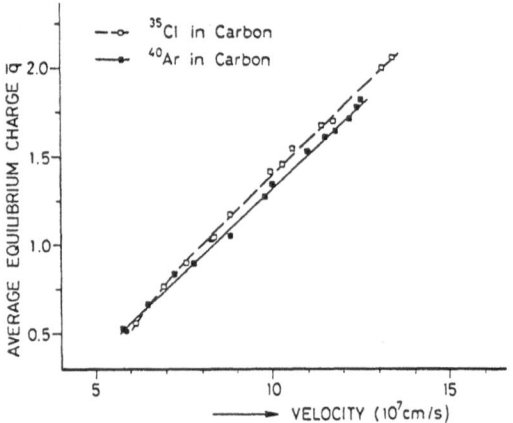

Fig. 8.26. Average equilibrium charge
of Ar and Cl ions emerging from carbon foils as a function of velocity

of Ar and Cl ions (70 to 350 keV). At the higher velocities the Cl does indeed yield
a slightly but measurably higher charge state (Fig.8.26), but the full effect due
to the additional anticipated higher L vacancy population is not observed, presumably because electron capture at the exit surface partially compensates for the additional charge. At lower velocities where surface capture is even more important,
the formation of Cl^- ions lowers the average charge on Cl to below the Ar value. Although capture at the surface is not well characterized it is easy to imagine that
outer-shell vacancies already present at emergence could be filled. However, it is
difficult to understand how surface events could much affect the ionization occurring
because of inner-shell vacancies. The Auger lifetimes are $\sim 10^{-14}$ s and the ions move
at $\sim 10^8$ cm s^{-1}. Thus the ionization should occur ~ 100 Å away from the surface where
electron capture from surface states is highly improbable.

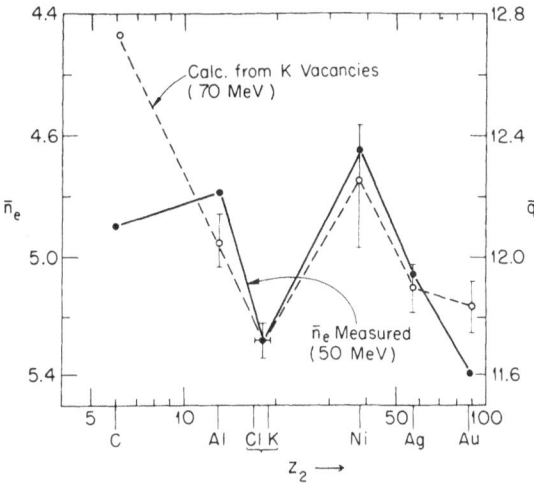

Fig. 8.27. Mean value of the number of remaining electrons on Cl ions emerging at 50 MeV from solid targets of varying Z. The dashed curve indicates differences expected from K vacancy fraction measured in these targets at 70 70 MeV (see Fig.8.22)

Noting the results of HOPKINS et al. [8.53] which showed an oscillatory behavior in the K vacancy fraction of 2 MeV/amu Cl ions in solid targets of varying Z_2 (Fig.8.22), DATZ et al. [8.57] sought a correlation with K vacancy fraction and emergent charge state distribution. The mean charge state,

$$\bar{q} = \sum_i \phi(q_i) q_i$$

of ^{35}Cl ion beams emerging at 54 MeV from foils of C, Al, KCl, Ni, Ag, and Au were determined, and the mean number of bound electrons $\bar{n}_e = Z - \bar{q}$ is plotted as a function of Z_2 in Fig.8.27. The dashed curve in Fig.8.27 shows a comparison with the effect which would be predicted based upon the established K vacancy fraction (HOPKINS et al. [8.53]) and a simple model of the inner-shell relaxation process.

Normalizing to the KCl point where the K vacancy fraction is negligible, they assessed the effect of a single Cl-K vacancy on the remaining number of bound electrons after emergence. First, of course, the K vacancy will reduce $\bar{n}_e$ by one unit. Secondly, the presence of a K vacancy in an emergent ion with L electrons present will lead to loss of a second electron by Auger with a probability $1-\omega_q$ where ω_q is the fluorescence yield for Cl^{q+}. Here they used an estimate $\omega_q \sim 0.2$ based upon computed values of ω for the various configurations of highly ionized Ar^{q+} (BHALLA [8.58]). Finally, since the equilibrium charge state is principally determined by capture and loss to the Cl L-shell, they consider the effect of a K vacancy on the processes occurring in L-shell. Here they assume that the principal effect is an increase in the effective Z at the L-shell radius by one unit, and using the semi-empirical formulation of NICOLAEV and DMITRIEV [8.18], they compute $\bar{q}'/Z$ for $Z=17$ and for $Z=18$ at the same velocity (1.5 MeV/amu); the result is $\bar{q}'/Z=0.738$ for $Z=17$

and $\bar{q}'/Z=0.731$ for $Z=18$ from which a change in n_e' of 0.4 unit of charge is expected. Thus the net expected change in $\bar{n}_e$ for a given K electron vacancy fraction $\phi(K)$ is:

$$\bar{n}_e = \phi(K)[1 + (1-\omega_q) - (\bar{n}_{e(Z+1)}' - \bar{n}_{e(Z)}')] = 1.4 \, \phi(K)$$

Using the values of $\phi(K)$ from Fig.8.22, they obtain the dashed curve shown in Fig.8.27. With the exception of the point at the lowest value of Z_2 (carbon), the agreement with the expectations of the model appears quite good, and indicates a beginning to the understanding of inner-shell vacancy-charge state correlations.

References

8.1 H.D. Betz, L. Grodzins: Phys. Rev. Lett. 25, 211 (1970)
8.2 V.S. Nicolaev, I.S. Dmitriev, Y.A. Tashaev, Y.A. Teplova, Y.A. Fainberg: J. Phys. B, Atom. Molec. Phys. 8, L58 (1975)
8.3 N.O. Lassen: Kgl. Dansk. Vid., Mat.-Fys. Medd. 26, No. 5 and No. 12 (1951)
8.4 P.E. Pierce, M. Blann: Phys. Rev. 173, 390 (1968)
8.5 M.D. Brown, C.D. Moak: Phys. Rev. B6, 90 (1972)
8.6 N. Bohr, J. Lindhard: Kgl. Dansk. Vid., Mat.-Fys. Medd. 28, No. 7 (1954)
8.7 W. Brandt, R. Laubert, M. Mourino, A. Schwartchild: Phys. Rev. Lett. 30, 358 (1973)
8.8 M.T. Robinson, O.S. Oen: Phys. Rev. 132, 2385 (1963)
8.9 J. Stark: Phys. Zschr. 13, 973 (1912)
8.10 J. Lindhard: Phys. Lett. 12, 126 (1964)
8.11 D.S. Gemmel: Rev. Mod. Phys. 46, 129 (1974)
8.12 S. Datz, B.R. Appleton, C.D. Moak: In *Channeling*, ed. by D.V. Morgan (John Wiley and Sons, New York 1974) pp. 153-180
8.13 M.T. Robinson: Phys. Rev. B4, 1461 (1971)
8.14 F.H. Eisen: Phys. Lett. 23, 401 (1966)
8.15 J.H. Barrett, B.R. Appleton, T.S. Noggle, C.D. Moak, J.A. Biggerstaff, S. Datz, R. Behrisch: In *Atomic Collisions in Solids V*, ed. by S. Datz, B.R. Appleton and C.D. Moak (Plenum Press, New York 1975) pp. 645-669
8.16 B.R. Appleton, C.D. Moak, T.S. Noggle, J.H. Barrett: Phys. Rev. Lett. 28, 1307 (1972)
8.17 H.D. Betz: Rev. Mod. Phys. 44, 465 (1972)
8.18 V.S. Nicolaev, I.S. Dmitriev: Phys. Lett. 28A, 277 (1968)
8.19 C.D. Moak, H.O. Lutz, L.B. Bridwell, L.C. Northcliffe, S. Datz: Phys. Rev. 176, 427 (1968)
8.20 H.O. Betz: Nucl. Inst. Methods 132, 19 (1976)
8.21 S. Datz: Nucl. Inst. Methods 132, 7 (1976)
8.22 H.O. Lutz, S. Datz, C.D. Moak, T.S. Noggle: Phys. Rev. Lett. 33A, 309 (1970)
8.23 S. Datz, F.W. Martin, C.D. Moak, B.R. Appleton, L.B. Bridwell: Rad. Effects 12, 163 (1972)
8.24 N. Bohr: Kgl. Danske Vid., Mat.-Fys. Medd. 28, No. 7 (1948)
8.25 P. Sigmund: In *Radiation Damage Processes in Materials*, ed. by C.H.S. Dupuy (Nordhoof, 1975) pp. 3-119
8.26 J. Lindhard, V. Nielsen, M. Scharff: Kgl. Danske Vid., Mat.-Fys. Medd. 36, No. 10 (1968)
8.27 H.A. Bethe: Ann. Phys. 5, 325 (1930)
8.28 W.H. Barkas, N.J. Dyer, H.H. Heckman: Phys. Rev. Lett. 11, 26 (1963)
8.29 H.H. Andersen, H. Simonsen, H. Sørenson: Nucl. Phys. A125, 171 (1969)
8.30 J.D. Jackson, R.L. McCarthy: Phys. Rev. B6, 4131 (1972)

8.31 J.C. Ashley, W. Brandt, R.H. Ritchie: Phys. Rev. B5, 2393 (1972); A8, 2402 (1973)
8.32 J. Lindhard: Nucl. Inst. Methods 132, 1 (1976)
8.33 J. Lindhard, M. Scharff, H.E. Schiøtt: Kgl. Danske Vid., Mat.-Fys. Medd. 33,
     No. 14 (1968)
8.34 O.B. Firsov: Zh. Eksper. Teor. Fys. 36, 1517 (1959); Engl. Trans.; Sov. Phys.
     JETP 9, 1076 (1959)
8.35 M. Brown, C.D. Moak: Phys. Rev. B6, 90 (1972)
8.36 W. Booth, I.S. Grant: Nucl. Phys. 63, 481 (1965)
8.37 L.C. Northcliffe, R.F. Schilling: Nucl. Data, Sect. A7, 233 (1970)
8.38 J.S. Forster, D. Ward, H.R. Andrews, G.C. Ball, G.J. Costa, W.G. Davies, I.V.
     Mitchell: Nucl. Inst. Methods 136, 349 (1976)
8.39 C.D. Moak, B.R. Appleton, J.A. Biggerstaff, M.D. Brown, S. Datz, T.S. Noggle,
     H. Verbeek: Nucl. Inst. Methods 132, 95 (1976)
8.40 J. Bøttiger, F. Basson: Radiation Effects 2, 105 (1969)
8.41 I.M. Cheshire, G. Dearnally, J.M. Poate: Proc. Roy. Soc. A311, 47 (1969)
8.42 J.S. Briggs, A.P. Pathak: Solid State Phys. 7, 1929 (1974)
8.43 S. Datz, P.F. Dittner, J. Gomez del Campo, P.D. Miller, J.A. Biggerstaff:
     Phys. Rev. Lett. 38, 1145 (1977)
8.44 H.O. Lutz, H.J. Stein, S. Datz, C.D. Moak: Phys. Rev. Lett. 28, 8 (1972)
8.45 R. Der, R. Fortner, T. Kavanagh, J. Garcia: Phys. Rev. Lett. 27, 1631 (1971)
8.46 R. Fortner: Phys. Rev. A10, 2218 (1974)
8.47 R. Fortner, J.D. Garcia: Phys. Rev. A12, 856 (1975)
8.48 H.D. Betz, F. Bell, H. Panke, G. Kalkoffen, M. Weltz, D. Evers: Phys. Rev.
     Lett. 33, 807 (1974)
8.49 L.C. Feldman, P.J. Silverman, R.J. Fortner: Nucl. Inst. Methods 132, 29 (1976)
8.50 J.R. Mowat, I.A. Sellin, D.J. Pegg, R.S. Peterson, M.D. Brown, J.R. Macdonald:
     Phys. Rev. Lett. 30, 289 (1973)
8.51 S. Datz, B.R. Appleton, J.R. Mowat, R. Laubert, R.S. Petersen, R.S. Thoe,
     I.A. Sellin: Phys. Rev. Lett. 33, 733 (1974)
8.52 R.L. Kaufman, K.A. Jamison, T.J. Gray, P. Richard: (to be published)
8.53 F. Hopkins, J. Sokolov, A. Little: (to be published)
8.54 A.R. Knudson, P.G. Burkhalter, D.J. Nagel: Phys. Rev. A10, 2118 (1974)
8.55 W.C. Turkenburg, B.G. Colenbrander, H.H. Kersten, F.W. Saris: Surface Science
     47, 272 (1975)
8.56 R. Fortner, J.D. Garcia: In *Atomic Collisions in Solids V*, ed. by S. Datz,
     B.R. Appleton, and C.D. Moak (Plenum Press, New York 1975) p. 469
8.57 S. Datz, C.D. Moak, P.F. Dittner, J. Gomez del Campo, P.D. Miller, J.A. Bigger-
     staff: Abstr. X Int. Conf. on Phys. of Electr. and Atomic Coll., Paris, July
     1977, p. 645
8.58 C.P. Bhalla: Phys. Rev. A8, 2877 (1973)

Subject Index

Lecture Notes in Physics

Selected titles

Editors: J. Ehlers, K. Hepp, R. Kippenhahn, H. A. Weidenmüller, J. Zittartz

Managing Editor: W. Beiglböck

Volume 55

Nuclear Optical Model Potential

Proceedings of the Meeting held in Pavia, April 8 and 9, 1976
Editors: S. Boffi, G. Passatore
62 figures, 6 tables. VI, 221 pages. 1976
ISBN 3-540-07864-9

Volume 56

Current Induced Reactions

International Summer Institute on Theoretical Particle Physics in Hamburg 1975
Editors: J. G. Körner, G. Kramer, D. Schildknecht
147 figures, 13 tables. V, 553 pages. 1976
ISBN 3-540-07866-5

Volume 61

Photonuclear Reactions I

International School on Electro- and Photonuclear Reactions, Erice, Italy 1976
Editors: S. Costa, C. Schaerf
362 figures, 62 tables. VII, 650 pages. 1977
ISBN 3-540-08139-9

Volume 62

Photonuclear Reactions II

International School on Electro- and Photonuclear Reactions, Erice, Italy 1976
Editors: S. Costa, C. Schaerf
174 figures, 13 tables. VII, 301 pages. 1977
ISBN 3-540-08140-2

Volume 66

A. H. Völkel

Fields, Particles and Currents

V, 354 pages. 1977
ISBN 3-540-08347-2

Volume 67

W. Drechsler, M. E. Mayer

Fiber Bundle Techniques in Gauge Theories

Lectures in Mathematical Physics at the University of Texas at Austin
Edited by A. Böhm, J. D. Dollard
IX, 248 pages. 1977
ISBN 3-540-08350-2

Springer-Verlag
Berlin
Heidelberg
New York

Beam-Foil Spectroscopy

Editor: S. Bashkin

91 figures. XIII, 318 pages. 1976
(Topics in Current Physics, Volume 1)
ISBN 3-540-07914-9

The purpose of this volume is to make available a variety of information which stems from the modern technique called Beam-Foil Spectroscopy. The book includes chapters written by outstanding experts in the fields of optical and electron spectra, the measurement of level lifetimes, theoretical calculations, applications of the data to astrophysics, and intrumentation. The experiments discussed range from rather low particle energies (some hundreds of keV) to the very high energies produced with tandem Van de Graaffs and the Berkeley Heavy-Ion Linear Accelerator. The book is intended to help physicists, astrophysicists, and chemists in understanding and using the beam-foil method. It is especially timely since new aspects of beam-foil spectroscopy are developing rapidly and a basic knowledge of the field is essential before one can take advantage of the recent advances.

Contents: S. Bashkin: Introduction. – S. Bashkin: Experimental Methods. – I. Martinson: Studies of Atomic Spectra by the Beam-Foil Method. – L. J. Curtis: Lifetime Measurements. – O. Sinanoğlu: Theoretical Oscillator Strengths of Neutral, Singly-Ionized, and Multiply-Ionized Atoms. – W. Wiese: Regularities of Atomic Oscillator Strengths in Isoelectronic Sequences. – W. Whaling: Applications to Astrophysics: Absorption Spectra. – L. J. Heroux: Applications of Beam-Foil Spectroscopy to the Solar Ultraviolet Emission Spectrum. – R. Marrus: Studies of Hydrogen-Like and Helium-Like Ions of High Z. – J. Macek, D. Burns: Coherence, Alignment and Orientation Phenomena in the Beam-Foil Light Source. – I. A. Sellin: The Measurement of Autoionizing Ion Levels and Lifetimes by Fast Projectile Electron Spectroscopy.

W. Nörenberg, H. A. Weidenmüller

Introduction to the Theory of Heavy-Ion Collisions

155 figures, 4 tables. IX, 273 pages. 1976
(Lecture Notes in Physics, Volume 51)
ISBN 3-540-07801-0

This volume compiles some basic tools of theoretical heavy-ion physics. The authors present essentially those facts they consider to be reasonably well-established and important for future development. They include many experimental data and also mention some numerical calculations. The level of presentation is intelligible for graduate students. The contents: Classical theory (elastic scattering and deeply inelastic collisions), gross properties and compound-nucleus formation (Q-values, grazing, yrast levels), nuclear scattering theory, elastic scattering of heavy-ions, coulomb excitation, inelastic scattering and transfer theory, statistical theory. The last chapter treats the interrelation of heavy-ion physics and atomic physics.

Contents: Introduction. – Classical theory of HI collisions. – Gross properties of HI reactions. Compound-nucleus formation. – Some elements of nuclear scattering theory. – Elastic scattering. – Coulomb excitation. – Inelastic scattering and transfer reactions. – Statistical theory. – Atomic effects in ion-atomic collisions. –

Springer-Verlag
Berlin
Heidelberg
New York